Quality Tools for Managing Construction Projects

Industrial Innovation Series

Series Editor

Adedeji B. Badiru

Department of Systems and Engineering Management
Air Force Institute of Technology (AFIT) – Dayton, Ohio

PUBLISHED TITLES

Carbon Footprint Analysis: Concepts, Methods, Implementation, and Case Studies, *Matthew John Franchetti & Defne Apul*

Computational Economic Analysis for Engineering and Industry, *Adedeji B. Badiru & Olufemi A. Omitaomu*

Conveyors: Applications, Selection, and Integration, *Patrick M. McGuire*

Global Engineering: Design, Decision Making, and Communication, *Carlos Acosta, V. Jorge Leon, Charles Conrad, and Cesar O. Malave*

Handbook of Industrial Engineering Equations, Formulas, and Calculations, *Adedeji B. Badiru & Olufemi A. Omitaomu*

Handbook of Industrial and Systems Engineering, *Adedeji B. Badiru*

Handbook of Military Industrial Engineering, *Adedeji B.Badiru & Marlin U. Thomas*

Industrial Control Systems: Mathematical and Statistical Models and Techniques, *Adedeji B. Badiru, Oye Ibidapo-Obe, & Babatunde J. Ayeni*

Industrial Project Management: Concepts, Tools, and Techniques, *Adedeji B. Badiru, Abidemi Badiru, & Adetokunboh Badiru*

Inventory Management: Non-Classical Views, *Mohamad Y. Jaber*

Kansei Engineering - 2 volume set

- Innovations of Kansei Engineering, *Mitsuo Nagamachi & Anitawati Mohd Lokman*
- Kansei/Affective Engineering, *Mitsuo Nagamachi*

Knowledge Discovery from Sensor Data, *Auroop R. Ganguly, João Gama, Olufemi A. Omitaomu, Mohamed Medhat Gaber, & Ranga Raju Vatsavai*

Learning Curves: Theory, Models, and Applications, *Mohamad Y. Jaber*

Modern Construction: Lean Project Delivery and Integrated Practices, *Lincoln Harding Forbes & Syed M. Ahmed*

Moving from Project Management to Project Leadership: A Practical Guide to Leading Groups, *R. Camper Bull*

Project Management: Systems, Principles, and Applications, *Adedeji B. Badiru*

Project Management for the Oil and Gas Industry: A World System Approach, *Adedeji B. Badiru & Samuel O. Osisanya*

Quality Management in Construction Projects, *Abdul Razzak Rumane*

Quality Tools for Managing Construction Projects, *Abdul Razzak Rumane*

Social Responsibility: Failure Mode Effects and Analysis, *Holly Alison Duckworth & Rosemond Ann Moore*

Statistical Techniques for Project Control, *Adedeji B. Badiru & Tina Agustiady*

STEP Project Management: Guide for Science, Technology, and Engineering Projects, *Adedeji B. Badiru*

Systems Thinking: Coping with 21st Century Problems, *John Turner Boardman & Brian J. Sauser*

Techonomics: The Theory of Industrial Evolution, *H. Lee Martin*

Triple C Model of Project Management: Communication, Cooperation, Coordination, *Adedeji B. Badiru*

FORTHCOMING TITLES

Communication for Continuous Improvement Projects, *Tina Agustiady & Adedeji B. Badiru*

Cellular Manufacturing: Mitigating Risk and Uncertainty, *John X. Wang*

Essentials of Engineering Leadership and Innovation, *Pamela McCauley-Bush & Lesia L. Crumpton-Young*

Handbook of Emergency Response: A Human Factors and Systems Engineering Approach, *Adedeji B. Badiru & LeeAnn Racz*

Sustainability: Utilizing Lean Six Sigma Techniques, *Tina Agustiady & Adedeji B. Badiru*

Technology Transfer and Commercialization of Environmental Remediation Technology, *Mark N. Goltz*

Quality Tools for Managing Construction Projects

Abdul Razzak Rumane

CRC Press is an imprint of the
Taylor & Francis Group, an **informa** business

CRC Press
Taylor & Francis Group
6000 Broken Sound Parkway NW, Suite 300
Boca Raton, FL 33487-2742

First issued in paperback 2021

Version Date: 20130220

ISBN 13: 978-1-03-209912-5 (pbk)
ISBN 13: 978-1-4665-5214-2 (hbk)

Publisher's Note
The publisher has gone to great lengths to ensure the quality of this reprint but points out that some imperfections in the original copies may be apparent.

Library of Congress Cataloging-in-Publication Data

Rumane, Abdul Razzak.
Quality tools for managing construction projects / author, Abdul Razzak Rumane.
pages cm. -- (Industrial innovation series)
Includes bibliographical references and index.
ISBN 978-1-4665-5214-2 (hardback)
1. Building--Quality control. 2. Building--Superintendence. I. Title.

TH438.2.R863 2013
624.068'4--dc23 2013003955

Visit the Taylor & Francis Web site at
http://www.taylorandfrancis.com

and the CRC Press Web site at
http://www.crcpress.com

Dedicated to my parents … for their prayers and love. My prayers are always for my father who always encouraged me. I wish he could have been here to see this book and to give me his blessing. My prayers and love are with my mother who always inspires me.

Contents

List of Illustrations

List of Tables

Foreword

Quality in design and construction means different things to different people.

To the architect and designers, it is the result of designing a beautiful and functional edifice that is relevant and meaningful years after construction is completed. To the builders, the contractors, and tradesmen who interpret and build from the designer's drawings, quality is a both a deliverable from the designers in understandable, biddable, and buildable documents, and a service in communication and timely response during construction. To the owner, quality is a deliverable from both the designer and builder in receiving a design and resulting building that is not only fit for purpose, but is a positive addition to its surroundings. Quality is also a service to the owner during the design and construction process from both designers and builders and after the project is complete through continued customer service that enhances the reputation of the owner in the community.

Therefore, dealing with such a multilayered and fungible intangible as quality during the design and construction process is difficult for all parties involved in the process. Furthermore, creating processes that bring about and manage quality, as well as provide metrics for ensuring that a quality outcome is an integral part of all activities, is important for the successful outcome of each project. Additionally, quality processes should be embedded not just at the project level, but within the entire enterprise, with designer, builder, and owner committed to integrating quality into all their business activities.

Dr. Rumane understands the tools and systems not just for managing quality, but for creating quality-focused project and enterprise processes that will ensure the quality of the deliverables and services during and after design and construction.

In this book, he has identified the quality tools and systems that have been successfully integrated into manufacturing and services, and then presented them in clear and easily adoptable methods for use in design and construction projects, as well as described how to incorporate them into all of the organizations that make up the design and construction industry, including designer, builder, and owner services.

Quality Tools for Managing Construction Projects focuses on the history of quality as a science and then the nature of construction projects, and how to successfully implement quality systems at every phase of a project. Following the construction cycle, Dr. Rumane delineates the quality tools and their application in construction projects, ending with the implementation of quality systems throughout the entire design and construction cycle.

Dr. Rumane is a practiced educator in the construction sciences, but he is also a student in continuously learning the new systems of quality including Lean, Six Sigma, and TRIZ. His descriptions of implementing these new tools, as well as traditional methodologies such as TQM and Testing and Commissioning, make this book a timeless reference for the quality and building professional.

Additionally, Dr. Rumane is actively engaged with professional quality societies such as ASQ, where I met him while I was chair of the Design and Construction Division. As one of the few registered architects within the society, I valued his focus on the entire lifecycle of construction, from design to operations, not just referencing discrete and separate activities like inspection during construction. His systems approach recognizes and supports the collaborative approach that modern design and construction projects need. Furthermore, Dr. Rumane is able to demonstrate that successful quality management is more than a series of handoffs between teams who have completed tasks. Designers should be involved with the operations, and builders should be involved in design, if only to leverage the feedback loop that makes up a robust quality system going back to W. Edwards Deming's Plan–Do–Check–Act Cycle.

Finally, *Quality Tools for Managing Construction Projects* is a book that will be valuable to all members of the design and construction team. As the public who utilizes the results of the designed and built environment, we all benefit from quality systems that focus on adding value and eliminating waste, as well as creating a more sustainable and vibrant economy, culture, and world.

Cliff Moser, AIA, MSQA, LEED AP
Past Chair, Design and Construction Division, ASQ

Acknowledgments

Share the knowledge with others is the motto of this book.

Numerous colleagues and friends extended help while preparing the book by arranging reference material; many thanks to all of them for their support.

I also want to extend my appreciation to the following; the publishers and authors whose writings are included in this book for offering their support by allowing me to reprint their material; reviewers from various professional organizations for their valuable input to improve my writing; members of ASQ Design & Construction Division, The Institution of Engineers (India), and Kuwait Society of Engineers for their support to bring out this book; Series Editor Dr. Adedeji B. Badiru and Senior Acquisitions Editor Cindy Renee Carelli of CRC Press and other staff for their support and contributions to making this construction related book a reality; and to Cliff Moser, past chair, Design and Construction Division, ASQ, for his nicely worded and thought-provoking foreword.

For their encouragement, I also want to acknowledge Raymond R. Crawford of Parsons Brinckerhoff; Dr. Ted Coleman for his good wishes and support all the time; Dr. N. N. Murthy of Jagruti Kiran Consultants; and engineer Ganesan Swaminathan of SSH International Consultants.

In addition, I am grateful to the many well wishers whose inspiration made it possible for me to complete this book.

The contributions of my son Ataullah, my daughter Farzeen, and daughter-in-law Ferha are well worth mentioning here. They encouraged me and helped me in my preparatory work to achieve the final product. I am indebted to my parents and family members for their continuous support, encouragement, good wishes, and prayers.

Finally, my special thanks to my wife, Noor Jehan, for her patience as she had to suffer a lot because of my busy schedule.

Most of the data discussed in this work are from the author's practical and professional experience, and are accurate to the best of author's knowledge and ability. However, if any discrepancies are observed in the presentation, I would appreciate these being communicated to me.

Abdul Razzak Rumane, PhD

About the Author

Abdul Razzak Rumane, PhD, is a certified consultant engineer in electrical engineering. He obtained a Bachelor of Engineering (Electrical) degree from Marathwada University (now Dr. Babasaheb Ambedkar Marathwada University) India in 1972, and received his PhD from Kennedy Western University (now Warren National University) in 2005. His dissertation topic was "Quality Engineering Applications in Construction Projects." Dr. Rumane's professional career exceeds forty years including ten years in manufacturing industries and about thirty years in construction projects. Presently, he is associated with Dar Alkuwait Alkhaleejia Consultation and Training, Kuwait, as Director, Construction Quality Management.

Dr. Rumane is associated with a number of professional organizations. He is a chartered fellow and Registered Senior Consultant of the Chartered Quality Institute (UK), fellow of the Institution of Engineers (India), and has an honorary fellowship of Chartered Management Association (Hong Kong). He is also a senior member of the Institute of Electrical and Electronics Engineers (USA), a senior member of American Society for Quality, a member of Kuwait Society of Engineers, a member of SAVE International (The Value Society), and is a certified AVS (Associate Value Specialist), and member of Project Management Institute. He is also an associate member of the American Society of Civil Engineers, a member of the London Diplomatic Academy, a member of the International Diplomatic Academy, and a member of the board of governors of the International Benevolent Research Forum.

As an accomplished engineer, Dr. Rumane has been awarded an honorary doctorate in engineering from Yorker International University. Albert Schweitzer International Foundation honored him with a gold medal for his "Outstanding contribution in the field of electrical engineering/consultancy in construction projects in Kuwait." The European Academy of Informatisation honored him with the "World Order of Science–Education–Culture" and the title of "Cavalier," and The Sovereign Order of the Knights of Justice, England honored him with its Meritorious Service Medal. He was selected as one of the Top 100 Engineers in 2009 of IBC

(International Biographical Centre, Cambridge, UK). He was also honorary chairman of the Institute of Engineers (India), Kuwait Chapter, for the years 2005 to 2007. Dr. Rumane is an author of the book titled *Quality Management in Construction Projects* published by CRC Press (a Taylor & Francis Group Company).

Introduction

Quality is a universal phenomenon that has been a matter of great concern throughout the recorded history of humankind. It has always been the intention of the builders and makers of products to ensure that these meet the customer's needs.

Construction projects have many participants including the owner, designer, contractor, and many other professionals from construction-related industries. Each of these participants is involved in implementing quality in construction projects. They all are both influenced by and dependent on each other, in addition to other players involved in the construction process. Therefore, construction projects have become more complex and technical; extensive efforts are required to reduce rework and costs associated with time, materials, and engineering.

Quality in construction projects is achieved through the application of various quality control principles, procedures, concepts, methods, tools, and techniques, and their applications to various activities/components/subsystems at different phases of the life cycle of a construction project to improve construction process to conveniently manage the project and make them more qualitative, competitive, and economical.

Quality tools and techniques are very important to develop a comprehensive project quality management system. Application of tools and techniques in construction projects has a great influence on the cost-effectiveness results of construction projects and achieving successful project performance. Quality management tools and techniques help in project planning, execution, monitoring, and control of a project, and evolve a project management system that makes project deliverables. They make it possible to:

- Meet defined scope
- Complete as per schedule
- Complete within budget

This book has been developed to provide significant information about usage and applications of various tools and techniques in different phases of a construction project. The book focuses on three quality management processes; namely, planning quality, performing quality assurance, and performing quality control.

For the sake of proper understanding, the book is divided into four chapters, and each chapter is divided into a number of sections covering quality related topics that have importance or relevance for understanding quality and project management concepts for construction projects.

Chapter 1 is an overview of quality in construction projects. It provides a brief introduction to quality history, a definition of quality, the birth of total quality, and the definition of quality in construction projects. It also discusses, in brief, the philosophies of "Quality Gurus" and total quality management (TQM). It includes information about project definition, types of construction projects, different phases (conceptual design, schematic design, design development, construction, testing, commissioning, and handing over) of the construction project life cycle and the subdivision of each phase into various elements/activities/subsystems having a functional relationship with the upper element of breakdown structure. A brief introduction to quality standards, quality management systems, and the development of an Integrated Quality Management System for development of a contractor's quality control plan is also discussed in this chapter.

Chapter 2 is about quality tools. It gives a brief description of various types of quality tools used in practice, by the construction industry, under sections such as Classic Tools of Quality, Management and Planning Tools, Process Analysis Tools, Process Improvement Tools, Innovation and Creation Tools, Lean Tools, Cost of Quality, Quality Function Deployment, and Six Sigma and TRIZ.

Chapter 3 is about quality tools in construction projects. It details different types of quality tools used for project development, project planning, project monitoring and control, quality management, risk management, procurement/contract management, safety management, quality assessment/measurement, training and development, and customer satisfaction.

Chapter 4 is about applications of quality tools in construction projects. It elaborates application of various tools and techniques for specific and appropriate usage at different phases of construction projects, starting from initiation of project to testing, commissioning, and handover of the project. This includes examples of detailed application of quality tools during project inception, concept design, schematic design, design development, construction documents, bidding and tendering, construction, and testing and handover. It also discusses application of quality during operation and maintenance, as well as assessment of quality.

Application of Six Sigma analytic tools in construction projects is elaborated in the appendices. These appendices include an example of each of the following:

1. Application of the DMADV tool to develop design for construction projects
2. Application of the DMADV tool to develop contactor's construction schedule
3. Application of the DMAIC tool in construction projects to develop quality management system for concrete structural works

This book, I am certain, will meet the requirements of construction professionals, students, and academics and satisfy their needs.

List of Abbreviations

AAMA	American Architectural Manufacturers Association
ACI	American Concrete Institute
ACMA	American Composite Manufacturers Association
AISC	American Institute of Steel Construction
ANSI	American National Standards Institute
API	American Petroleum Institute
ARI	American Refrigeration Institute
ASCE	American Society of Civil Engineers
ASHRAE	American Society of Heating, Refrigeration and Air-conditioning Engineers
ASTM	American Society of Testing Materials
ASQ	American Society for Quality
BMS	Building Management System
BREEAM	Building Research Establishment Environmental Assessment Method
BSI	British Standard Institute
CEN	European Committee for Standardization
CIBSE	Chartered Institution of Building Services Engineers
CIE	International Commission on Illumination
CII	Construction Industry Institute
CSC	Construction Specifications, Canada
CSI	Construction Specification Institute
CTI	Cooling Tower Industry
DIN	Deutsches Institute für Normung
EIA	Electronic Industry Association
EN	European Norms
FIDIC	Fédération International des Ingéneurs-Counseils
HQE	Higher Quality Environmental
ICE	Institute of Civil Engineers (UK)
IEC	International Electrotechnical Commission
IEEE	Institute of Electrical and Electronics Engineers
IP	Ingress Protection
ISO	International Organization for Standardization
LEED	Leadership in Energy and Environmental Design
NEC	National Electric Code
NEMA	National Electrical Manufacturers Association (USA)

NFPA	National Fire Protection Association
NWWDA	National Wood, Window and Door Association
PMI	Project Management Institute
PMBOK	Project Management Book of Knowledge
QS	Quantity Surveyor
RFID	Radio Frequency Identification
SDI	Steel Door Institute
TIA	Telecommunications Industry Association
UL	Underwriters Laboratories

SYNONYMOUS

Owner	Client, employer
Consultant	Architect/engineer (a/e), designer, design professionals, consulting engineers, supervision professional
Engineer	Resident project representative
Engineer's representative	Resident engineer
Contractor	Constructor, builder
Quantity surveyor	Cost estimator, contract attorney, cost engineer, cost and works superintendent

1

Overview of Quality in Construction Projects

1.1 Quality History

Quality issues have been of great concern throughout the recorded history of human beings. During the New Stone Age, several civilizations emerged, and some 4000–5000 ago, considerable skills in construction had been acquired. The pyramids in Egypt were built approximately during 2589–2566 BC. The king of Babylonia (1792–1750 BCE) had codified the law, and according to which, during the Mesopotamian era, builders were responsible for maintaining the quality of buildings and were given the death penalty if any of their constructed buildings collapsed and its occupants were killed. Records exist to show that the extension of Greek settlements around the Mediterranean after 200 BCE featured temples and theatres were built using marble. India had strict standards for working in gold in the fourth century BCE.

During the Middle Ages, guilds took the responsibilities for quality control upon themselves. Guilds and governments carried out quality control while consumers carried out informal quality inspection during every age of humanity.

The guilds' involvement in quality was extensive. All craftsmen living in a particular area were required to join the corresponding guild and were responsible for controlling the quality of their own products. If any of the items was found defective, then the craftsman was forced to discard the faulty items. The guilds also punished members who turned out shoddy products. Guilds maintained inspections and audits to ensure that artisans followed the quality specifications. Guild hierarchy consisted of three categories of workers: apprentice, the journeyman, and the master. Guilds had established specifications for input materials, manufacturing processes, and finished products, as well as methods of inspection and test. Guilds were active in managing the quality during Middle Ages until the Industrial Revolution reduced their influence.

The Industrial Revolution began in Europe in the mid-eighteenth century and gave birth to factories. The goals of the factories were to increase productivity and reduce costs. Prior to the Industrial Revolution, items were produced by individual craftsman for individual customers, and it was possible for workers to control the quality of their own products. Working conditions then were more conducive for professional pride. Under the factory system, the tasks needed to produce a product were divided up among several or many factory workers. In this system, large groups of workmen performed a similar type of work, and each group worked under the supervision of a foreman who also took on the responsibility to control the quality of the work performed. Quality in the factory system was ensured through skilled workers, and the quality audit was done by inspectors.

The broad economic result of the factory system was mass production at low cost. The Industrial Revolution changed the situation dramatically with the introduction of this new approach to manufacturing.

In the early nineteenth century, the approach to manufacturing in the United States tended to follow the craftsmanship model used in the European countries.

In the late nineteenth century, Fredrick Taylor's system of Scientific Management was born. Taylor's goal was to increase production. He achieved this by assigning planning to specialized engineers, and the execution of the job was left to the supervisors and workers. Taylor's emphasis on increasing production had a negative effect on quality. With this change in the production method, inspection of finished goods became the norm rather than inspection at every stage. To remedy the quality decline, factory managers created inspection departments having their own functional bosses. These departments were known as quality control departments.

The beginning of the twentieth century marked the inclusion of process in quality practices. During World War I, the manufacturing process became more complex. Production quality was the responsibility of quality control departments. The introduction of mass production and piecework created quality problems, as workmen were interested in increasing their earnings by producing more, which in turn led to bad workmanship. This led factories to introduce full-time quality inspectors, which marked the real beginning of inspection quality control and thus the beginning of quality control departments headed by superintendents. Walter Shewhart introduced statistical quality control in processes. His concept was that quality is not relevant for the finished product, but for the process that created the product. Shewhart's approach to quality was based on continuous monitoring of process variation. The statistical quality control concept freed the manufacturer from a time-consuming 100% quality control system because it accepted that variation is tolerable up to certain control limits. Thus quality control focus shifted from the end of the line to the process.

The systematic approach to quality in industrial manufacturing started during the 1930s when the cost of scrap and rework attracted attention. With the impact of mass production, which was required during World War II, it became necessary for manufacturing units to introduce a more stringent form of quality control. Called Statistical Quality Control (SQC), SQC made a significant contribution in that it provided a sampling inspection rather than a comprehensive inspection. This type of inspection, however, did lead to a lack of realization of the importance of engineering to product quality.

The concept and techniques of modern quality control were introduced in Japan immediately after World War II. The statistical and mathematical techniques, sampling tables, and process control charts emerged during this period.

From the early 1950s to the late 1960s, quality control evolved into quality assurance, with its emphasis on problem avoidance rather than problem detection. The quality assurance perspective suffered from a number of shortcomings as its focus was internal. Quality assurance was generally limited to those activities that were directly under the control of the organization; important activities such as transportation, storage, installation, and service were typically either ignored or given little attention. The quality assurance concept pays little or no attention to the competition's offerings. This resulted in integration of the quality actions on a companywide scale and application of quality principles in all the areas of business from design to delivery instead of confining the quality activities to production activities. This concept was called Total Quality Control and was popularized by Armand V. Feigenbaum, a quality guru from the United States.

From the foregoing brief overview and many other writings about the history of quality, it is evident that the quality system in its different forms has moved through distinct quality eras such as:

1. Quality Inspection
2. Quality Control
3. Quality Assurance
4. Total Quality

Introduction and promotion of companywide quality control led to a revolution in management philosophy. To help sell their products in international markets, the Japanese took some revolutionary steps to improve quality:

1. Upper-level managers personally took charge of leading the revolution.
2. All levels and functions received training in the quality disciplines.
3. Quality improvement projects were undertaken on a continuing basis at a revolutionary pace.

Thus, the concept of quality management started after Word War II, broadening into the development of initiatives that attempted to engage all employees in the systematic effort for quality. Quality emerged as a dominant thinking, becoming an integral part of an overall business system focused on customer satisfaction, which became known as Total Quality Management (TQM), with its three constitutive elements:

1. Total: Organizationwide
2. Quality: Customer Satisfaction
3. Management: Systems of Managing

TQM was stimulated by the need to compete in the global market, where higher quality, lower cost, and more rapid development are essential to market leadership. TQM was/is considered to be a fundamental requirement for any organization to compete, let alone lead its market. It is a way of planning, organizing, and understanding each activity of the process and removing all the unnecessary steps routinely followed in an organization. TQM is a philosophy that makes quality values the driving force behind leadership, design, planning, and improvement in activities. It acknowledges quality as a strategic objective and focuses on continuous improvement of products' processes, services, and cost, to compete in the global market by minimizing rework and maximizing profitability to achieve market leadership and customer satisfaction. It is a way of managing people and business processes to ensure customer satisfaction. TQM involves everyone in the organization in the effort to increase customer satisfaction and achieve superior performance of the products or services through continuous quality improvement. TQM helps in the following:

- Achieving customer satisfaction
- Continuous improvement
- Developing teamwork
- Establishing a vision for the employees
- Setting standards and goals for the employees
- Building motivation within the organization
- Developing a corporate culture

The TQM approach was developed immediately after World War II. There are prominent researchers and practitioners whose works have dominated the quality movement. Their ideas, concepts, and approaches in addressing specific quality issues have become part of the accepted wisdom in the field of quality, resulting in a major and lasting impact on the business. These persons are known as quality gurus. They all emphasize involvement of organizational management in quality efforts. These philosophers (gurus) are:

1. Philip B. Crosby
2. W. Edwards Deming
3. Armand V. Feigenbaum

4. Kaoru Ishikawa
5. Joseph M. Juran
6. John Oakland
7. Shigeo Shingo
8. Genichi Taguchi

Their approaches to quality emphasize customer satisfaction, management leadership, teamwork, continuous improvement, and minimizing defects. The common features of their philosophies can be summarized as follows:

1. Quality is conformance to the customer's defined needs.
2. Senior management is responsible for quality.
3. Continuously improve process, product, and services through application of various tools and procedures to achieve a higher level of quality.
4. Establish performance measurement standards to avoid defects.
5. Take a team approach by involving every member of the organization.
6. Provide training and education for everyone in the organization.
7. Establish leadership to help employees perform a better job.

TQM was considered a fundamental requirement for any industry to compete, let alone lead, in its market. The TQM methods were used until the end of the twentieth century. The concept and culture of the Integrated Quality Management System (IQM) emerged at the beginning of the twenty-first century. The IQM system integrates all the relevant systems in it for competitive advantages. Table 1.1 summarizes periodic changes in quality systems and the birth of IQMs.

1.2 Construction Projects

A project is a temporary endeavor undertaken to create a unique product or service. Temporary means that every project has a definite beginning and a definite end. Unique means that the product or service is different in some distinguishing way from all similar products or services. Projects are often critical components of the performing organization's business strategy. Examples of projects include:

- Developing a new product or service
- Effecting a change in structure, staffing, or style of an organization

TABLE 1.1

Periodic Changes in Quality System

Period	System
Middle Ages (1200–1799)	Guilds—Skilled craftsman were responsible for controlling their own products.
Mid-18th century (Industrial Revolution)	Establishment of factories. Increase in productivity. Mass production. Assembly lines. Several workers were responsible for producing a product. Production by skilled workers and quality audit by inspectors.
Early 19th century	Craftsmanship model of production.
Late 19th century (1880s)	Fredrick Taylor and Scientific Management. Quality management through Inspection.
Beginning of 20th century (1920s)	Walter Shewhart introduced Statistical Process Control. Introduction of full-time Quality Inspection and quality control department. Quality management.
1930s	Introduction of the sampling method.
1950s	Introduction of Statistical Quality Process in Japan.
Late 1960s	Introduction of Quality Assurance.
1970s	Total Quality Control. Quality Management.
1980s	Total Quality Management (TQM).
Beginning of 21st century	Integrated Quality Management (IQM).

Source: Abdul Razzak Rumane. (2010). *Quality Management in Construction Projects*. CRC Press, Boca Raton, FL. Reprinted with permission of Taylor & Francis Group.

- Designing a new transportation vehicle/aircraft
- Developing or acquiring a new or modified information system
- Running a campaign for political office
- Implementation of a new business procedure or process
- Constructing a building or facilities

The duration of a project is finite; projects are not ongoing efforts, and the project ceases when its declared objectives have been attained. Some of the characteristics of projects, for example, are the following:

1. They are implemented by people.
2. They are constrained by limited resources.
3. They are planned, executed, and controlled.

Based on various definitions, the project can be defined as follows:

> A project is a plan or program performed by people who have been assigned resources to achieve an objective within a finite duration.

Construction has a history of several thousand years. The first shelters were built from stone or mud and the materials collected from the forests to provide protection against cold, wind, rain, and snow. These buildings were primarily for residential purposes, although some may have had a commercial function.

During the New Stone Age, people introduced dried bricks, wall construction, metal working, and irrigation. Gradually, people developed the skills to construct villages and cities, and considerable skills in building were acquired. This can be seen from the great civilizations in different parts of the world some 4000–5000 years ago. During the Greek settlements, which dates back to about 2000 BCE, buildings were made of mud using timber frames. Later, temples and theaters were built from marble. Some 1500–2000 years ago, Rome became the leading center of world culture, which extended to construction.

Marcus Pollo (1 BCE) was a military and civil engineer from Rome who published books. These were the world's first major publications on architecture and construction, and they dealt with building materials, the styles and design of building types, the construction process, building physics, astronomy, and building machines.

During the Medieval Age (476–1492), improvements in agriculture, artisanal productivity, and exploration, as well as a consequent broadening of commerce took place, and in the late Middle Ages building construction became a major industry. Craftsmen were given training and education so that they could develop skills and raise their status. At this time, guilds were responsible for managing quality.

The fifteenth century ushered in a renaissance or renewal in architecture, building, and science. Significant changes occurred during the seventeenth century and thereafter, owing to the increasing transformation of construction and the urban habitat.

The scientific revolution of the seventeenth and eighteenth centuries gave birth to the great Industrial Revolution of the eighteenth century. After some delay, construction followed these developments in the nineteenth century.

In the first half of the twentieth century, the construction industry became an important sector throughout the world, employing many workers. During this period, skyscrapers, long-span dams, shells, and bridges were developed to satisfy new requirements; these developments marked the continuing progress of construction techniques. The provision of services such as heating, air conditioning, electrical lighting, mains water, and elevators in buildings became common. The twentieth century saw the transformation of the construction and building industry into a major economic sector. During the second half of the twentieth century, the construction industry began to industrialize, introducing mechanization, prefabrication, and system building. The design of building services systems changed considerably in the last twenty years of the twentieth century. It became the responsibility of the designer to conform to health, safety, and environmental regulations while designing any building.

Building and Commercial, the traditional A/E type of construction projects, account for an estimated 25 percent of the annual construction volume.

Building construction is a labor-intensive endeavor. Every construction project has some elements that are unique. No two construction or R&D projects are alike. Though it is clear that construction projects are usually more routine than R&D projects, some degree of customization is a characteristic of these projects.

Figure 1.1 illustrates several types of projects.

Construction projects comprise a cross section of many different participants. These participants both influence and depend on each other

<table>
<tr><td>1</td><td colspan="4">Process Type Projects</td></tr>
<tr><td>1.1</td><td colspan="4">Liquid chemical plants</td></tr>
<tr><td>1.2</td><td colspan="4">Liquid/solid plants</td></tr>
<tr><td>1.3</td><td colspan="4">Solid process plants</td></tr>
<tr><td>1.4</td><td colspan="4">Petrochemical plants</td></tr>
<tr><td>1.5</td><td colspan="4">Petroleum refineries</td></tr>
<tr><td>2</td><td colspan="4">Non-Process Type Projects</td></tr>
<tr><td>2.1</td><td>Power plants</td><td colspan="3" rowspan="4"></td></tr>
<tr><td>2.2</td><td>Manufacturing plants</td></tr>
<tr><td>2.3</td><td>Support facilities</td></tr>
<tr><td>2.4</td><td>Miscellaneous (R&D) projects</td></tr>
<tr><td>2.5</td><td>Civil construction projects</td><td rowspan="4">Categories of civil construction projects and Commercial A/E projects</td><td>Residential construction</td><td>Family homes, multi unit town houses, garden apartments, high-rise apartments, villas</td></tr>
<tr><td>2.6</td><td>Commercial A/E projects</td><td>Building construction (institutional and commercial)</td><td>Schools, universities, hospitals, commercial office complexes, theaters, government buildings, warehouses, recreation centers, holiday resorts, neighborhood centers</td></tr>
<tr><td colspan="2" rowspan="2"></td><td>Industrial construction</td><td>Petroleum refineries, petroleum plants, Power plants, heavy manufacturing plants</td></tr>
<tr><td>Heavy engineering</td><td>Dams, tunnels, bridges, highways, railways, airports, urban rapid transit system, ports, harbors, water treatment and distribution, sewage and storm water collection, treatment, and disposal system, power lines, and communication networks</td></tr>
</table>

FIGURE 1.1
Types of projects.

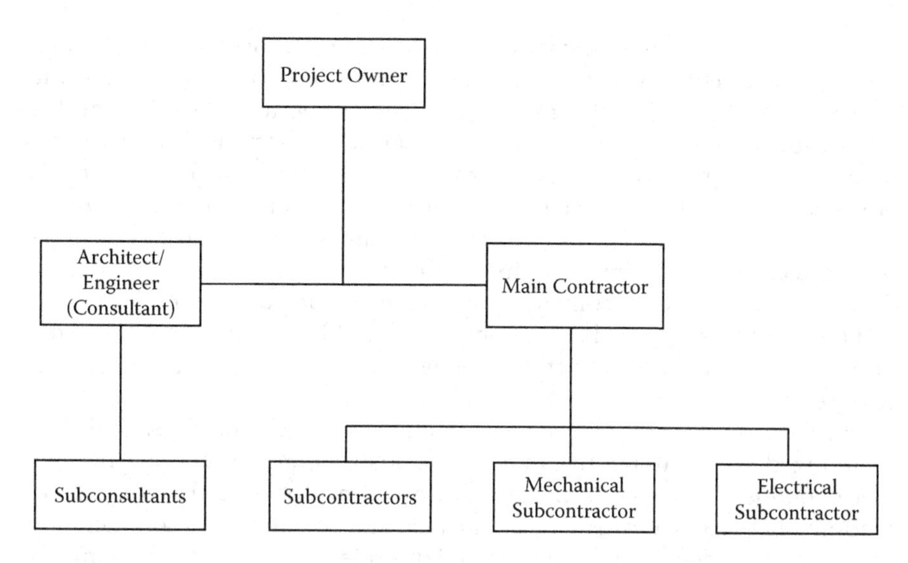

FIGURE 1.2
Concept of traditional projects organization.

in addition to other players involved in the construction process. Figure 1.2 illustrates the concept of traditional construction project organization.

Traditional construction projects involve three main groups:

1. Owner—A person or an organization that initiates and sanctions a project. He/she requests the need for the facility and is responsible for arranging the financial resources for creation of the facility.
2. Designer (A/E)—Architects or engineers or consultants are the entities appointed by the owners and who are accountable for converting the owner's conception and needs into a specific facility with detailed directions through drawings and specifications, within the economic objectives. They are responsible for the design of the project and in certain cases supervision of the construction process.
3. Contractor—A construction firm engaged by the owner to complete the specific facility by providing the necessary staff, workforce, materials, equipment, tools, and other accessories to the satisfaction of the owner/end user in compliance with the contract documents. The contractor is responsible for implementing the project activities and achieving the owner's objectives.

Construction projects are executed based on a predetermined set of goals and objectives. In traditional construction projects, the owner heads the team, designating a project manager. The project manager is a person/member of the owner's staff or an independently hired person/firm who has the overall or principal responsibility for the management of the project as a whole.

In certain cases, owners engage a professional firm, called the Construction Manager, who is trained in the management of construction processes, to assist in developing bid documents and overseeing and coordinating the project for the owner. The basic construction management concept is that the owner assigns a contract to a firm that is knowledgeable and capable of coordinating all the aspects of the project to meet the intended use of the project by the owner. In construction management type of construction projects, the consultant (Architect/Engineer) prepares complete design drawings and contact documents, then the project is put for competitive bid and the contact is awarded to the competitive bidder (contractor). The owner hires a third party (Construction Manager) to oversee and coordinate the construction.

Construction projects are mainly capital investment projects. They are customized and nonrepetitive in nature. Construction projects have become more complex and technical, and the relationships and the contractual grouping of those who are involved are also more complex and contractually varied. The products used in construction projects are expensive, complex, immovable, and long-lived. Generally, a construction project is composed of building materials (civil), electromechanical items, finishing items, and equipment. These are normally produced by other construction-related industries/manufacturers. These industries produce products as per their own quality management practices complying with certain quality standards or against specific requirements for a particular project. The owner of the construction project or his representative has no direct control over these companies unless her/his representative/appointed contractor commits to buying their product for use in their facility. These organizations may have their own quality management program. In manufacturing or service industries, the quality management of all in-house manufactured products is performed by the manufacturer's own team or under the control of the same organization having jurisdiction over their manufacturing plants at different locations. Quality management of vendor-supplied items/products is carried out as stipulated in the purchasing contract as per the quality control specifications of the buyer.

1.3 Quality Definition for Construction Projects

Quality has different meanings to different people. The definition of quality relating to manufacturing, processes, and service industries is as follows:

- Meeting the customer's need
- Customer satisfaction
- Fitness for use

- Conforming to requirements
- Degree of excellence at an acceptable price

The International Organization for Standardization (ISO) defines quality as "the totality of characteristics of an entity that bears on its ability to satisfy stated or implied needs."

However, the definition of quality for construction projects is different from that of manufacturing or service industries, as the product is not repetitive, but a unique piece of work with specific requirements.

Quality in construction projects is not only the quality of the product and equipment used in the construction of facility, but it is the total management approach to completing the facility. The quality of construction depends mainly on the control of construction, which is the primary responsibility of the contractor.

Quality in manufacturing passes through a series of processes. Material and labor are input through a series of processes out of which a product is obtained. The output is monitored by inspection and testing at various stages of production. Any nonconforming product identified is either repaired, reworked, or scrapped, and appropriate steps are taken to eliminate problem causes. Statistical process control methods are used to reduce the variability and to increase the efficiency of processes. In construction projects, the scenario is not the same. If anything goes wrong, the nonconforming work is very difficult to rectify, and remedial action is sometimes not possible.

Quality management in construction projects is different from that of manufacturing. Quality in construction projects is not only the quality of products and equipment used in the construction, it is the total management approach to completing the facility as per the scope of work to customer/owner satisfaction within the budget and within the specified schedule to meet the owner's defined purpose. The nature of the contracts between the parties plays a dominant part in the quality system required from the project, and the responsibility for achieving them must therefore be specified in the project documents. The documents include plans, specifications, schedules, bill of quantities, and so on. Quality control in construction typically involves ensuring compliance with minimum standards of material and workmanship in order to ensure the performance of the facility according to the design. These minimum standards are contained in the specification documents. For the purpose of ensuring compliance, random samples and statistical methods are commonly used as the basis for accepting or rejecting work completed and batches of materials. Rejection of a batch is based on nonconformance or violation of the relevant design specifications.

Based on the foregoing, the quality of construction projects can be defined as follows: Construction project quality is the fulfillment of the owner's needs as per the defined scope of work within the budget and specified schedule to satisfy the owner's/user's requirements. These three components can be called the Construction Project Trilogy, which is illustrated in Figure 1.3.

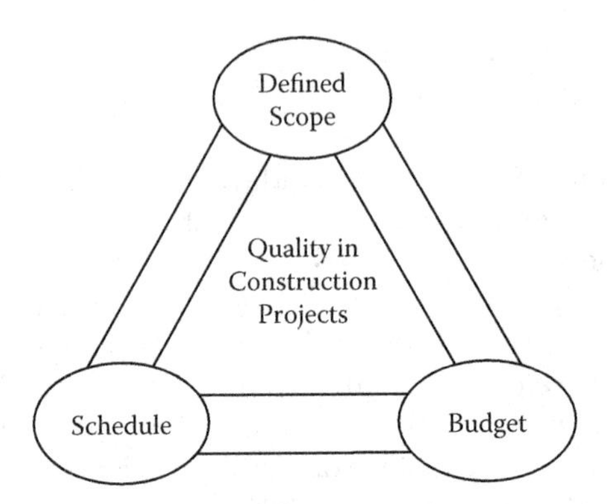

FIGURE 1.3
Construction Project Trilogy.

Thus, quality of construction projects can be evolved as follows:

1. Properly defined scope of work.
2. Owner, project manager, design team leader, consultant, and construction manager are responsible for implementing quality.
3. Continuous improvement can be achieved at different levels as follows:
   a. Owner—Specify the latest needs.
   b. Designer—Specification to include the latest quality materials, products, and equipment.
   c. Constructor—Use the latest construction equipment to build the facility.
4. Establishment of performance measures:
   a. Owner—(I) To review and ensure that the designer has prepared contract documents that satisfy his needs
      II. To check the progress of work to ensure compliance with the contract documents
   b. Consultant—(I) As a consultant designer, to include the owner's requirements explicitly and clearly defined in the contract documents
      II. As a supervision consultant, supervise the contractor's work as per contract documents and the specified standards
   c. Contractor—To construct the facility as specified and use the materials, products, and equipment that satisfy the specified requirements
5. Team Approach—Every member of the project team should be familiar with the principles of TQM, recognizing that TQM is a collaborative effort. Everybody should participate in all the functional

areas to improve the quality of the project works. They should know that achieving project quality is a collective effort by all the participants.

6. Training and Education—The consultant and contractor should have customized training plans for their management, engineers, supervisors, office staff, technicians, and laborers.
7. Establish Leadership—Organizational leadership should be established to achieve the specified quality. Encourage and help the staff and laborers to understand the quality to be achieved for the project.

These definitions when applied to construction projects relate to the contract specifications or owner/end user requirements to be constructed in such a way that construction of the facility is suitable for the owner's use or it meets the owner's requirements. Quality in construction is achieved through complex interaction of many participants in the facility's development process.

The quality plan for construction projects is part of the overall project documentation consisting of:

1. Well-defined specifications for all the materials, products, components, and equipment to be used to construct the facility
2. Detailed construction drawings
3. Detailed work procedure
4. Details of the quality standards and codes to be compiled
5. Cost of the project
6. Manpower and other resources to be used for the project
7. Project completion schedule

Participation and involvement of all three parties at different levels of the construction phases is required to develop a quality system and apply quality tools and techniques. With the application of various quality principles, tools, and methods by all the participants at different stages of a construction project, rework can be reduced, resulting in savings in the project cost and making the project qualitative and economical. This will ensure completion of construction and make the project optimally qualitative, competitive, and economical. Table 1.2 illustrates the quality principles of construction projects.

1.4 Construction Project Life Cycle

Most construction projects are custom oriented, having a specific need and a customized design. It is always the owner's desire that his project should be unique and better. Further, it is the owner's goal and objective that

TABLE 1.2

Principles of Quality in Construction Projects

Principle	Construction Project's Principle
Principle 1	Owner, consultant, contractor are fully responsible for application of quality management system to meet defined scope of work in the contract documents.
Principle 2	Consultant is responsible to provide owner's requirements explicitly, clearly defining them in the contract documents.
Principle 3	Method of payments (work progress, material, equipment, etc.) to be clearly defined in the contract documents. Rate analysis of BOQ or BOM item to be agreed before signing of contract.
Principle 4	Contract documents should include a clause to settle any dispute arising during the construction stage.
Principle 5	Contractor should study all the documents during tendering/bidding stage and submit his proposal, taking into consideration all the requirements specified in the contract documents.
Principle 6	Contractor shall follow an agreed-upon quality assurance and quality control plan. Consultant shall be responsible for overseeing compliance with contract documents and specified standards.
Principle 7	Contractor is responsible for providing all the resources, manpower, material, equipment, etc., to build the facility per specifications.
Principle 8	Contractor shall follow the submittal procedure specified in the contract documents.
Principle 9	Each member of project team should participate in all the functional areas to continuously improve project quality.
Principle 10	Contractor is responsible for constructing the facility as specified and using the material, products, equipment, and methods that satisfy the specified requirements.
Principle 11	Contractor to build the facility as stipulated in the contract documents, plan and specifications within budget and on schedule to meet Owner's objectives.
Principle 12	Contractor should perform the works per agreed-upon construction program and hand over the project per the contracted schedule.

the facility is completed on time. The expected time schedule is important from both financial and acquisition of the facility by the owner/end user.

The system life cycle is fundamental to the application of systems engineering. The systems engineering approach to construction projects helps one understand the entire process of project management. This approach allows for management and control of activities at different levels of various phases to ensure timely completion of the project with economical use of resources to make the construction project most qualitative, competitive, and economical.

Systems engineering starts from the complexity of the large-scale problem as a whole and moves toward structural analysis and partitioning processes until the questions of interest are answered. This process of decomposition is called a Work Breakdown Structure (WBS). The WBS is a hierarchical representation of system levels. Similar to a family tree, the WBS consists of a number of levels, starting with the complete system at level 1 at the top and progressing downward through as many levels as necessary to obtain elements that can be conveniently managed.

Benefits of systems engineering applications are:

- Reduction in cost of system design and development, production/construction, system operation and support, system retirement, and material disposal
- Reduction in system acquisition time
- More visibility and reduction in the risks associated with the design decision-making process

It is difficult to generalize the concepts of the project life cycle to the system life cycle. However, considering that there are innumerable processes that make up a construction project, the technologies and processes applied to systems engineering can also be applied to construction projects. The number of phases depends on the complexity of the project. The duration of each phase may vary from project to project. Generally, construction projects have five common phases. These are as follows:

1. Conceptual Design
2. Schematic Design
3. Design Development
4. Construction
5. Testing, Commissioning, and Handover

Each phase can further be subdivided using the WBS principle to reach a level of complexity where each element/activity can be treated as a single unit that can be conveniently managed. WBS represents a systematic and logical breakdown of the project phase into its components (activities). It is constructed by dividing the project into major elements with each of these being divided into subelements. This is done until a breakdown is achieved in terms of manageable units of work for which responsibility can be defined. WBS involves envisioning the project as a hierarchy of goals, objectives, activities, subactivities, and work packages. The hierarchical decomposition of activities continues until the entire project is displayed as a network of separately identified and nonoverlapping activities, Each activity will be single purposed, of a specific time duration, and manageable; its time and cost estimates easily derived, deliverables clearly understood, and responsibility for its completion clearly assigned. The Work Breakdown Structure helps in

- Effective planning by dividing the work into manageable elements that can be planned, budgeted, and controlled
- Assignment of responsibility for work elements to project personnel and outside agencies
- Development of control and information systems

WBS facilitates the planning, budgeting, scheduling, and control activities for the project manager and his team. By application of WBS, the construction phases are further divided into various activities. This division will improve the control and planning of the construction project at every stage before a new phase starts. The components/activities of construction project life-cycle phases divided on the basis of the WBS principle are listed below:

1. Conceptual Design
    - Identification of Need and Objectives (TOR: Terms of Reference)
    - Identification of Project Team
    - Data Collection
    - Identification of Alternatives
    - Time Schedule
    - Financial Implications/Resources
    - Development of Concept Design
2. Schematic Design
    - General Scope of Works/Basic Design
    - Regulatory/Authorities Approval
    - Schedule
    - Budget
    - Contract Terms and Conditions
    - Value Engineering Study
3. Design Development
    - Detail Design of the Works
    - Regulatory/Authorities Approval
    - Contract Documents and Specifications
    - Detailed Plan
    - Budget
    - Estimated Cash Flow
    - Tender/Bid Documents
4. Construction
    - Mobilization
    - Execution of Works

- Planning and Scheduling
- Management of Resources/Procurement
- Monitoring and Control
- Quality
- Inspection

5. Testing, Commissioning, and Handover
   - Testing
   - Commissioning
   - Regulatory/Authorities Approval
   - As-Built Drawings/Records
   - Technical Manuals and Documents
   - Training of User's Personnel
   - Hand Over Facility to Owner/End User
   - Move-in-Plan
   - Substantial Completion

Table 1.3 illustrates subdivided activities/components of the construction project life cycle.

These activities may not be strictly sequential; however, the breakdown allows implementation of project management functions more effectively at different stages.

1.5 Quality Standards

A quality system is a framework for quality management. It embraces the organizational structure, procedure, and processes needed to implement quality management. The adequacy of a quality system, quality of products, services, and process are judged by their compliance to specified/relevant standards. Standards have important economic and social repercussions. They are useful to industrial and business organizations of all types, to government and other regulatory bodies, to conformity assessment professionals, to suppliers, to customers of products and services in both the public and private sector, and to people in general in their role as customers and users. Standards provide governments with a technical base for health, safety, and environmental legislation.

A standard is simply a definition of how something should be.

Standards are documents used to define acceptable conditions or behaviors and to provide a baseline for assuring that conditions or behaviors meet the acceptable criteria. In most cases, standards define minimum criteria; world-class quality is, by definition, beyond the standard level of performance.

TABLE 1.3

Construction Project Life Cycle

Conceptual Design	Schematic Design	Design Development	Construction	Testing, Commissioning, and Handover
• Identification of Need and Objectives (TOR)	• General Scope of Work/Basic Design	• Detail Design of the Works	• Mobilization	• Testing
• Identification of Project Team	• Regulatory Approval	• Regulatory/Authorities Approval	• Execution of Works	• Commissioning
• Data Collection	• Schedule	• Contract Documents and Specifications	• Planning and Scheduling	• Regulatory/Authorities Approval
• Identification of Alternatives	• Budget	• Detail Plan	• Management of Resources/Procurement	• Move-in-Plan
• Time Schedule	• Contract Terms and Conditions	• Budget	• Monitoring and Control	• As-built Drawings/Records
• Financial Implications/Resources	• Value Engineering Study	• Estimated Cash Flow	• Quality	• Technical Manuals and Documents
• Development of Concept Design		• Tender/Bidding	• Inspection	• Training of User's Personnel
				• Handover of Facility to Owner/End User
				• Substantial Completion

Source: Abdul Razzak Rumane. (2010). *Quality Management in Construction Projects*. CRC Press, Boca Raton, FL. Reprinted with permission of Taylor & Francis Group.

Standards can be written or unwritten, voluntary or mandatory. Unwritten quality standards are generally not acceptable.

Standard setting is one of the first issues in developing a quality assurance system, and increasingly organizations are relying on readily available standards rather than developing their own. Each standard should be:

- Clearly written in simple language that is unambiguous
- Convenient to understand
- Specific in setting out precisely what is expected
- Measurable, so that the organization can know whether or not it is being met
- Achievable, that is, the organization must have the resources available to meet the standard
- Constructible

There are many organizations that produce standards; some of the best-known organizations in the quality field are:

1. International Organization for Standardization (ISO)
2. International Electrotechnical Commission (IEC)
3. American Society for Quality (ASQ)
4. American National Standards Institute (ANSI)
5. American Society for Testing and Materials (ASTM)
6. American Standards for Mechanical Engineering (ASME)
7. Institute of Electrical and Electronic Engineers (IEEE)
8. European Committee for Standardization (CEN)
9. European Committee for Electrotechnical Standardization (CENELEC)
10. American Society of Heating, Refrigerating, and Air Conditioning Engineers (ASHRAE)
11. National Fire Protection Association (NFPA)
12. British Standard Institute

Standards produced by these organizations/institutes are recognized worldwide. These standards are referred to in the contract documents created by the designers to specify products, systems, or services to be used in a project. They are also used to specify the installation method to be followed or the fabrication works to be performed during the construction process.

Apart from these, there have been many other national and international quality system standards. These various standards have commonalities and historical linkages. However, in order to facilitate international trade, delegates from 25 countries met in London in 1946 to create a new international organization. The objective of this organization was to facilitate international coordination and unification of industrial standards. Thus, the new organization, International Organization for Standardization (ISO), officially began operation on 23 February 1947.

ISO is a network of national standards institutes of 164 countries (as of December 2012), on the basis of one member per country, with a Central Secretariat in Geneva, Switzerland, that coordinates the system.

ISO is the world's largest developer and publisher of international standards. It is a nongovernmental organization that forms a bridge between the public and private sectors. ISO has more than 19,000 international standards. Of all the standards produced by ISO, the ones that are most widely known are those of ISO 9000 and ISO 14000 series. ISO 9000 has become an international reference for quality requirements in business-to-business dealings, and ISO 14000 looks set to achieve at least as much, if not more, in helping organizations to meet their environmental commitments. ISO 9000 and ISO 14000 families are known as "generic management system standards."

The ISO 9000 family is primarily concerned with "quality management." This means what an organization does to fulfill:

- The customers' quality requirements
- Applicable regulatory requirements, while aiming to enhance customer satisfaction
- Continual improvement of its performance in pursuit of the objectives

The ISO 14000 family deals primarily with "environmental management." This means what organizations do to:

- Minimize harmful effects on the environment caused by its activities
- Achieve continual improvement of its environmental performance

1.6 Quality Management System

ISO 9000 quality system standards are tested frameworks for taking a systematic approach to managing the business process so that the organizations turn out products or services conforming to the customer's satisfaction. The typical ISO quality management system is structured on four levels, usually portrayed as a pyramid.

On top of the pyramid is the quality policy, which sets out what management requires its staff to do in order to ensure the success of the quality management system. Beneath the policy is the quality manual, which details the work to be done. Beneath the quality manual are work instructions or procedures. The number of manuals containing work instructions or procedures is determined based on the size and complexity of the organization. The procedures mainly address the following questions:

- What is to be done?
- How is it done?
- How does one know that it has been done properly (for example, by inspecting, testing, or measuring)?
- What is to be done if there are problems (for example, failure)?

The bottom level of the hierarchy contains forms and records that are used to capture the history of routine events and activities.

The ISO 9000 quality management system requires documentation that includes a quality manual and quality procedures, as well as work instructions and quality records. All documentation (including quality records) must be controlled according to a document control procedure. The structure of the quality management system depends largely on the management structure in the organization.

ISO 9001:2008 identifies certain minimum requirements that all quality management systems must meet to ensure customer satisfaction. ISO 9001:2008 specifies requirements for the quality management system when an organization:

- Needs to demonstrate its ability to consistently provide products that meets customer needs and applicable regulatory requirements
- Aims to enhance customer satisfaction through the effective application of the system, including processes for continual improvement of the system and the assurance of conformity to customer needs and applicable regulatory requirements.

A quality system has to cover all the activities leading to creation of the final product or service. The quality system depends entirely on the scope of operation of the organization and particular circumstances, such as the number of employees, type of organization, and the physical size of the premises of the organization. The quality manual is the document that identifies and describes the quality management system:

1. Identify the process (activities and necessary elements) needed for the quality management system
2. Determine the sequence and interaction of these processes and how they fit together to accomplish quality goals
3. Determine how these processes are effectively operated and controlled
4. Measurement, monitoring, and analysis of these processes and action necessary to implement to correct the process and achieve continual requirement

5. All information that is available to support the operation and monitoring of the process
6. Options to help create the right management system

ISO 9001:2000 requirements fall into the following five sections:

1. Quality Management System
2. Management Responsibility
3. Resource Management
4. Product Realization
5. Measurement Analysis and Improvement

In the construction industry, a contractor may be working at any time on a number of projects of a varied nature. These projects have their own contract documents to implement project quality that require the contractor to submit the contractor's quality control plan. This plan ensures that the specific requirements of the project are considered to meet the client's requirements. Therefore, while preparing a quality management system at the corporate level, the organization has to take into account tailor-made requirements for the projects and the manual should be prepared accordingly.

1.7 Integrated Quality Management

An Integrated Quality Management System (IQMS) is the integration and proper coordination of functional elements of quality to achieve efficiency and effectiveness in implementing and maintaining an organization's quality management system to meet customer requirements and satisfaction. An IQMS consists of any element or activity that has an effect on quality. Customer satisfaction is the goal of quality objectives.

During the past three decades, many programs have been implemented for organizational improvement. In 1980s, programs such as statistical process control, various quality tools, and TQM were implemented. In the 1990s, the ever-popular ISO 9000 came into being, which resulted in improved productivity, cost reduction, improved time, improved quality, and customer satisfaction.

With globalization and competition, it became necessary for organizations to improve continuously to achieve the highest performance and competitive advantage.

In the 1980s, the major challenge facing most organizations was to improve quality. In the 1990s, it was to improve faster by restructuring and reengineering all operations.

In today's global competitive environment, organizations are facing many challenges due to increased customer demand for higher performance requirements at a competitive cost. They are finding that their survival in the competitive market is increasingly in doubt. To achieve competitive advantage, effective quality improvement is critical.

Processes and systems are essential for the performance and expansion of any organization. ISO 9000 is an excellent tool to develop a strong foundation for good processes and systems. The ISO 9000 quality management system is accepted worldwide, and ISO 9000 certification has global recognition.

An Integrated Quality Management System (IQMS) is developed by merging recommendations and specifications from ISO 9000 (Quality Management System), ISO 14000 (Environmental Management System), and OHSAS 18000 (Occupational Health and Safety Management) together with other contract documents. If an organization has a certified Quality Management System (ISO 9000), it can build an IQMS system by adding environmental, health, safety, and other requirements of management system standards.

The benefits of implementing an IQMS are:

- Reduced duplication and therefore cost
- Improved resource allocation
- Standardized process
- Elimination of conflicting responsibilities and relationships
- Consistency
- Improved communication
- Reduced risk and increased profitability
- Facilitation of training development
- Simplification of document maintenance
- Reduced record keeping
- Ease of managing legal and other requirements

Construction projects are unique and nonrepetitive in nature and have their own quality requirements, which can be developed by integration of project specifications and the organization's quality management system. Normally, quality management system manuals consist of procedures to develop a project quality control plan, taking into consideration contract specifications. This plan is called the Contractor's Quality Control Plan (CQCP). Certain projects specify that value engineering studies be undertaken during the construction phase. The contractor is required to include these studies while developing the CQCP. This plan can be termed an Integrated Quality Management System (IQMS) for construction projects. The contractor has to implement a quality system to ensure that the construction is carried out in accordance with the specification details and approved COQP. Figure 1.4 illustrates the logic flow diagram for the development of an IQMS for construction projects.

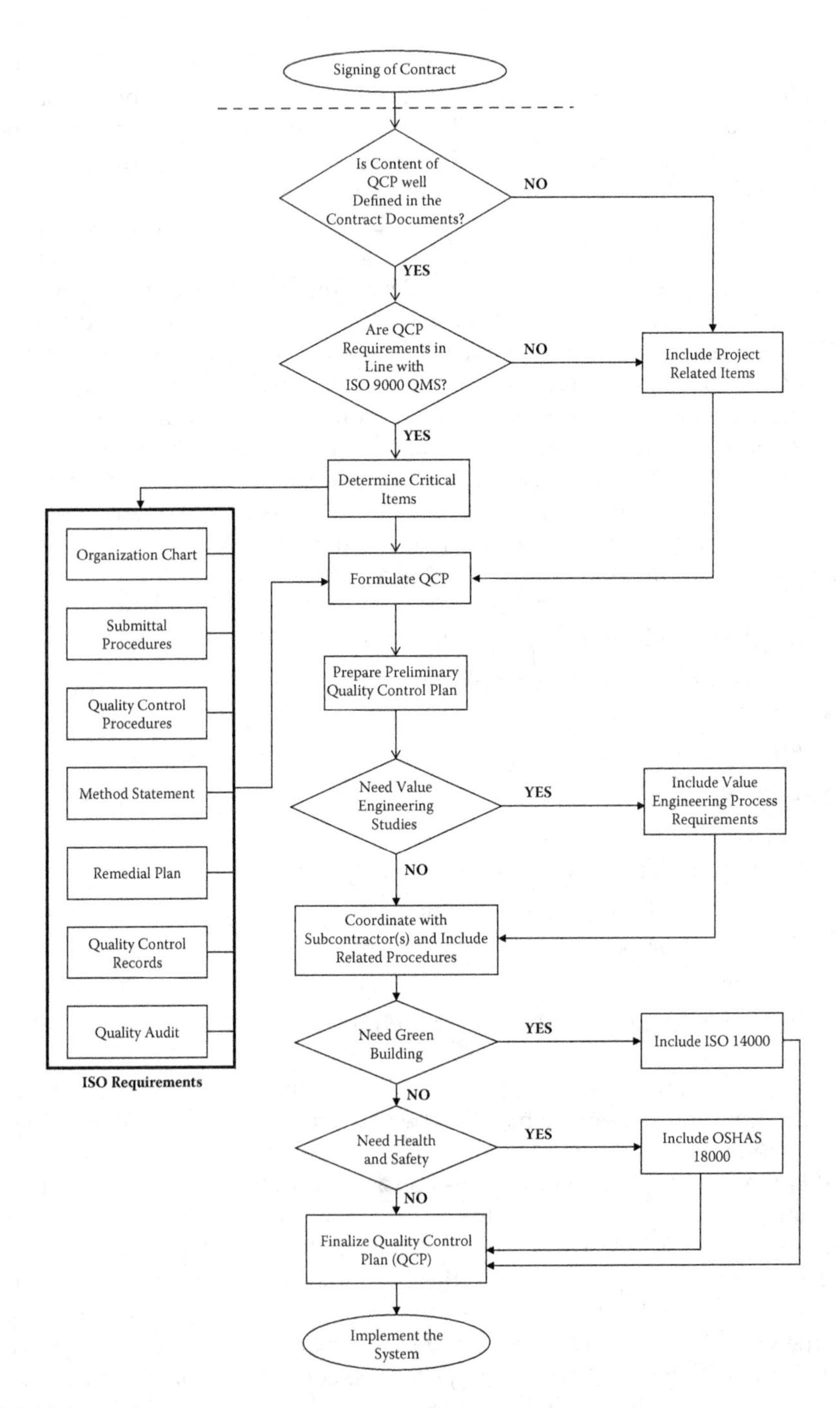

FIGURE 1.4
Logic flow diagram for development of IQMS. (Source: Abdul Razzak Rumane (2010). *Quality Management in Construction Projects*. CRC Press, Boca Raton. With permission.)

2

Quality Tools

2.1 Introduction

Quality tools are the charts, check sheets, diagrams, graphs, techniques, and methods used to create an idea, engender planning, analyze the cause, analyze the process, foster evaluation, and create a wide variety of situations for continuous quality improvement. Applications of tools enhance chances of success, help maintain consistency and accuracy, and improve efficiency and process improvement.

In practice, there are several types of tools, techniques, and methods that are used as quality improvement tools and have a variety of applications in manufacturing and the process industry. However, not all of these tools are used in construction projects because these projects are customized and nonrepetitive. Some of the most commonly used quality management tools in the construction industry are listed under the following broad categories:

1. Quality classic tools
2. Management and planning tools
3. Process analysis tools
4. Process improvement tools
5. Innovation and creative tools
6. Lean tools
7. Cost of quality
8. Quality function deployment
9. Six Sigma
10. TRIZ

A brief description of these tools is given below.

2.2 Quality Classic Tools

Quality classic tools have a long history. These tools are listed in Figure 2.1.

All of these tools have been in use since World War II. Some of them date back to earlier than 1920. The approaches include both quantitative and qualitative aspects, which taken together, focus on companywide quality.

The remainder of this section provides brief definitions of these quality tools (values shown in the figures are only provided for example purposes).

2.2.1 Cause and Effect Diagram

A cause and effect diagram is also called an Ishikawa diagram, after its developer Kaoru Ishikawa, or a fishbone diagram. It is used to identify possible causes and effects in processes. It is used to explore all the potential or real causes that result in a single output. The causes are organized and displayed in a graphical manner to their level of importance or details. The diagram is a graphical display of multiple causes with a particular effect. The causes are organized and arranged mainly into four categories. These are

Sr. No.	Name of Quality Tool	Usage
Tool 1	Cause and effect diagram	To identify possible causes and effects in processes.
Tool 2	Check sheet	To provide a record of quality. How often does it occur?
Tool 3	Control chart	A device in Statistical Process Control to determine whether or not the process is stable.
Tool 4	Flowchart	Used for graphical representation of a process in sequential order.
Tool 5	Histogram	Graphs used to display frequencies of various ranges of values of a quantity.
Tool 6	Pareto chart	Used to identify the most significant cause or problem.
Tool 7	Pie chart	Used to show classes or groups of data in proportion to the whole set of data.
Tool 8	Run chart	Used to show measurement against time in a graphical manner with a reference line to show the average of the data.
Tool 9	Scatter diagram	Used to determine whether there is a correlation between two factors.
Tool 10	Stratification	Used to show the pattern of data collected from different sources.

FIGURE 2.1
Quality classic tools.

1. Machine
2. Manpower
3. Material
4. Method

The effect or problem being investigated is shown at the end of the horizontal arrow. Potential causes are shown as labeled arrows entering the main cause arrow. Each arrow may have a number of other arrows entering it, as the principal cause or factors are reduced to their subcauses. Figure 2.2 illustrates an example cause and effect diagram for rejection of executed (installed) false ceiling works by the supervision engineer for not complying with contract specifications.

2.2.2 Check Sheet

A check sheet is a structured list, prepared from the collected data to indicate how often each item occurs. It is an organized way of collecting and structuring data. The purpose of a check sheet is to collect the facts in the most efficient manner. Data is collected and ordered by adding tally or

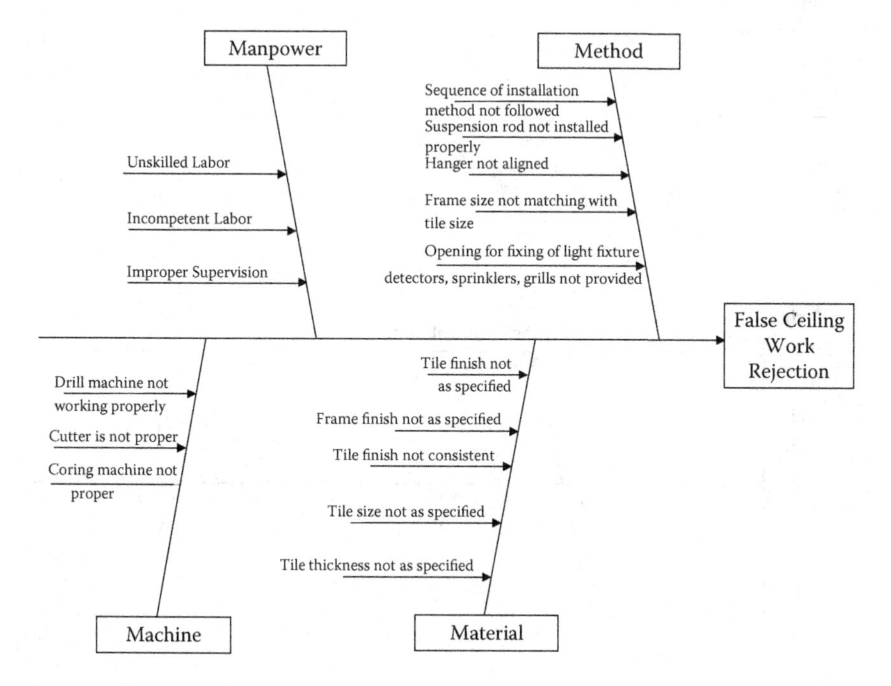

FIGURE 2.2
Cause and effect diagram.

check marks against predetermined categories of items or measurements. Figure 2.3 illustrates a check sheet for a checklist approval record for wire bundles.

2.2.3 Control Chart

The control chart is the fundamental tool of statistical process control. It is a graph used to analyze how a process behaves over time and to show whether it is stable or is being affected by a special cause of variation and creating an out-of-control condition. It is used to determine whether the process is stable or varies between predictable limits. It can be employed to distinguish between the existence of a stable pattern of variation and the occurrence of an unstable pattern. With control charts, it is easy to see both special and common cause variation in a process. There are many types of control charts. Each is designed for a specific kind of process or data. Figure 2.4 illustrates the control chart for the distribution of air handling unit (AHU) air.

Approval Record for Wire Bundles

	Approved	Not Approved	Total	% Not Approved
1.5 mm^2 Wire	48	2	50	4
2.5 mm^2 Wire	80	5	85	6
4.0 mm^2 Wire	24	1	25	4
6.0 mm^2 Wire	15	0	15	0
10.0 mm^2 Wire	9	1	10	1

FIGURE 2.3
Check sheets.

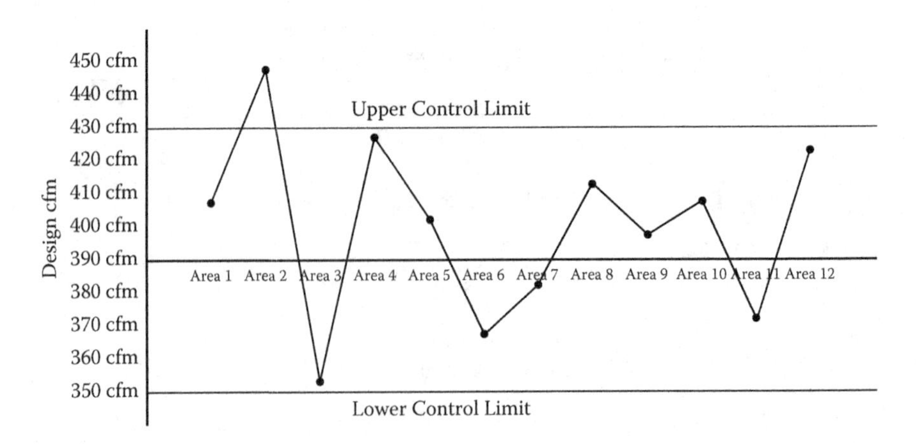

FIGURE 2.4
Control chart.

2.2.4 Flowchart

The flowchart is a pictorial tool that is used for representing processes in sequential order. A flowchart uses graphic symbols to depict the nature and flow of the steps in a process. It helps one see whether the steps of the process are logical, uncover problems or miscommunications, define the boundaries of a process, and develop a common base of knowledge about a process. The flow of steps is indicated with arrows connecting the symbols. Flowcharts can be applied at all stages of the project life cycle. Figure 2.5 illustrates the flowchart for the contractor's staff approval in construction projects.

2.2.5 Histogram

The histogram is a pictorial representation of frequency distribution of data. It is created by grouping the data points into cells and displays how frequently

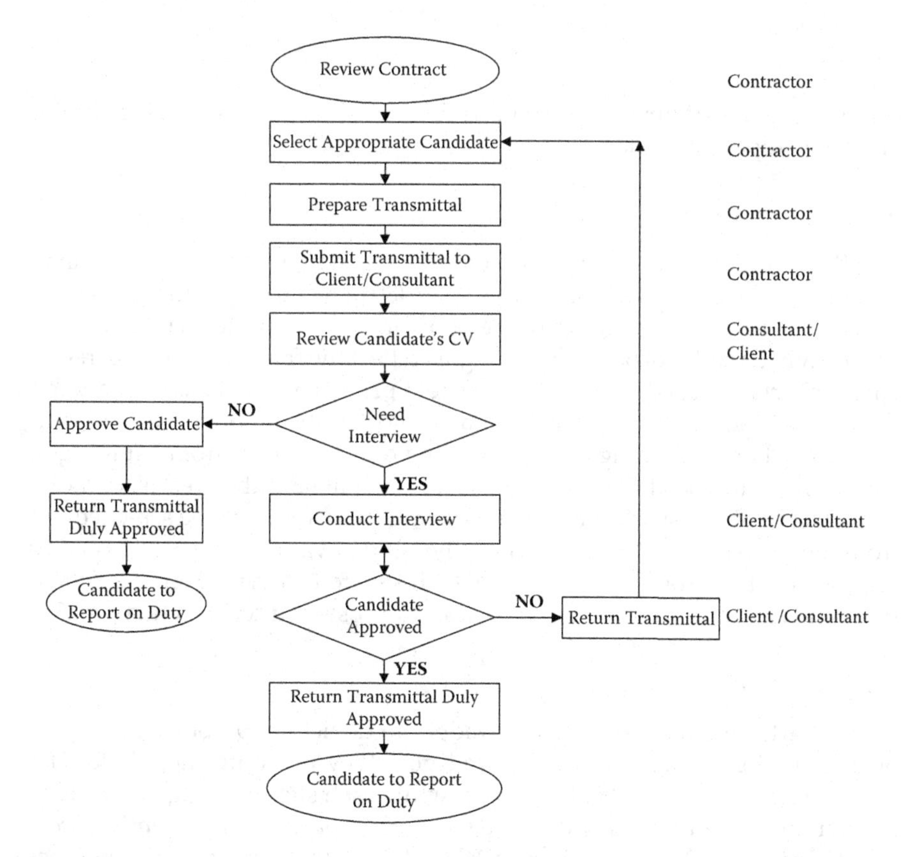

FIGURE 2.5
Flowchart.

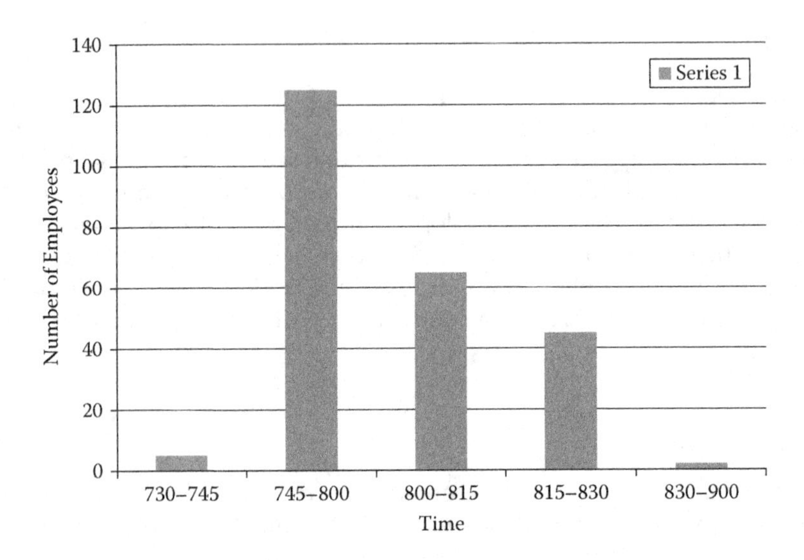

FIGURE 2.6
Histogram.

different values occur in the data set. Figure 2.6 illustrates a histogram for employee reporting time.

2.2.6 Pareto Chart

The Pareto chart is named after Vilfredo Pareto, a nineteenth century Italian economist who postulated that a large share (80 percent) of wealth is owned by a small (20 percent) percentage of the population. The Pareto chart is a graph chart having a series of bars whose heights reflect the frequency of occurrence. Pareto charts are used to display the Pareto principle in action, by arranging data so that the few vital factors that cause most of the problems reveal themselves. The bars are arranged in descending order of height from left to right. Pareto charts are used to identify those factors that have the greatest cumulative effect on the system, and thus less significant factors can be screened out from the process. The Pareto chart can be used at various stages in a quality improvement program to determine which step to take next. Figure 2.7 illustrates the Pareto chart showing the division of costs of a construction project.

2.2.7 Pie Chart

A pie chart is a circle divided into wedges to depict the proportions of data or information in order to understand how they make up the whole. The entire pie chart represents all the data, while each slice or wedge represents a different class or group within the whole. The portions of the entire circle or pie sum up to 100 percent. Figure 2.8 illustrates the breakup of the contractor's site staff at a construction project site.

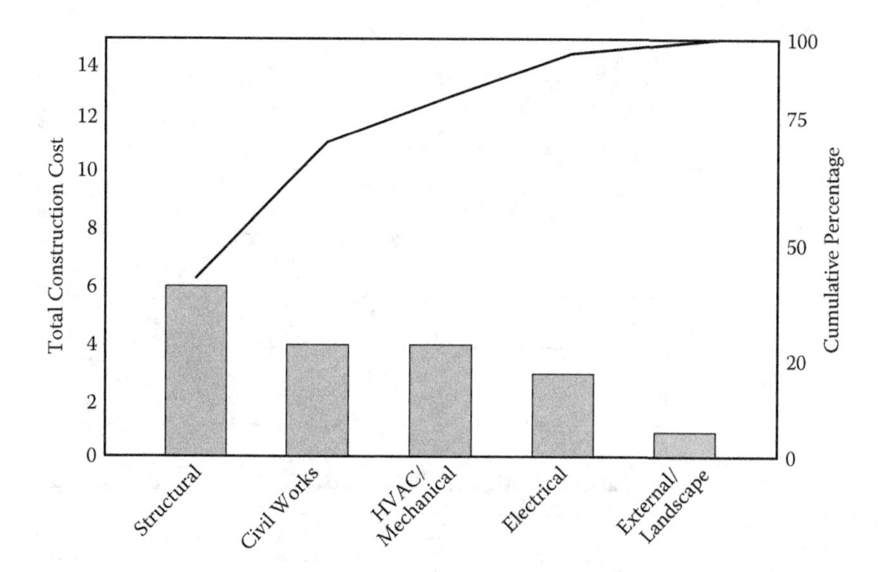

FIGURE 2.7
Pareto chart.

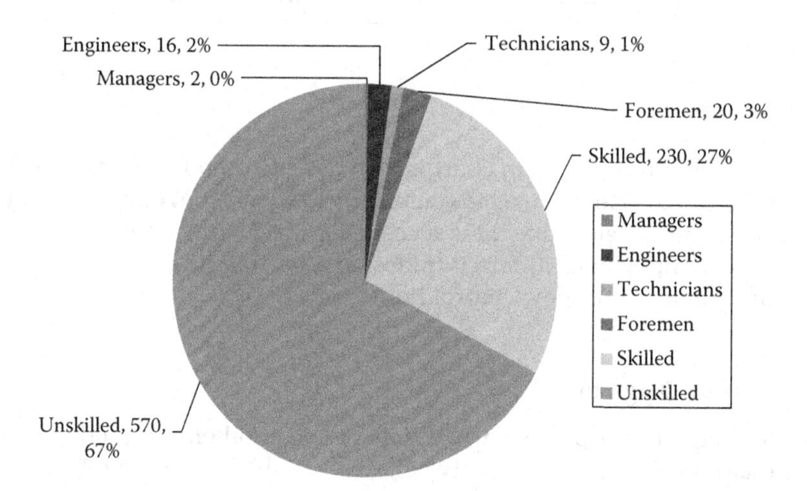

FIGURE 2.8
Pie chart.

2.2.8 Run Chart

A run chart is a graph plotted by showing measurement (data) against time. Run charts are used to show the trends or changes in a process variation over time over the average. They also can be used to determine if the pattern can be attributed to common causes of variation, or if special causes of

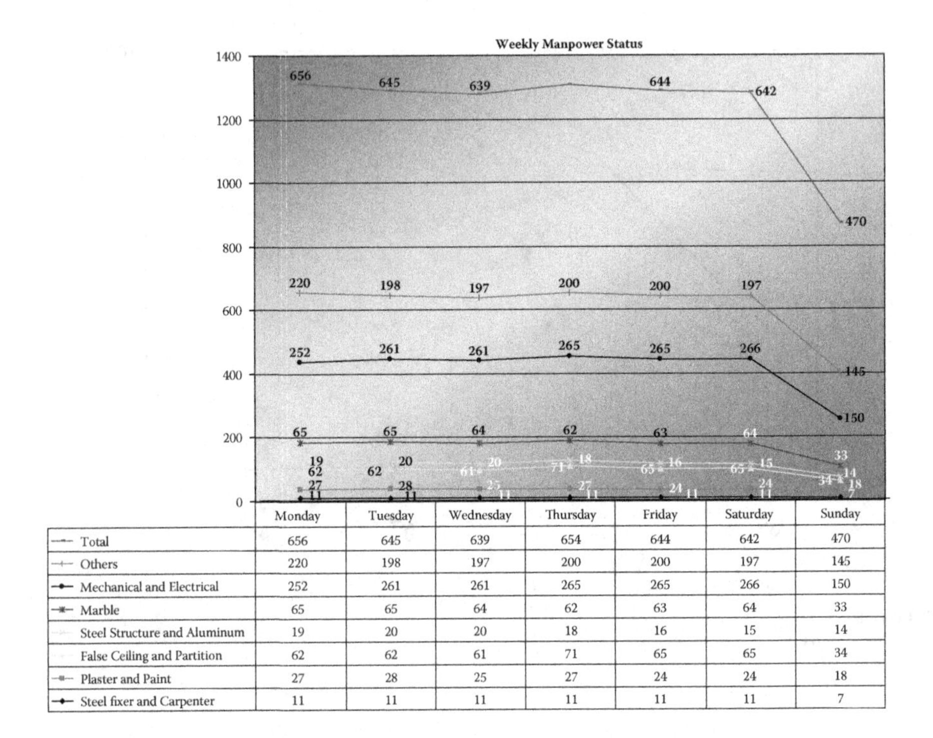

	Monday	Tuesday	Wednesday	Thursday	Friday	Saturday	Sunday
Total	656	645	639	654	644	642	470
Others	220	198	197	200	200	197	145
Mechanical and Electrical	252	261	261	265	265	266	150
Marble	65	65	64	62	63	64	33
Steel Structure and Aluminum	19	20	20	18	16	15	14
False Ceiling and Partition	62	62	61	71	65	65	34
Plaster and Paint	27	28	25	27	24	24	18
Steel fixer and Carpenter	11	11	11	11	11	11	7

FIGURE 2.9
Run chart.

variation were present. A run chart is also used to monitor process performance. Run charts can be used to track improvements that have been put in place, checking to determine their success. Figure 2.9 illustrates a run chart for weekly manpower of different trades on a project. It is similar to a control chart but does not show control limits.

2.2.9 Scatter Diagram

A scatter diagram is a plot of one variable versus another. It is used to investigate the possible relationship between two variables that both relate to the same event. It helps to know how one variable changes with respect to the other. It can be used to identify potential root causes of problems and to evaluate cause and effect relationships. Figure 2.10 illustrates a scatter diagram for beam quantity of various lengths.

2.2.10 Stratification

Stratification is a graphical representation of data collected from different data. Figure 2.11 illustrates a stratification diagram for cable drums.

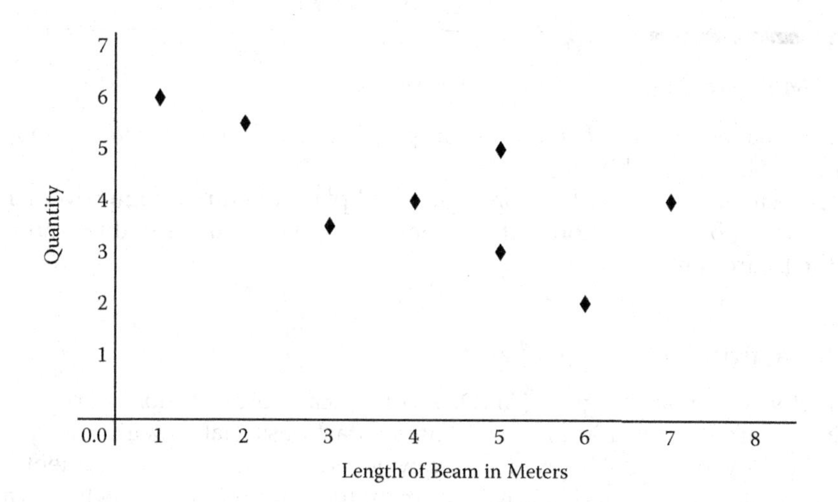

FIGURE 2.10
Scatter diagram.

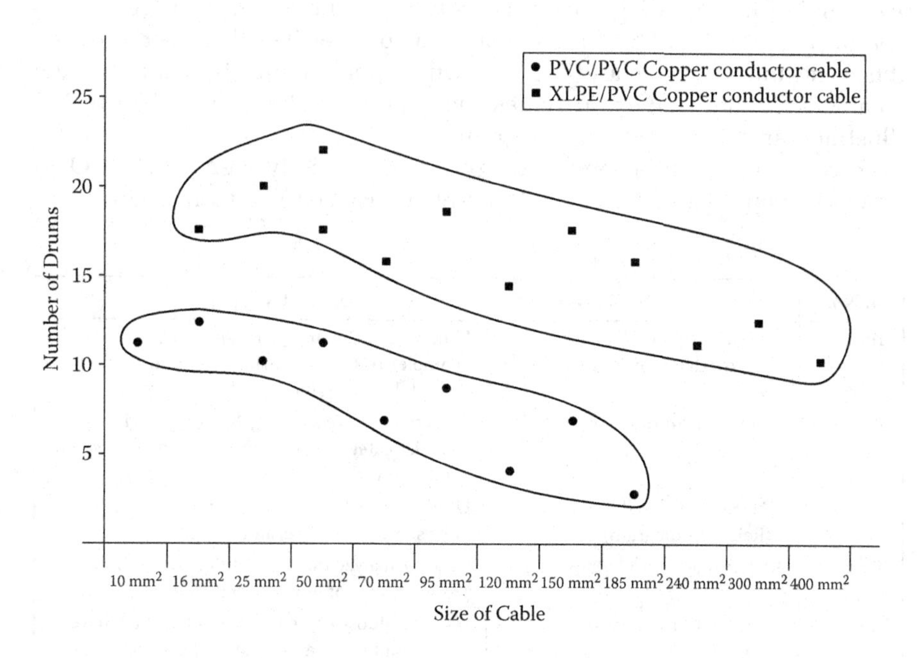

FIGURE 2.11
Stratification.

2.3 Management and Planning Tools

Seven management tools are most popular after quality classic tools. These tools are listed in Figure 2.12.

These tools are focused on managing and planning quality improvement activities. A brief definition of these quality tools is as follows (values shown in the figures are indicative only).

2.3.1 Activity Network Diagram

An activity network diagram (AND) is a graphical representation chart showing the interrelationship among activities (task) associated with a project. The activity network diagram was developed by the U.S. Department of Defense. It was first used as a management tool for military projects. It was adapted as educational tool for business managers. In an activity network diagram, each activity is represented by one and only one arrow in the network, and is associated with an estimated time to perform the activity. An activity network diagram analyzes the sequences of tasks necessary to complete the project. The direction of the arrow specifies the order in which the events must occur. The event represents a point in time that indicates the completion of one or more activities and beginning of new ones. Figure 2.13 illustrates an activity network diagram.

There are two kinds of network diagrams, the "Activity-on-Arrow" (A-O-A) network diagram and the "Activity-on-Node" (A-O-N) network diagram.

Sr. No.	Name of Quality Tool	Usage
Tool 1	Activity network diagram (Arrow diagram)/Critical Path Method	Used when scheduling or monitoring task is complex or lengthy or time consuming and has schedule constraints.
Tool 2	Affinity diagram	Used to organize a large group of items into smaller categories that are easier to understand and deal with.
Tool 3	Interrelationship digraph (Relations diagram)	Used to show logical relationships between ideas, process, cause, and effect.
Tool 4	Matrix diagram	Used to analyze the correlations between two or more groups of information.
Tool 5	Prioritization matrix	Used to choose one or two options which have important criteria from several options.
Tool 6	Process decision program chart	Used to help contingency plan.
Tool 7	Tree diagram	Used to break down or stratify ideas in progressively more detailed steps.

FIGURE 2.12
Management and planning tools.

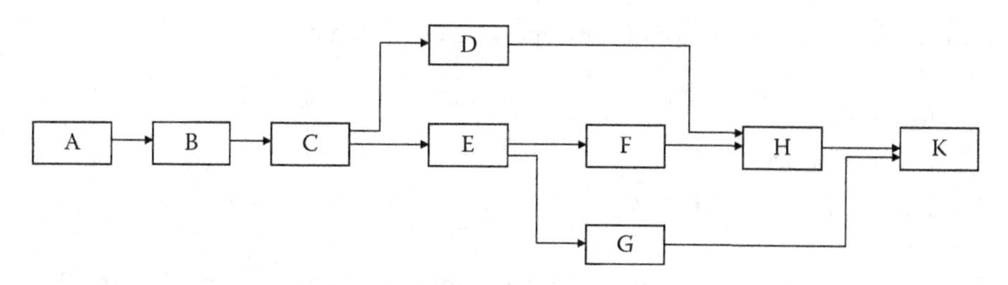

FIGURE 2.13
Activity network diagram.

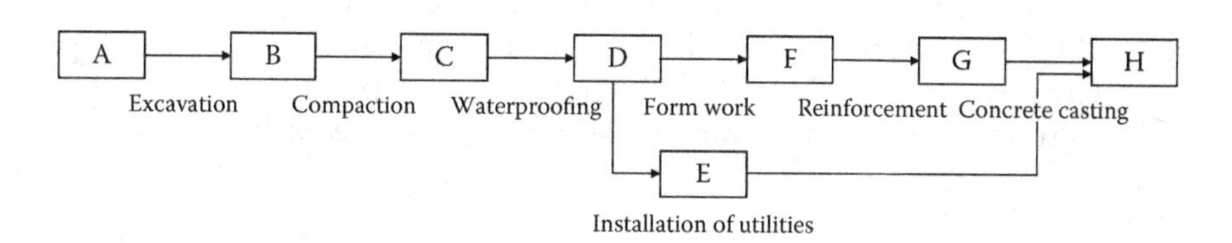

FIGURE 2.14
Arrow diagram.

Arrow diagrams or activity-on-arrows (A-O-A) uses a diagramming method to represent the activities on arrows and connect them at nodes (circles) to show the dependencies. With the A-O-A method, the detailed information about each activity is placed on an arrow or as footnotes at the bottom.

Figure 2.14 illustrates the arrow diagramming method for concrete foundation work.

Activities originating from a certain event cannot start until the activities terminating the same event have been completed. This is known as precedence relations. These relationships are drawn using the precedence diagramming method (PDM). The PDM technique is also referred to as "Activity-on-Node" (A-O-N) because it shows the activities in a node (box) with arrows showing dependencies. An A-O-N network diagram has the activity information written in small boxes, which are the nodes of the diagram. Arrows connect the boxes to show the logical relationships between pairs of activities.

In a networking diagram, all activities are related in some direct way and may be further constrained by indirect relationships. The following are direct logical relationships or dependencies among project-related activities:

1. Finish-to-Start: Activity A must finish before activity B can begin.
2. Start-to-Start: Activity A must begin before activity B can begin.
3. Start-to-Finish: Activity A must begin before activity B can finish.
4. Finish-to-Finish: Activity A must finish before activity B can finish.

Apart from these, there are other dependencies such as:

1. Mandatory
2. Discretionary
3. External

Figure 2.15 illustrates dependency relationship diagrams, and Figure 2.16 illustrates the precedence diagramming method.

An activity network diagram (AND) or arrow diagram is a tool used for detailed planning, for analyzing the schedule during execution, and for controlling a complex or large-scale project. A network diagram uses nodes and arrows. Date information is added to each activity node.

The Critical Path Method (CPM) chart is an expanded activity network diagram showing the estimated time to complete each activity and

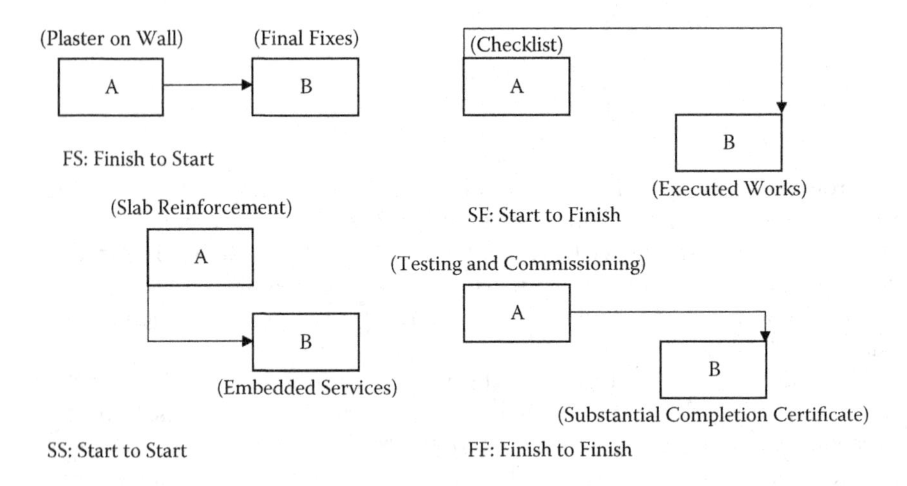

FIGURE 2.15
Dependency relationship.

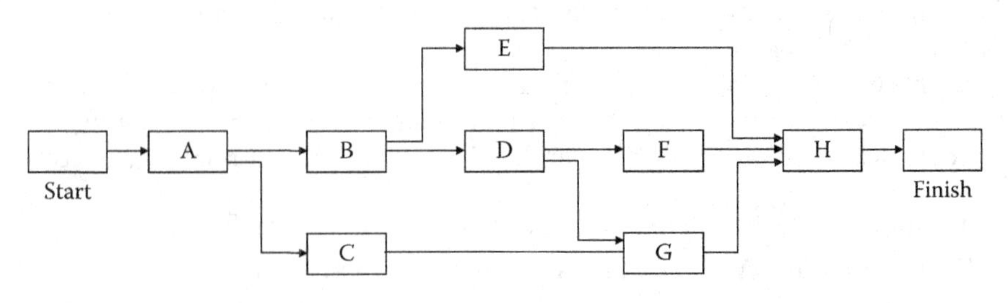

FIGURE 2.16
Precedence diagramming method.

connecting these activities based on the task to be performed. The critical path is a sequence of interrelated predecessor/successor activities that determines the minimum completion time for a project. The duration of the critical path is the sum of the activities' durations along the path. The activities in the critical path have the least scheduling flexibility. Any delays along the critical path would imply that additional time would be required to complete the project. There may be more than one critical path among all the project activities, so completion of the entire project could be affected due to a delaying activity along any of the critical paths. Figure 2.17 illustrates the activity relationship for a substation project, and Figure 2.18 shows the Critical Path Method (CPM) diagram for construction of the substation.

An activity network diagram is also known as the Program Evaluation and Review Technique (PERT). PERT is used to schedule, organize, and coordinate tasks within a project.

PERT planning involves the following nine steps:

1. Identify specific activity.
2. Identify milestones of each activity.
3. Determine the proper sequence of each activity.
4. Construct a network diagram.
5. Estimate the time required to complete each activity. The three point estimation method using the following formula can be used to determine an approximate estimated time for each activity:

   Estimated time = (Optimistic + 4*Most likely + pessimistic)/6

6. Compute the Early Start (ES), Early Finish (EF), Late Start (LS), and Late Finish (LF) times for each activity in the network.
7. Determine the critical path.
8. Identify the critical path for possible schedule compression.
9. Evaluate the diagram for milestones and target dates in the overall project.

Figure 2.19 illustrates the Gantt chart.

2.3.2 Affinity Diagram

The affinity diagram is a tool that gathers a large group of ideas/items and organizes them into a smaller grouping based on their natural relationships. An affinity diagram is a refinement of brainstorming ideas into smaller groups, which can be dealt with more easily and satisfy team members.

Activity Number	Description of Activity	Duration in Days	Preceding Activity
1	Start	0	
2	Mobilization	21	1
3	Preparation of site	15	2
4	Staff approval	15	2
5	Material approval	15	2,4
6	Shop Drawing approval	15	2,5
7	Procurement (Structural work)	15	5
8	Procurement (Pipes, Ducts, Sleeves)	7	5
9	Procurement (H.T Switchgear, Transformer)	60	5
10	Procurement (Civil, MEP, Furnishing)	30	5
11	Excavation	4	3,6
12	Blinding concrete	4	5,7,11
13	Raft foundation	7	5,6,7,12
14	Utility services (Embedded)	4	8,12
15	Concrete (Floor)	1	6,13,14
16	Trenches	7	13
17	Embedded services/ducts in trench	2	8,13
18	Concrete (Transformer area)	1	16,17
19	Walls and Columns	7	15,18
20	Form work for slab	3	6,7,19
21	Reinforcement	2	7,20
22	Embedded services	1	8,20
23	Concrete (Roof slab)	1	21,22
24	Masonry work	14	6,10,23
25	Installation of equipment	7	9,18,24
26	Installation of electromechanical items	14	6,10,23,24
27	Installation of ventilation system	4	10,23,24
28	Finishes	7	10,24
29	Installation of final fixes	3	10,28
30	Furnishing	2	10,28
31	Testing of equipment	2	25
32	Testing of HVAC, Firefighting, Electrical System	4	25,26,27
33	Handing Over	1	29,30,31,32
34	End	0	

Note: The duration is indicative only, to help make clear the sequencing and interrelationship of activities.

FIGURE 2.17
Activities to construct substation building.

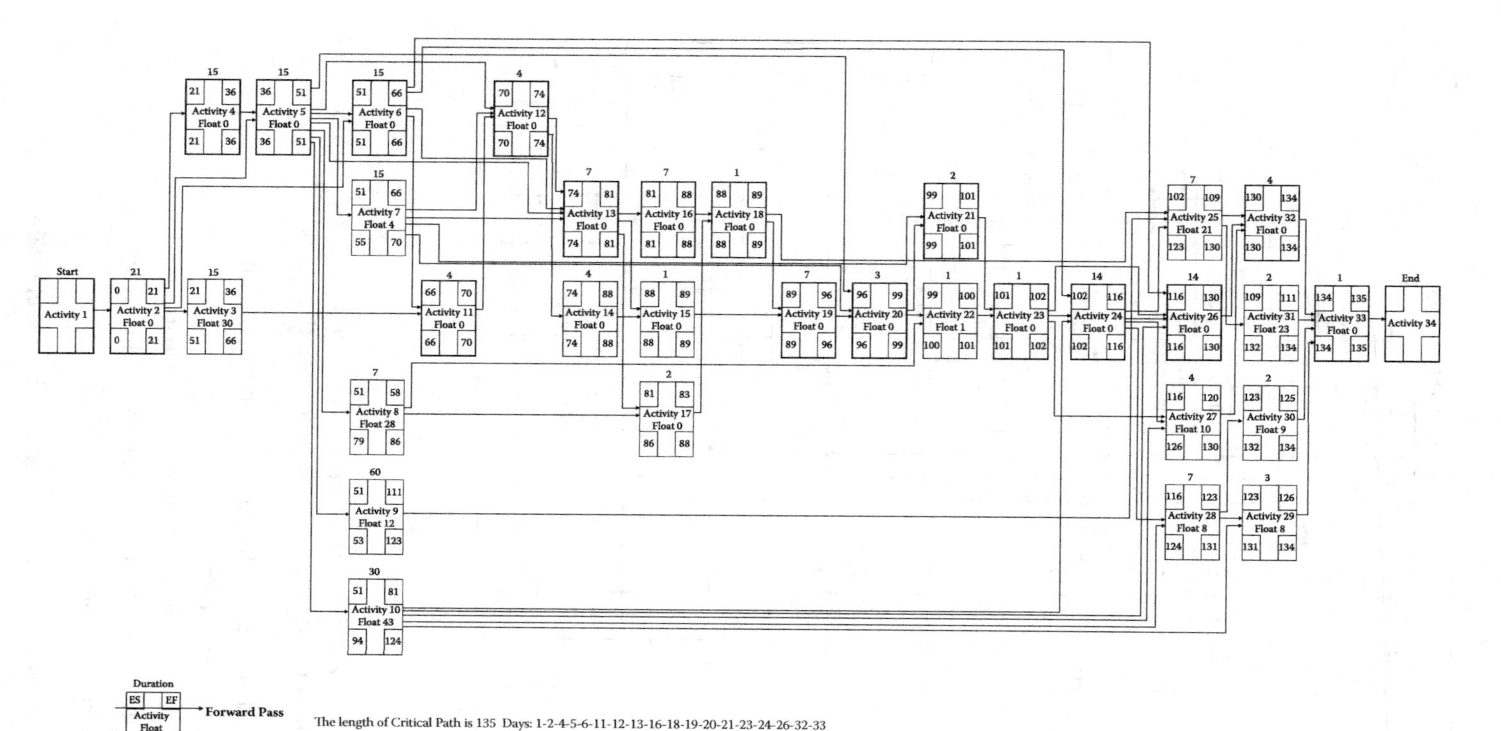

FIGURE 2.18
Critical Path Method.

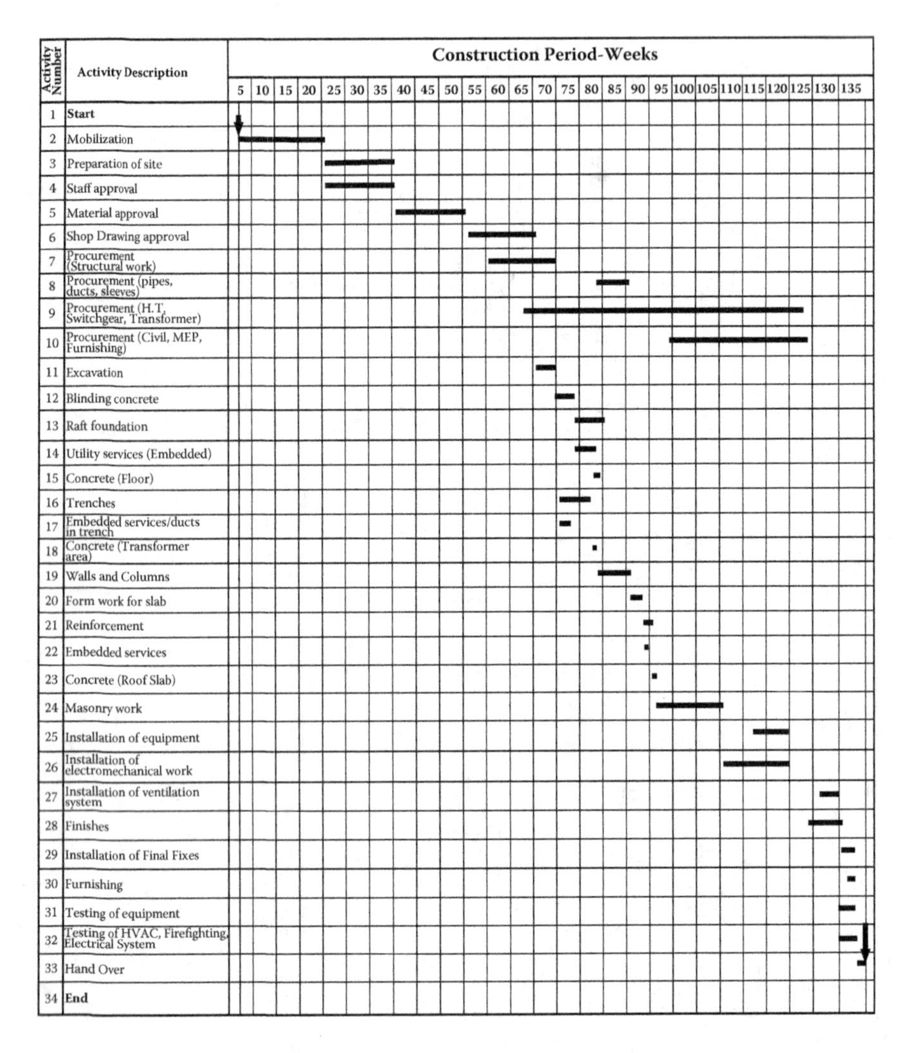

FIGURE 2.19
Gantt chart.

The affinity process is often used to group ideas generated by brainstorming. An affinity diagram is created per the following five steps:

1. Generate ideas and list the ideas without criticism.
2. Display the ideas in a random manner.
3. Sort the ideas and place them into multiple groups.
4. Continue until smaller groups satisfy all the members.
5. Draw the affinity diagram.

Figure 2.20 illustrates an affinity diagram for concrete slabs.

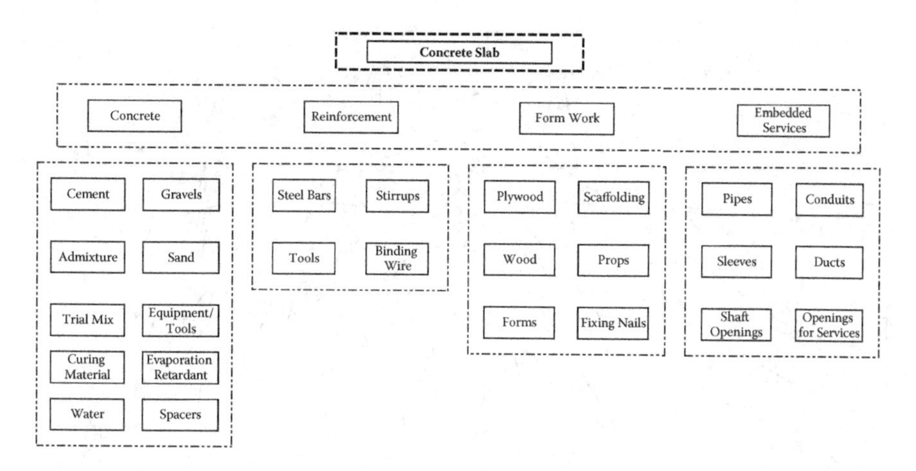

FIGURE 2.20
Affinity diagram.

2.3.3 Interrelationship Digraph

An interrelationship digraph (di is for directional) is an analysis tool that allows the team members to identify logical cause and effect relationships between ideas. It is drawn to show all the different relationships between factors, areas, or process. The digraph makes it easy to pick out the factors in a situation that are the ones driving many of the other symptoms or factors. While affinity diagrams organize and arrange the ideas into groups, interrelationship digraphs identify problems to define the ways in which ideas influence one another. An interrelation digraph is used to identify cause and effect relationships with the help of directional arrows among critical issues. The number of arrows coming into the node determine Outcome (Key Indicator), while the outgoing arrows determine the Cause (Driver) of the issue.

Figure 2.21 illustrates an interrelationship digraph for causes of bridge collapse.

2.3.4 Matrix Diagram

A matrix diagram is constructed to analyze systematically the correlations between two or more group of items or ideas. The matrix diagram can be shaped in the following ways:

1. L shaped
2. T shaped
3. X shaped
4. C shaped
5. Inverted Y shaped
6. Roof shaped

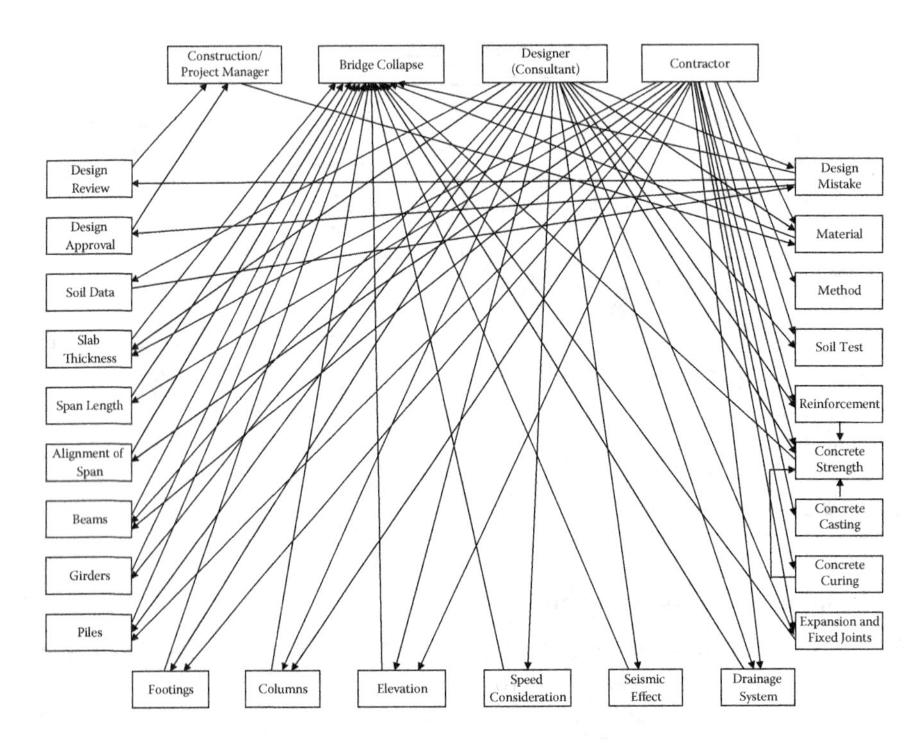

FIGURE 2.21
Interrelationship digraph.

Each shape has its own purpose:

1. An L-shaped matrix is used to show interrelationships between two groups or processes.
2. A T-shaped matrix is used to show the relation between three groups. For example, consider there are three groups A, B, and C. In a T-shaped matrix group, A and B are each related to group C whereas group A and B are not related to each other.
3. An X-shaped matrix is used to show the relationship among four groups. Each group is related to two other groups in a circular fashion.
4. A C-shaped matrix interrelates three groups of processes or ideas in three dimensions.
5. An inverted Y-shaped matrix is used to show the relation between three groups. Each group is related with other groups in a circular fashion.
6. A roof-shaped matrix relates one group of items to itself. This matrix is used with L- or T-shaped matrices.

Customer Requirements of Distribution Boards				
Serial Number	Component Details	Customer		
		A	B	C
1	Isolator	1	1	-
2	Molded Case Circuit Breaker (MCCB)	-		1
3	HRC Fuse	1	1	-
4	Earth Leakage Circuit Breaker	2	1	2
5	Miniature Circuit Breaker (MCB)	18	12	18
6	Single Bus Bar	-	1	-
7	Double Bus Bar	2	-	2
8	Enclosure with Lock	Yes	No	Yes
9	Surface Mounting	Yes	No	Yes
10	Flush Mounting	-	Yes	

FIGURE 2.22
L-shaped matrix.

Manufacturing Plant	Products				
Customer					
International				#	#
European Manufacturer	≠	#	#		
Local Plant	#	≠	≠		
# Large Capacity ≠ Small Capacity	600 KVA Transformer	1000 KVA Transformer	1250 KVA Transformer	1600KVA Transformer	2000KVA Transformer
ABC Company	#	≠	≠		
XYZ Company	≠	#	#		
Others				#	#

FIGURE 2.23
T-shaped matrix.

Figures 2.22, 2.23, and 2.24 illustrate an L-shaped matrix, a T-shaped matrix, and a roof-shaped matrix.

2.3.5 Prioritization Matrix

The prioritization matrix assists in choosing between several options in order of importance and priority. It helps decision makers determine the order of importance considering the relative merit of each of the activities or goal being considered. The prioritization matrix focuses the attention of team members on those key issues and options that are more important for the organization or project.

Figure 2.25 illustrates the prioritization matrix.

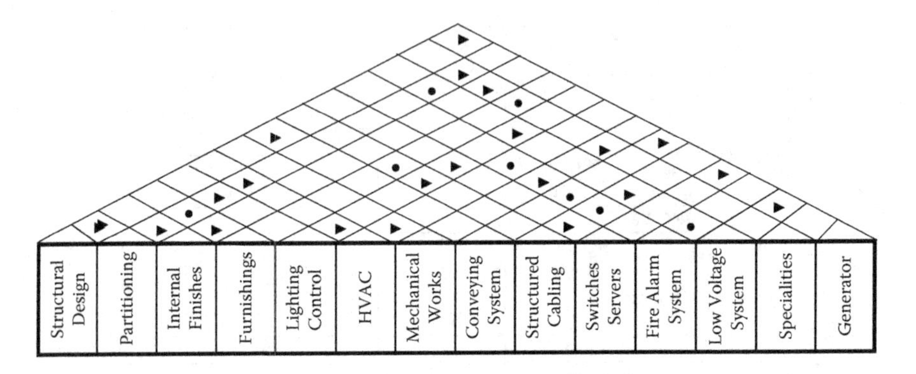

FIGURE 2.24
Roof-shaped matrix.

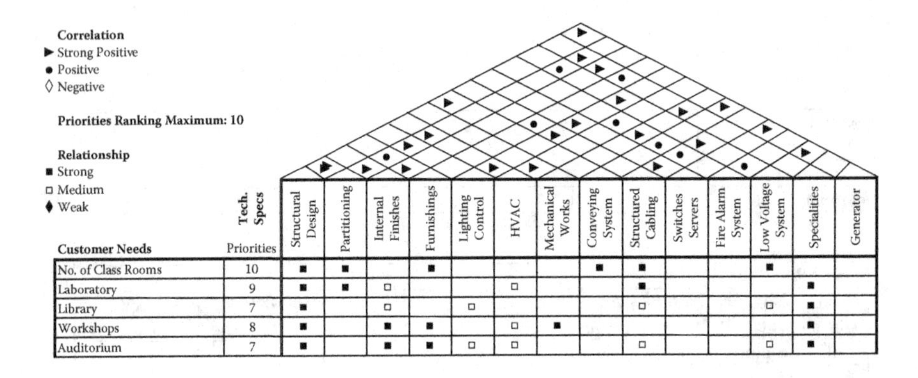

FIGURE 2.25
Prioritization matrix.

2.3.6 Process Decision Program

The process decision program is a technique used to help prepare contingency plans. The process decision program systematically identifies what might go wrong in a project plan or project schedule. It describes specific actions to be taken to prevent the problems from occurring in the first place, and to mitigate or avoid the impact of the problems if they do occur.

Figure 2.26 illustrates the process decision program for submission of contract documents.

2.3.7 Tree Diagram

Tree diagrams are used to break down or stratify ideas progressively into more detailed steps. A tree diagram breaks broader ideas into specific details and helps make the decision to select an alternative easier. It is used

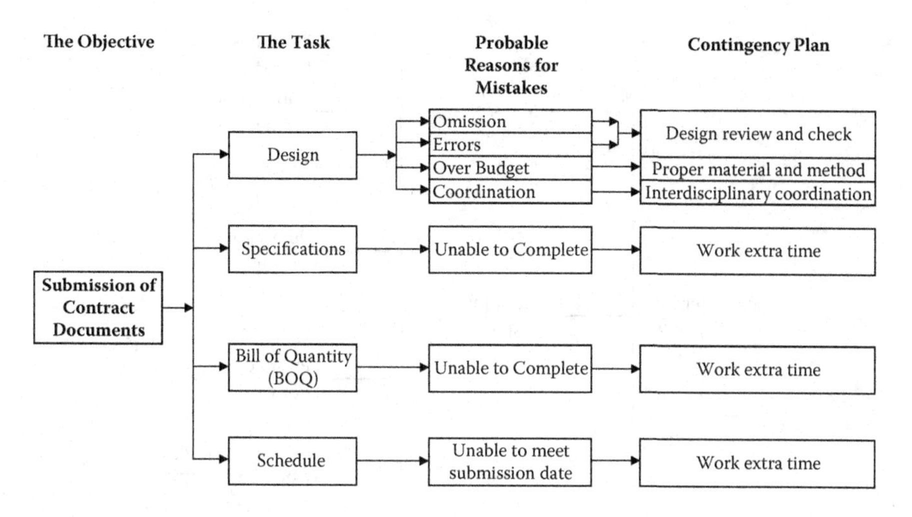

FIGURE 2.26
Process decision diagram.

to figure out all the various tasks that must be undertaken to achieve a given objective.

Figure 2.27 shows a tree diagram for water in a storage tank.

2.4 Process Analysis Tools

Figure 2.28 shows process analysis tools.

2.4.1 Benchmarking

Benchmarking is the process of measuring the actual performance of the organization's products, processes, and services and comparing it with the best-known industry standards to assist the organization in improving the performance of their products, processes, and services. Benchmarking involves analyzing an existing situation, identifying and measuring factors critical to the success of the product or services, comparing them with other businesses, analyzing the results, and implementing an action plan to achieve better performance. The following is the process for benchmarking:

1. Collect internal and external data on work, process, method, product characteristics, and system selected for benchmarking.
2. Analyze data to identify performance gaps and determine cause and differences.

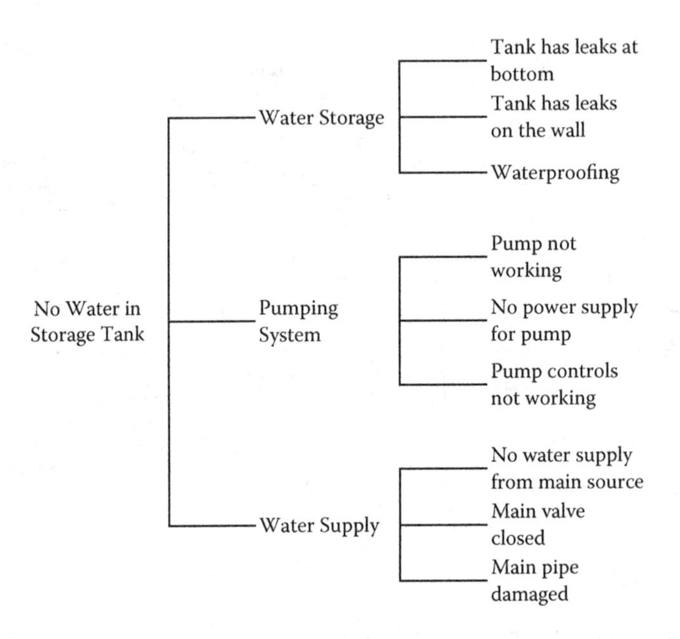

FIGURE 2.27
Tree diagram.

Sr. No.	Name of Quality Tool	Usage
Tool 1	Benchmarking	To identify best practices in the industry and improve the process or project.
Tool 2	Cause and effect	To identify possible cause and its effect in the process.
Tool 3	Cost of quality	To identify hidden or indirect cost affecting the overall cost of product/project.
Tool 4	Critical to quality	To identify quality features or characteristics most important to the client.
Tool 5	Failure Mode and Effects Analysis (FMEA)	To identify and classify failures according to their effects.
Tool 6	5 Why Analysis	Used to analyze and solve any problem where the root cause is unknown.
Tool 7	5W2H	The questions used to understand why the things happen the way they do.
Tool 8	Process mapping/Flowcharting	It is a technique used for designing, analyzing, and communicating work processes.

FIGURE 2.28
Process analysis tools.

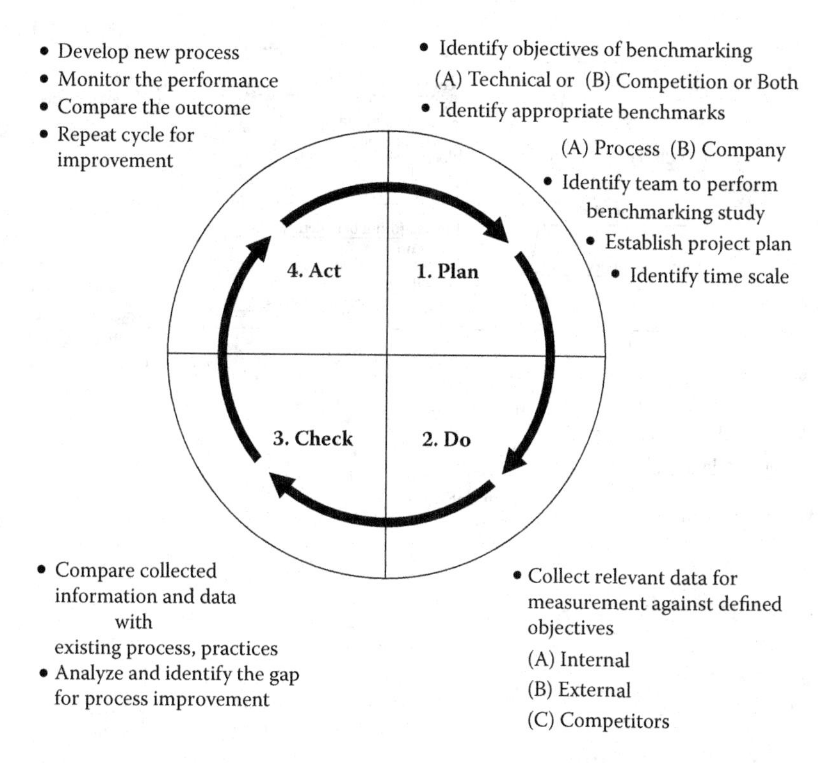

FIGURE 2.29
Benchmarking.

3. Prepare an action plan to improve the process in order to meet or exceed the best practices in the industry.
4. Search for the best practices among market leaders, competitors, and noncompetitors that lead to their superior performance.
5. Improve performance by implementing these practices.

Figure 2.29 shows the benchmarking process.

2.4.2 Cause and Effect Diagram

Cause and effect diagrams are classic quality control tools. They are used to analyze the cause and effect of defects or nonconformance, and the effect on the process due to these causes.

Figure 2.30 shows a cause and effect diagram for rejection of masonry work.

2.4.3 Cost of Quality

Cost of quality is discussed in detail in Section 2.8.

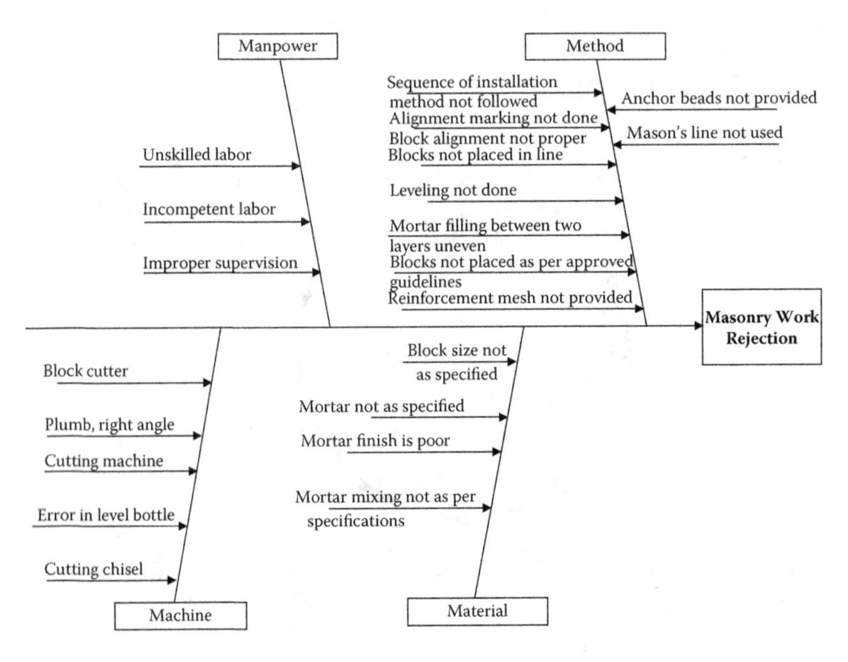

FIGURE 2.30
Cause and effect.

2.4.4 Critical to Quality

Critical to quality is a significant step in the design process of a product or service to identify the customer's/client's expectation and to fulfill their needs and requirements.

2.4.5 Failure Mode and Effects Analysis (FMEA)

Failure mode and effects analysis identifies all the possible failures in the design of a product, process, and service and their effects. Its aim is to reduce the risk of failure and improve the process.

Figure 2.31 illustrates the failure mode and effects analysis process, and Figure 2.32 illustrates an example form used to record FMEA readings.

2.4.6 5 Whys Analysis

A 5 whys analysis is used to analyze and solve any problem where the root cause is unknown.

Figure 2.33 illustrates a 5 Whys analysis chart for burning of cable.

2.4.7 5W2H

5W2H is about asking the questions to understand a process or problem.

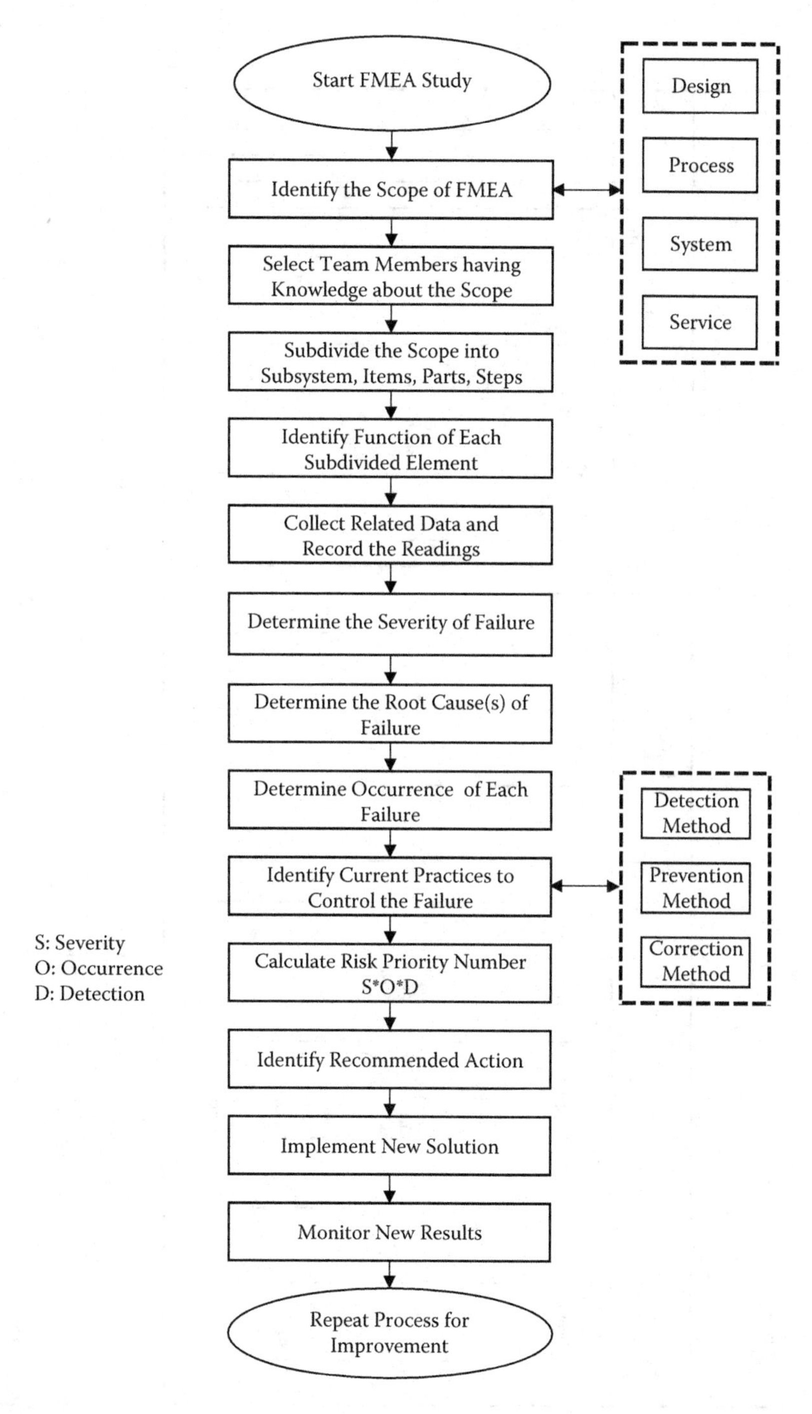

FIGURE 2.31
Failure mode and effects analysis process.

Failure Mode and Effect Analysis

Product Name:

Drawing Reference:
Revision:

EXAMPLE ANALYSIS

Team Members
1
2
3
4

Operation Number	Scope Description	Subdivided Elements	Failure Mode	Effects of Failure	Severity Rating	Cause of Failure	Occurrence Rating	Current Practice of Controls			Detection Rating	Current Status of Product				Recommended Corrective Action	Action By	Action Taken Date	Revise Status			
								Detection	Prevention	Correction		S	O	D	RPN				S	O	D	RPN
1	Emergency generator system	Operation	Generator failed to start	1. No lights 2. Life support equipment stopped functioning 3. No power supply for IT System 4. Water supply pumps stopped 5. No power supply for lift 6. Fire mode operation equipment will not operate 7. HVAC system stopped	7 10 9 6 4	1. No signal from ATS 2. Automatic starting system failed 3. Low Battery voltage for starter motor 4. Circuit breaker in off position 5. No diesel in day tank		1. Through BMS 2. No regular check 3. Manual 4. Manual	1. Not in practice	1. Manual						1. Regular check of starting system. 2. Check starter regularly. 3. Check diesel level regularly. Interface level indicator with BMS. 4. Check breaker position regularly. Interface with BMS.	Maintenance Engineer					

Legend:-
RPN: Risk Priority Number
S: Severity
O: Occurrence
D: Dtetection

ATS: Automatic Transfer Switch
BMS: Building Management System

FIGURE 2.32
FMEA form.

Serial Number	Why	Related Analyzing Question
1	Why	Why did the cable burn?
2	Why	Why did the earth leakage relay not trip?
3	Why	Why did the circuit breaker not trip?
4	Why	Why was poor cable insulation not noticed?
5	Why	Why was undersize rating of breaker with respect to current-carrying capacity of cable not noticed?

FIGURE 2.33
5 Whys analysis for cable burning.

Serial Number	Why	Related Analyzing Question
1	Why	Why did the slab collapse?
2	What	What is the reason for the collapse?
3	Who	Who is responsible?
4	Where	Where is the mistake?
5	When	When did the slab collapse?
6	How many	How many persons were affected (injured or died)?
7	How much	How much was the loss in terms of cost and time?

FIGURE 2.34
5W2H analysis for slab collapse.

The 5 Ws are

1. Why
2. What
3. When
4. Where
5. Who

And the 2 Hs are

1. How
2. How much

Figure 2.34 illustrates 5W2H for slab collapse.

2.4.8 Process Mapping/Flowcharting

Process mapping/flowcharting is a graphical representation of workflow giving a clear understanding of a process or services of parallel processes.

Process mapping/flowcharting are techniques that can be employed to not only produce a visual representation of the production processes but processes related to other departments.

Figure 2.35 illustrates a process mapping/flowcharting diagram for approval of a variation order.

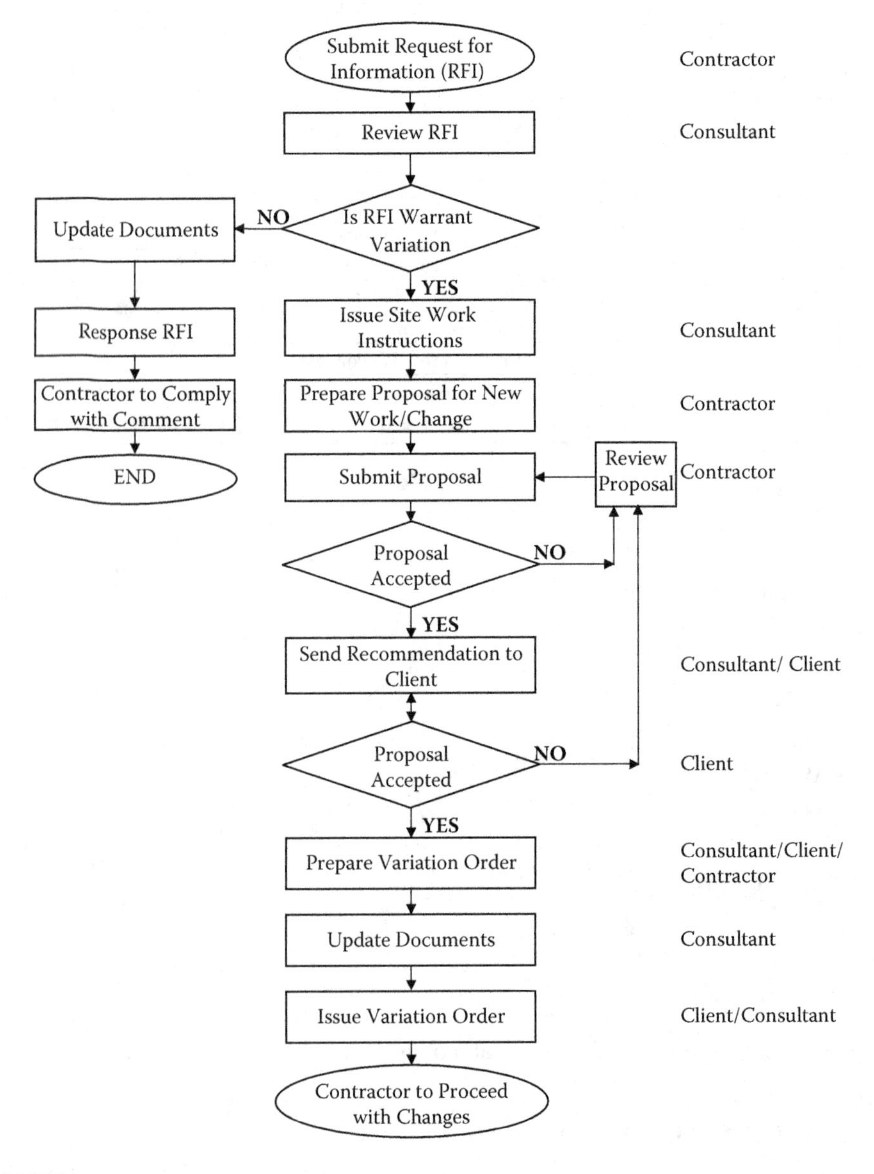

FIGURE 2.35
Process mapping/flowcharting.

2.5 Process Improvement Tools

Figure 2.36 shows process improvement tools.

2.5.1 Root Cause Analysis

Root cause analysis is used to analyze root causes of problems. The analysis is generally performed by using an Ishikawa diagram or a cause and effect diagram.

Figure 2.37 shows root cause analysis for rejection of executed marble work.

2.5.2 PDCA Cycle

Plan-Do-Check-Act (PDCA) is mainly used for continuous improvement. It consists of a four-step model for making changes. The PDCA cycle model can be developed as a process improvement tool to reduce the cost of quality.

Figure 2.38 illustrates the PDCA cycle for preparation of shop drawings.

2.5.3 SIPOC Analysis

This analysis is used to identify a Supplier–Input–Process–Output–Customer relationship. The purpose of a SIPOC analysis is to show the process flow by defining and documenting the suppliers, inputs, process steps, outputs, and customers.

Figure 2.39 illustrates a SIPOC analysis for an electrical panel.

2.5.4 Six Sigma-DMAIC

Six Sigma is discussed in detail in Section 2.10.

Sr. No.	Name of Quality Tool	Usage
Tool 1	Root cause analysis	Used to identify root causes of the problem
Tool 2	PDCA cycle	Used to plan for improvement followed by putting into action
Tool 3	SIPOC analysis	Used to identify the supplier–input–process–output–customer relationship
Tool 4	Six Sigma DMAIC	Used as an analytic tool for improvement
Tool 5	Failure Mode and Effects Analysis (FMEA)	Used to identify and classify failures according to effects and prevent or reduce failure
Tool 6	Statistical process control	Used to study how the process changes over time

FIGURE 2.36
Process improvement tools.

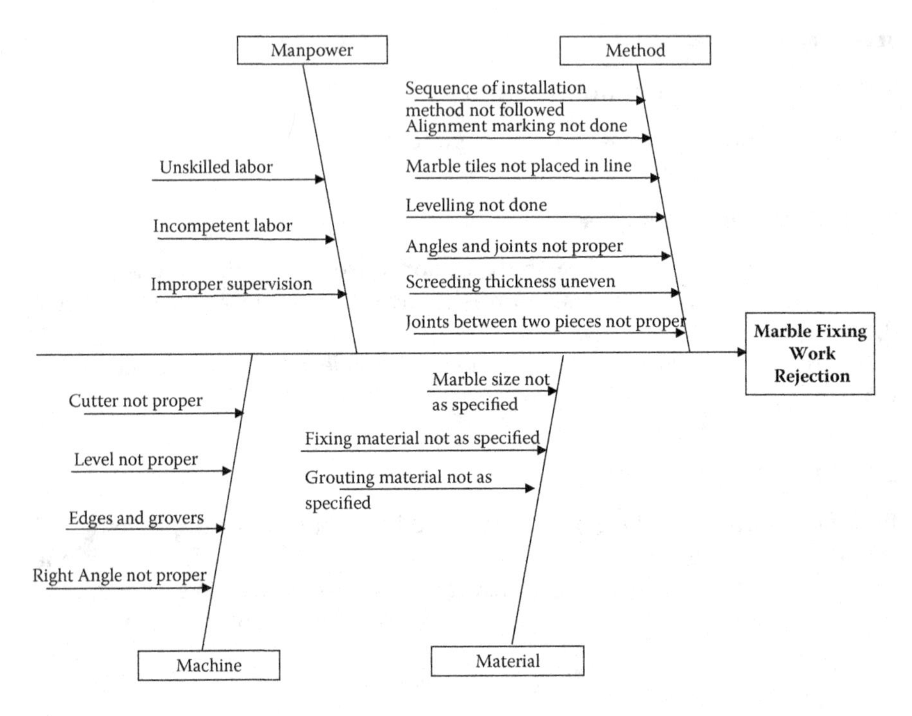

FIGURE 2.37
Root cause analysis.

2.5.5 Failure Mode and Effects Analysis (FMEA)

Failure mode and effects analysis is also used as a Process Improvement Tool. It identifies all the possible failures in the design of a product, process, and service and their effects. Its aim is to reduce the risk of failure and improve the process. Figure 2.31 illustrates the failure mode and effects analysis process.

2.5.6 Statistical Process Control

Statistical process control (SPC) is a quantitative approach based on the measurement of process control. Dr. Walter A. Shewhart developed the control charts as early as 1924. SPC charts are used for identification of common cause and special (or assignable) cause variations, and in assisting diagnosis of quality problems. SPC charts reveal whether a process is "in control," that is, stable and exhibiting only random variation, or "out of control" and needing attention. The control chart is one of the key tools of SPC. It is used to monitor processes that are not in control, using measured ranges. There are two types of process control charts:

1. Variable charts
2. Attribute charts

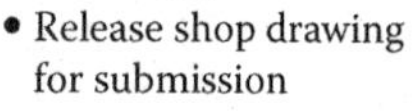
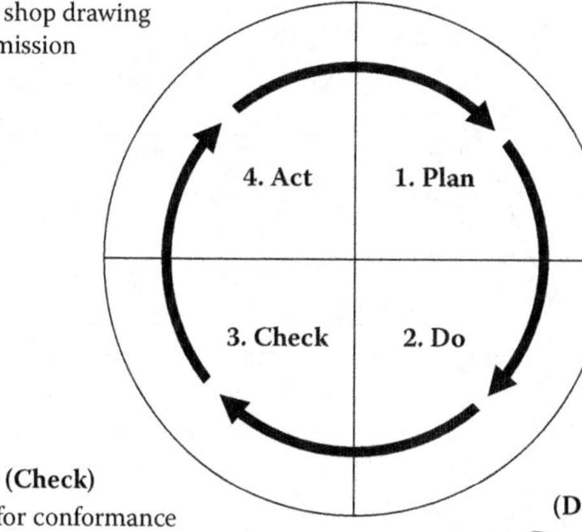

FIGURE 2.38
PDCA cycle.

(Who are suppliers?)	(What the suppliers are providing?)	(What is the process?)	(What is the output of process?)	(Who are the customers?)
Supplier	**Inputs**	**Process**	**Outputs**	**Customer**
Electrical Panel Builder/ Assembler	Main Low Tension Panel	Electrical Installation Work	Electrical Distribution Network	Power supply for project
	Main Switchboards			
	Distribution Boards			
	Starter Panels			
	Control Panels			

FIGURE 2.39
SIPOC analysis for electrical panel.

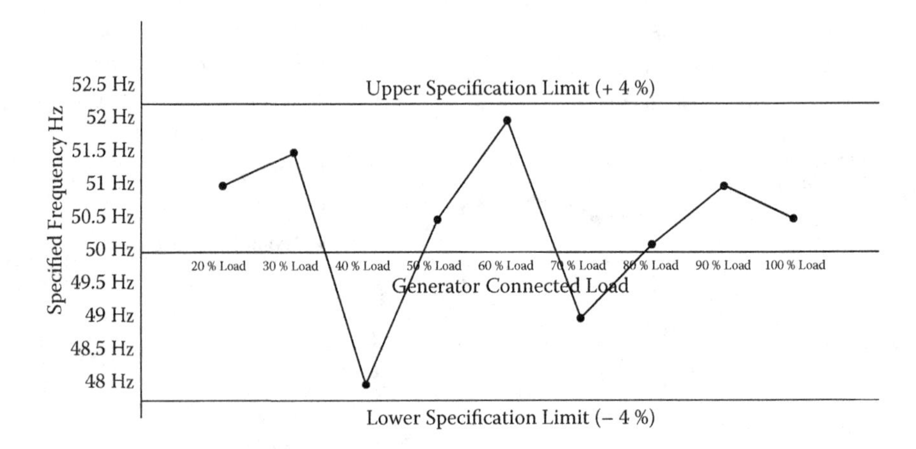

FIGURE 2.40
Statistical process control.

Variable charts relate to variable measurement such as length, width, temperature, weight, etc.

Attribute charts relate to the characteristics possessed (or not possessed) by the process or the product.

Figure 2.40 shows SPC charts for generator frequency.

2.6 Innovation and Creative Tools

Figure 2.41 shows innovation and creative tools.

2.6.1 Brainstorming

Brainstorming is listing all the ideas put forth by a group in response to a given question or problem. It is the process of creating ideas by storming some objective. In 1939, a team led by advertising executive Alex Osborn coined the term *brainstorm*. According to Osborn, to brainstorm means to use the brain to storm a creative problem. Classical brainstorming is the most well-known and often-used technique for quick idea generation. It is based on the fundamental principles of deferment of judgment and that quantity breeds quality. It involves questions such as

- Does the item have any design features that are not necessary?
- Can two or more parts be combined together?
- How can we cut down the weight?
- Are there nonstandard parts that can be eliminated?

Sr. No.	Name of Quality Tool	Usage
Tool 1	Brainstorming	Used to generate multiple ideas.
Tool 2	Delphi Technique	Used to get ideas from select group of experts.
Tool 3	5W2H	The questions used to understand why the things happen the way they do.
Tool 4	Mind mapping	Used to create a visual representation of many issues that can help one gain a better understanding of the situation.
Tool 5	Nominal group technique	Used to enhance brainstorming by ranking the most useful ideas.
Tool 6	Six Sigma DMADV	Used primarily for the invention and innovation of modified or new products, services, or processes.
Tool 7	TRIZ	Used to provide systematic methods and tools for analysis and innovative problem solving.

FIGURE 2.41
Innovation and creative tools.

There are four rules for successful brainstorming:

1. Criticism is ruled out.
2. Freewheeling is welcomed.
3. Quantity is required.
4. Contribution and improvement are sought.

A classical brainstorming session involves the following basic steps:

- *Preparation:* The participants are selected, and a preliminary statement of the problem is circulated.
- *Brainstorming:* A warm-up session with simple unrelated problems is conducted, the relevant problem and the four rules of brainstorming are presented, and ideas are generated and recorded using checklists and other techniques if necessary.
- *Evaluation:* The ideas are evaluated relative to the problem.

Generally, a brainstorming group should consist of four to seven people, although some suggest a larger group.

Figure 2.42 illustrates the brainstorming process.

2.6.2 Delphi Technique

The Delphi technique is intended to determine a consensus among experts on a subject. The goal of the Delphi technique is to pick the brains of experts in the subject area, treating them as contributors to create ideas. It is a method

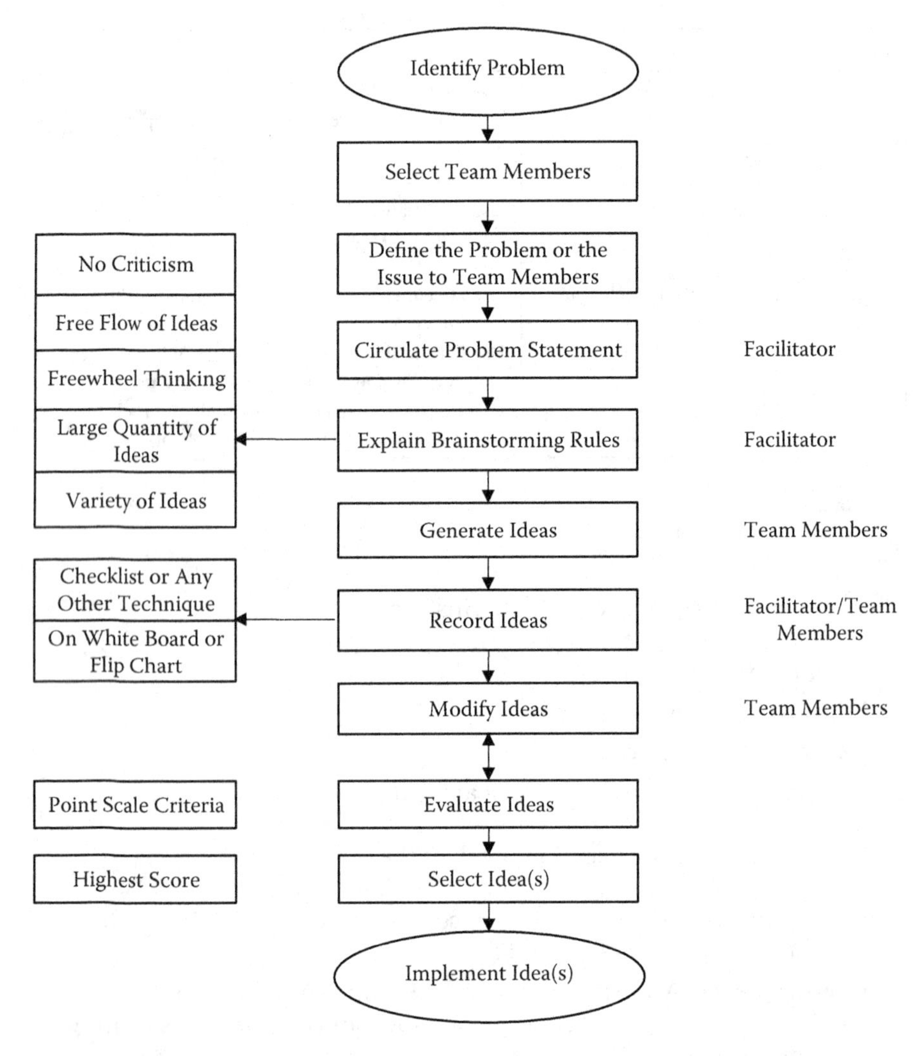

FIGURE 2.42
Brainstorming.

for consensus building by using questionnaires and obtaining responses from the panel of experts in the selected subjects. The Delphi technique employs multiple iterations designed to develop consensus opinion about the specific subject. The selected expert group answers questions posed by the facilitator. The responses are summarized and further circulated for group comments to reach a consensus. The iteration/feedback process allows the team members to reassess their initial judgment and change or modify the earlier suggestions.

Figure 2.43 shows the Delphi technique process.

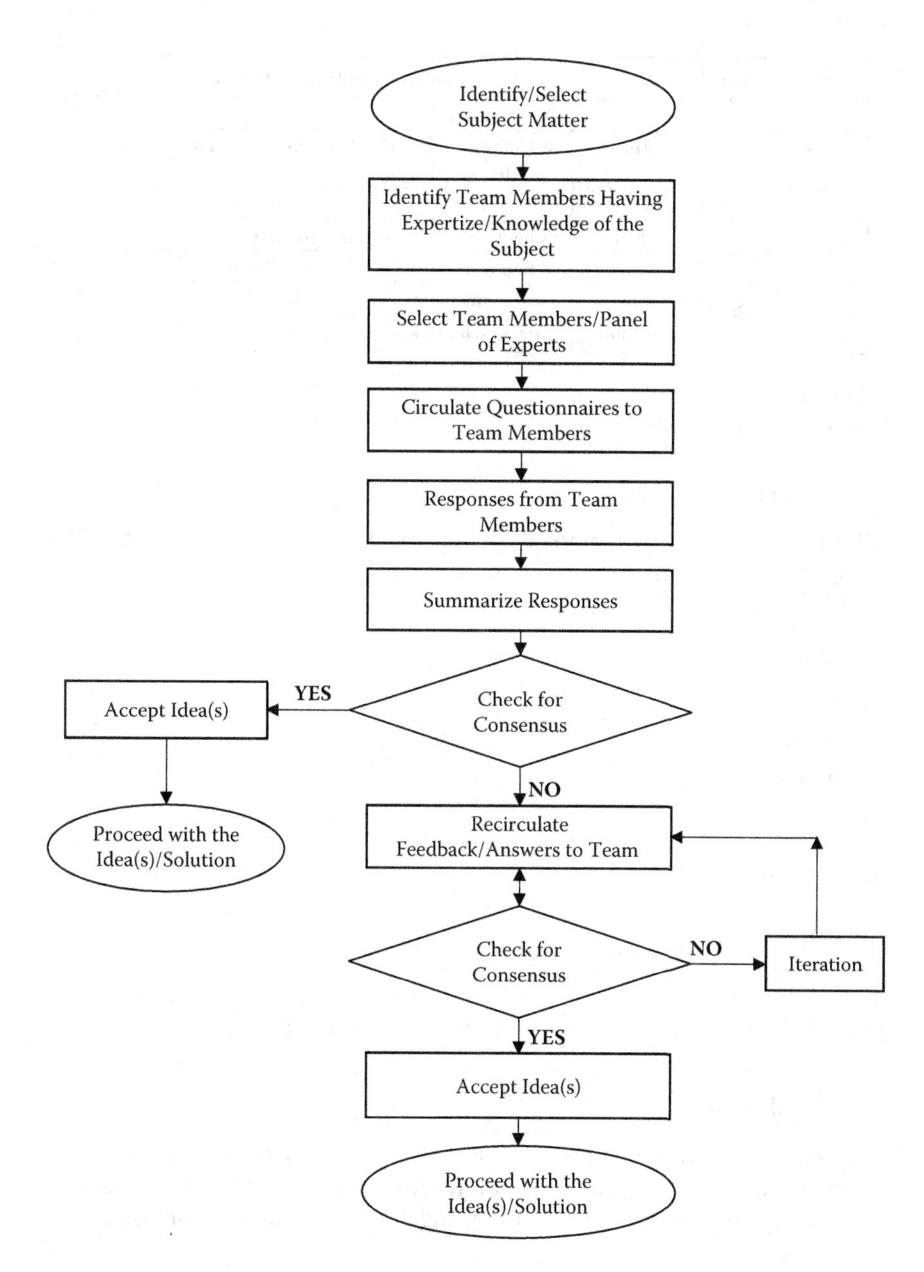

FIGURE 2.43
Delphi technique.

Serial Number	Why	Related Analyzing Question
1	Why	Why a new product?
2	What	What advantage will it have over other similar products?
3	Who	Who will be the customers for this product?
4	Where	Where can we market the product?
5	When	When will the product be ready for sale?
6	How many	How many pieces will be produced/sold per year?
7	How much	How much market share we will get for this product?

FIGURE 2.44
5W2H analysis for new product.

2.6.3 5W2H

5W2H is also used as an innovative and creative tool and involves asking questions to understand a process or problem.

The 5 Ws are

1. Why
2. What
3. When
4. Where
5. Who

And 2 Hs are

1. How
2. How much

Figure 2.44 shows 5W2H for development of a new product.

2.6.4 Mind Mapping

Mind mapping is a graphical representation of ideas so that one can gain a better understanding of the situation and create the solution or improve the task. Figure 2.45 shows mind mapping applied to improve site safety.

2.6.5 Nominal Group Technique

The nominal group technique (NGT) involves a structural group meeting designed to incorporate individual ideas and judgments into a group consensus. By correctly applying the NGT, it is possible for groups of people (preferably 5 to 10) to generate alternatives or other ideas for improving the competitiveness of the firm. The technique can be used to obtain

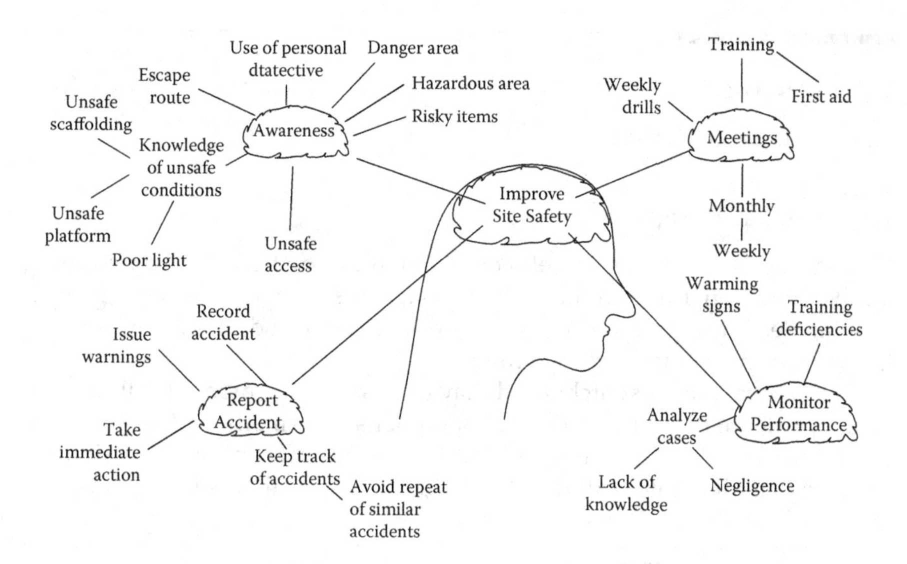

FIGURE 2.45
Mind mapping.

group thinking (consensus) on a wide range of topics. The technique, when properly applied, draws on the creativity of the individual participants, while reducing two undesirable effects of most group meetings:

1. The dominance of one or more participants
2. The suppression of conflicting ideas

The basic format of an NGT session is as follows:

- Individual silent generation of ideas
- Individual round-robin feedback and recording of the ideas
- Group's clarification of each idea
- Individual voting and ranking to prioritize ideas
- Discussion of group consensus results

The NGT session begins with an explanation of the procedure and a statement of question(s), preferably written by the facilitator.

2.6.6 Six Sigma-DMADV

Six Sigma is discussed in detail in Section 2.10.

2.6.7 TRIZ

TRIZ is discussed in detail in Section 2.11.

2.7 Lean Tools

Figure 2.46 shows Lean tools.

2.7.1 Cellular Design

Cellular design consists of self-contained units dedicated to performing all the operational requirements to accomplish sequential processing. With cellular design, individual cells can be fabricated and assembled to give the same performance and save time.

An electrical mains switchboard may consists of a number of cells assembled together to perform the desired operations. This method of design helps in easy maneuvering and assembly at the workplace for proper functioning.

Figure 2.47 illustrates cellular design for an electrical panel.

Sr. No.	Name of Quality Tool	Usage
Tool 1	Cellular Design	A self-contained unit dedicated to performing all the operational requirements to accomplish sequential processing.
Tool 2	Concurrent Engineering	Used for product cycle reduction time. It is a systematic approach for creating a product design that simultaneously considers all elements of the product life cycle.
Tool 3	5S	Used to eliminate waste that results from improper organization of work area.
Tool 4	Just in time (JIT)	Used to reduce inventory levels, improve cash flow, and reduce space requirements for storage of material.
Tool 5	Kanban	Used to signal that more material is required to be ordered. It is used to eliminate waste from inventory.
Tool 6	Kaizen	Used for continually eliminating waste from manufacturing processes by combining the collective talent of the company.
Tool 7	Mistake proofing	Used to eliminate the opportunity for error by detecting the potential source of error.
Tool 8	Outsourcing	Contracting out certain works, processes, and services to specialists in the discipline area.
Tool 9	Poka-Yoke	Used to detect the abnormality or error, fix or correct the error, and take action to prevent the error.
Tool 10	Single minute exchange of die (SMED)	Used to reduce setup time for changeover to new process.
Tool 11	Value stream mapping	Used to establish flow of material or information and eliminate waste and add value.
Tool 12	Visual management	Addresses both visual display and control. It exposes waste elimination/prevention.
Tool 13	Waste reduction	Focuses on reducing waste.

FIGURE 2.46
Lean tools.

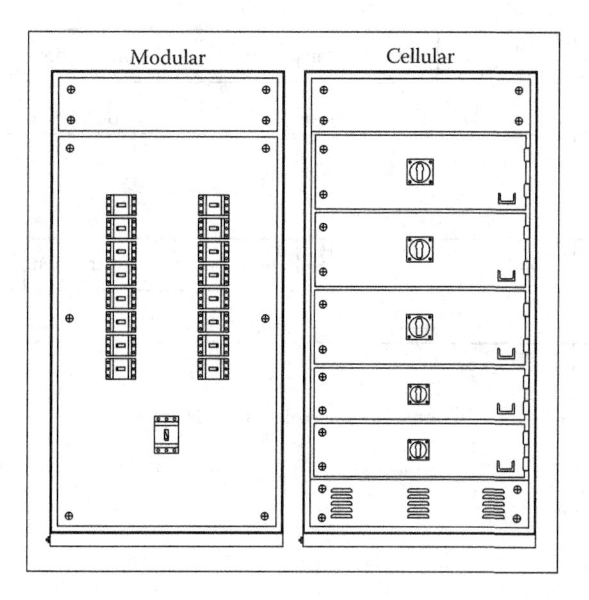

FIGURE 2.47
Cellular design.

2.7.2 Concurrent Engineering

The product life cycle begins with need and extends through concept design, preliminary design, detail design, production or construction, product use, phase out, and disposal. Concurrent engineering is defined as a systematic approach to creating a product design that simultaneously considers all the elements of the product life cycle, thus reducing the product life cycle time. It is used to expedite the development and launch of a new product. In construction projects, construction can simultaneously start while the design is under development.

Figure 2.48 shows concurrent engineering for a construction project life cycle.

2.7.3 5S

5S is a systematic approach for improvement of quality and safety by organizing a workplace. It is a methodology that advocates

- What should be kept
- Where it should be kept
- How it should be kept

5S is a Japanese concept of housekeeping that refers to five Japanese words starting with the letter S. Table 2.1 shows 5S for construction projects.

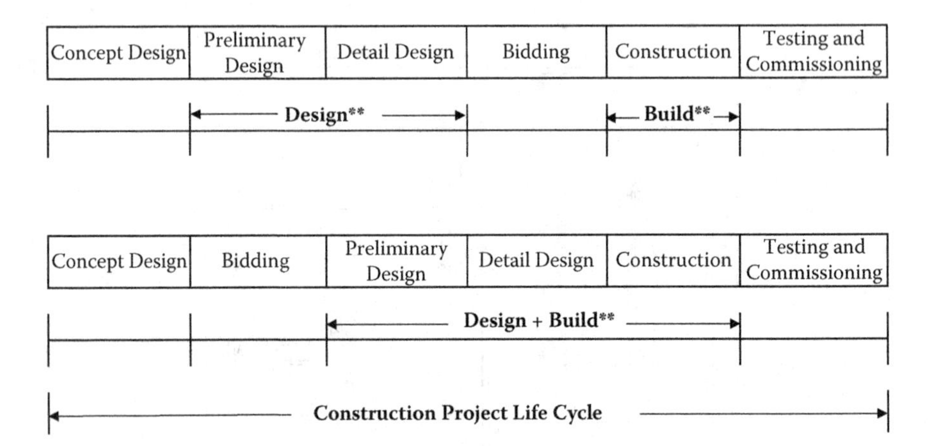

FIGURE 2.48
Concurrent engineering.

2.7.4 Just in Time (JIT)

This Lean tool is used to reduce inventory levels, improve cash flow, and reduce storage space requirements for material. For example:

1. Concrete blocks can be received at the site just before the start of block work and can be stacked near the work area where masonry work is in progress.
2. Chiller can be received at the site and directly placed on the chiller foundation without storing it in the storage yard.

2.7.5 Kanban

Kanban is used to signal that more material is required to be ordered. It is used to eliminate waste from inventory and inventory control, thus avoiding the extra storage required for a large inventory. In construction projects, electrical wires for circuiting can be ordered to be received on site when the wire-pulling work is under way. Similarly, concrete blocks and false ceiling tiles can be ordered and received as and when required.

2.7.6 Kaizen

Kaizen is used for continually improving through small changes to eliminate waste from the manufacturing process by combining the collective talent of every employee of the company.

TABLE 2.1

5S for Construction Projects

Sr. No.	5S	Related Action
1	Sort	• Determine what is to be kept in the open and what in the shed • Allocate area for each type of construction equipment and machinery • Allocate area for electrical tools • Allocate area for hand tools • Allocate area for construction material/equipment to be used/installed in the project • Allocate area for hazardous, inflammable material • Allocate area for chemicals, paints • Allocate area for spare parts for maintenance
2	Set in Order	• Keep/arrange equipment in such a way that their maneuvering/movement shall be easy • Vehicles to be parked in the yard in such a way that frequently used vehicles are parked near the gate • Frequently used equipment/machinery to be located near the workplace • Set boundaries for different type of equipment and machinery • Identify and arrange tools for easy access • Identify and store material/equipment per relevant division/section of contract documents • Identify and store materials in accordance with their usage per construction schedule • Determine items that need special conditions • Mark/tag the items/materials • Display route map and location • Put the materials in sequence per their use • Frequently used consumables to be kept near workplace • Label on the drawer with list of contents • Keep shuttering materials in one place • Determine inventory level of consumable items

3	Sweeping	• Clean site on daily basis by removing:
		  • cut pieces of reinforced bars
		  • cut pieces of plywood
		  • left-out concrete
		  • cut pieces of pipes
		  • cut pieces of cables and wires
		  • used welding rods
		• clean equipment and vehicles
		• check electrical tools after return by the technician
		• attend to breakdown report

(*Continued*)

TABLE 2.1 (*Continued*)

5S for Construction Projects

Sr. No.	5S	Related Action
4	Standardize	• Standardize the store by allocating separate areas for material used by different divisions/sections • Standardize areas for long lead items • Determine a regular schedule for cleaning the workplace • Make available standard tool kit/box for a group of technicians • Inform everyone of their responsibilities and related areas where the things are to be placed and will be available • Standardize the store for consumable items • Inform suppliers/vendors in advance the place for delivery of material
5	Sustain	• Follow the system until the end of the project

Source: Abdul Razzak Rumane. (2010). *Quality Management in Construction Projects*. CRC Press, Boca Raton, FL. Reprinted with permission of Taylor & Francis Group.

2.7.7 Mistake Proofing

Mistake proofing is used to eliminate the opportunity for error by detecting the potential source of error. Mistakes are generally categorized as follows:

1. Information
2. Mismanagement
3. Omission
4. Selection

Figure 2.49 shows a mistake proofing chart for eliminating design error.

2.7.8 Outsourcing

Outsourcing is contracting out certain work, processes, and services to a specialist in a particular discipline or area. For example, in construction projects, the following is a list of some of the work that is outsourced (subcontracted):

1. Structural concrete
2. Waterproofing work
3. HVAC work
4. Fire suppression work
5. Water supply piping
6. Electrical work

Serial Number	Items	Points to be Considered to Avoid Mistakes
1	Information	1. Terms of Reference (TOR) 2. Client's preferred requirements matrix 3. Data collection 4. Regulatory requirements 5. Codes and standards 6. Historical data 7. Organizational requirements
2	Mismanagement	1. Compare production with actual requirements 2. Interdisciplinary coordination 3. Application of different codes and standards 4. Drawing size of different trades/specialist consultants
3	Omission	1. Review and check design with TOR 2. Review and check design with client requirements 3. Review and check design with regulatory requirements 4. Review and check design with codes and standards 5. Check for all required documents
4	Selection	1. Qualified team members 2. Available material 3. Installation methods

FIGURE 2.49
Mistake proofing for eliminating design errors.

7. Security system
8. Low voltage works
9. Landscape work

2.7.9 Poka Yoke

Poke Yoke is a quality management concept developed by Shigeo Shino to prevent human errors occurring in the production line. The main objective of Poka Yoke is to achieve zero defects.

2.7.10 Single Minute Exchange of Die (SMED)

SMED is used to reduce setup time for changeover to a new process. For example, a spare circuit breaker of a similar rating can be used as an immediate replacement for a damaged circuit breaker in the electrical distribution board

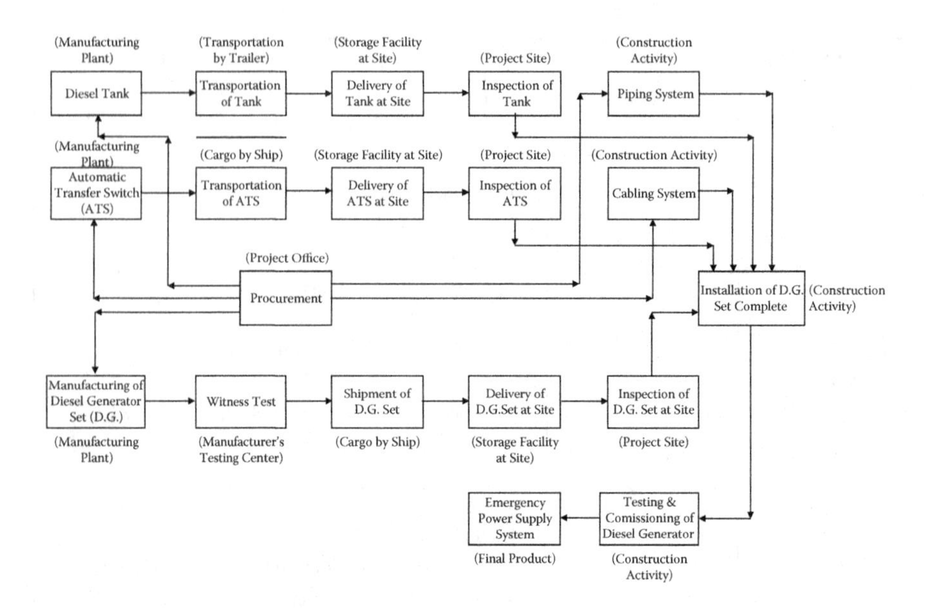

FIGURE 2.50
Value stream mapping.

to avoid breakdown of the electrical supply for a long duration. Subsequently, a new circuit breaker can be fixed in place of the spare breaker.

2.7.11 Value Stream Mapping

Value stream mapping is used to establish a flow of material or information and eliminate waste and add value. It is also used to identify areas for improvement.

Figure 2.50 shows a value stream mapping diagram for an emergency power system.

2.7.12 Visual Management

Visual Management addresses both visual display and control. It exposes waste elimination/prevention. Visual displays present information, while visual control focuses on a need to act.

2.7.13 Waste Reduction

Waste reduction focuses on reducing waste. The following are the general types of waste:

1. Defective parts
2. Delays, waiting

3. Excess inventory
4. Misused resources
5. Overproduction
6. Processing
7. Transportation
8. Untapped resources
9. Wasted motion

2.8 Cost of Quality

2.8.1 Introduction

Quality has an impact on the costs of products and services. The cost of poor quality is the annual monetary loss of products and processes that are not achieving their quality objective. The main components of the cost of low quality are

1. Cost of conformance
2. Cost of nonconformance

Table 2.2 illustrates the elements of the cost of quality.

2.8.2 Categories of Costs

Costs of poor quality are the costs associated with providing poor-quality products or services. These costs would not be incurred if things had been done right to achieve the quality objective. There are four categories of costs:

1. Internal failure costs—The costs associated with defects found before the customer receives the product or service. These also include the cost of failure to meet customer satisfaction and needs and the cost of inefficient processes.
2. External failure costs—The cost associated with defects found after the customer receives the product or service. It also includes lost opportunity for sales revenue.
3. Appraisal costs—The costs incurred to determine the degree of conformance with quality requirements.
4. Prevention costs—The costs incurred to keep failure and appraisal costs down to a minimum.

TABLE 2.2

Cost of Quality

Cost of Compliance	Cost of Noncompliance
• Quality planning	• Scrap
• Process control planning	• Rework
• Quality training	• Corrective action
• Quality audit	• Additional material/inventory cost
• Design review	• Expedition
• Product design validation	• Customer complains
• Work procedure	• Product recalls
• Method statement	• Warranty
• Process validation	• Maintenance service
• Field testing	• Field Repairs
• Third party inspection	• Rectification of returned material
• Receiving inspection	• Re-inspection or re-test
• Prevention action	• Downgrading
• In-process inspection	• Loss of business
• Outside endorsement	
• Calibration of equipment	
• Laboratory acceptance testing	

These cost categories allow the use of quality cost data for a variety of purposes. Quality costs can be used for measurement of progress, for analyzing the problem, or for budgeting. By analyzing the relative size of the cost categories, the company can determine if its resources are properly allocated.

2.8.3 Quality Cost in Construction*

Quality of construction is defined as

1. Scope of work
2. Time
3. Budget

Cost of quality refers to the total cost incurred during the entire life cycle of a construction project in preventing nonconformance with owner requirements (defined scope). There are certain hidden costs that may not directly

* *Source:* Abdul Razzak Rumane (2010), *Quality Management in Construction Projects.* CRC Press, Boca Raton, FL. Reprinted with permission from Taylor & Francis Group.

affect the overall cost of the project; however, they may add to the cost of the consultant/designer for completing the design within the stipulated schedule to meet owner requirements and conform to all the regulatory codes/standards, and for the contractor to construct the project within the stipulated schedule, meeting all the contract requirements. Rejection or nonapproval of the executed/installed works by the supervisor due to noncompliance with the specifications will cause the contractor loss in terms of

- Material
- Manpower
- Time

The contractor will have to rework or rectify the work, which will need additional resources and extra time. This may disturb the contractor's work schedule and affect the execution of other activities. The contractor has to emphasize a Zero Defect policy, particularly for concrete works. To avoid rejection of works, the contractor has to take the following seven measures:

1. Execution of works per approved shop drawings using approved material
2. Following the approved method of statement of work or the manufacturer's recommended method of installation
3. Conducting continuous inspection during the construction/installation process
4. Employing a properly trained workforce
5. Maintaining good workmanship
6. Identifying and correcting deficiencies before submitting the checklist for inspection and approval of work
7. Coordinating requirements of other trades, for example, if any opening is required in the concrete beam for crossing of a service pipe

The timely completion of a project is one of the objectives to be achieved. To avoid a delay in the completion schedule, proper planning and scheduling of construction activities is necessary. Since construction projects involve many participants, it is essential that the requirements of all the participants be fully coordinated. This will ensure execution of activities as planned, resulting in timely completion of the project.

Normally, the construction budget is fixed at the inception of a project. Any variations during the construction process must be avoided, as it may take time to get approval for additional budget, resulting in project delay. Quality costs related to construction projects can be summarized as follows:

Internal Failure Costs

- Rework
- Rectification
- Rejection of checklist
- Corrective action

External Failure Costs

- Breakdown of installed system
- Repairs
- Maintenance
- Warranty

Appraisal Costs

- Design review/preparation of shop drawings
- Preparation of composite/coordination drawings
- On-site material inspection/test
- Off-site material inspection/test
- Pre-checklist inspection

Prevention Costs

- Preventive action
- Training
- Work procedures
- Method statement
- Calibration of instruments/equipment

2.9 Quality Function Deployment (QFD)*

Quality function deployment (QFD) is a technique of translating customer requirements into technical requirements. It was developed in Japan by Dr. Yoji Akao in the 1960s to transfer the concepts of quality control from the manufacturing process into the new product development process. QFD is referred to as Voice of the Customer, which helps in identifying and

* *Source:* Abdul Razzak Rumane (2010), *Quality Management in Construction Projects*. CRC Press, Boca Raton, FL. Reprinted with permission from Taylor & Francis Group.

developing customer requirements through each stage of product or service development. It is a development process that utilizes a comprehensive matrix involving project team members.

QFD involves constructing one or more matrices containing information related to others. The assembly of several matrices showing the correlation with one another is called the *house of quality*. The house of quality matrix is the most recognized form of QFD. QFD is being applied virtually in every industry and business, such as aerospace, communication, software, transportation, manufacturing, service industries, and the construction industry. The house of quality is made up of the following six major components:

1. WHATS
2. HOWS
3. Correlation matrix (roof)–technical requirements
4. Interrelationship matrix
5. Target value
6. Competitive evaluation

WHATS is the first step in developing the house of quality. It is a structured set of needs/requirements ranked in terms of priority, and the levels of importance are specified quantitatively. It is generated by using questions such as:

- What are the types of finishes needed for the building?
- What type of air conditioning system is required for the building?
- What type of communication system is required for the building?
- What type of flooring material is required?
- Does the building need a security system?

HOWS is the second step, in which project team members translate the requirements (WHATS) into technical design characteristics (Specifications), which are listed across the columns. The correlation matrix identifies technical interaction or the physical relationship among the technical specifications. The interrelationship matrix illustrates a team member's perception of the interrelationship between the owner's requirements and technical specifications. The bottom part allows for technical comparison between possible alternatives, target values for each technical design characteristic, and performance measurement. The right side of the house of quality is used for planning purposes. It shows customer perceptions observed in a market survey.

The QFD technique can be used to translate the owner's need/requirements into development of a set of technical requirements during conceptual design.

Figure 2.51 shows the house of quality for a hospital building.

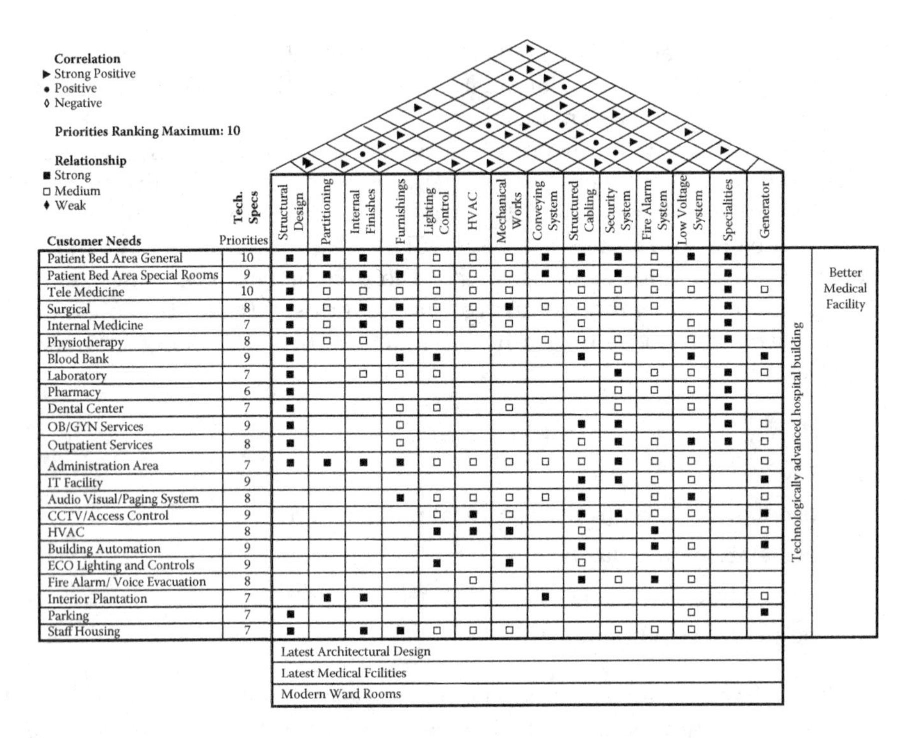

FIGURE 2.51
House of quality for hospital building.

2.10 Six Sigma*

2.10.1 Introduction

Six Sigma is, basically, a process quality goal. It is a process quality technique that focuses on reducing variation in processes and preventing deficiencies in a product. In a process that has achieved Six Sigma capability, the variation is small compared to the specification limits.

Sigma is a Greek letter (σ) that stands for standard deviation. Standard deviation is a statistical way to describe how much variation exists in a set of data, a group of items, or a process. The standard deviation is the most useful measure of dispersion. Six Sigma means that for a process to be at the Six Sigma level, the specification limits should be at least 6σ from the average point. So the total spread between the upper specification (control) limit and the lower specification (control) limit should be 12σ. With Motorola's Six Sigma program, no more than 3.4 defects per million fall outside the

* *Source:* Abdul Razzak Rumane (2010), *Quality Management in Construction Projects.* CRC Press, Boca Raton, FL. Reprinted with permission from Taylor & Francis Group.

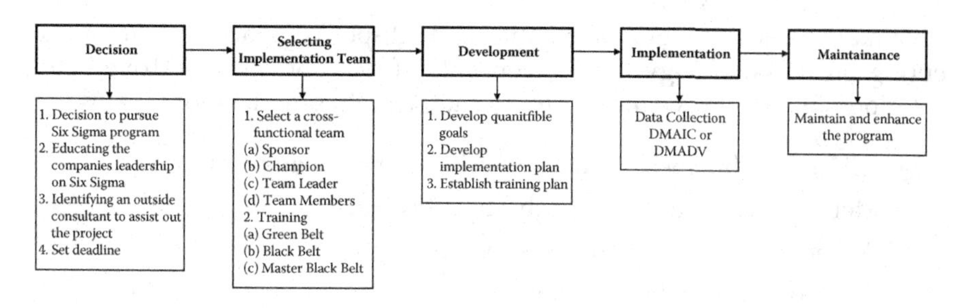

FIGURE 2.52
Six Sigma roadmap.

specification limits with a process shift of not more than 1.5σ from the average or mean. Six Sigma started as a defect reduction effort in manufacturing and was then applied to other business processes for the same purpose.

Six Sigma is a measurement of "goodness" using a universal measurement scale. It provides a relative way to measure improvement. "Universal" means sigma can measure anything from coffee mug defects to missed chances to close a sales deal. It simply measures how many times a customer's requirements were not met (a defect), given a million opportunities. Sigma is measured in defects per million opportunities (DPMO). For example, a level of sigma can indicate how many defective coffee mugs were produced when one million mugs were manufactured. Levels of sigma are associated with improved levels of goodness. To reach a level of Three Sigma, you can only have 66,811 defects, given a million opportunities. A level of Five Sigma only allows 233 defects. Minimizing variation is a key focus of Six Sigma. Variation leads to defects, and defects lead to unhappy customers. To keep customers satisfied, loyal, and coming back, you have to eliminate the sources of variation. Whenever a product is created or a service performed, it needs to be done the same way every time, no matter who is involved. Only then will you truly satisfy the customer. Figure 2.52 illustrates the Six Sigma roadmap.

2.10.2 Six Sigma Methodology

Six Sigma is an overall business improvement methodology that focuses an organization on

- Understanding and managing customer requirements
- Aligning key business process to achieve these requirements
- Utilizing rigorous data analysis to minimize variation in these processes
- Driving rapid and sustainable improvement in business process by reducing defects, cycle time, impact to the environment, and other undesirable variations
- Timely execution

As a management system, Six Sigma is a high-performance system for executing business strategy. It uses concepts of fact and data to drive better solutions. Six Sigma is a top-down solution to help organizations

- Align their business strategy to critical improvement efforts
- Mobilize teams to attack high-impact projects
- Accelerate improved business results
- Govern efforts to ensure that improvements are sustained

The Six Sigma methodology also focuses on

- Leadership principles
- Integrated approach to improvement
- Engaged teams
- Analytic tools
- Hard-coded improvements

2.10.2.1 Leadership Principles

The Six Sigma methodology has four leadership principles:

1. Align
2. Mobilize
3. Accelerate
4. Govern

A brief description of these leadership principles follows:

1. Align—The leadership should ensure that all improvement projects are in line with the organization's strategic goals.

   Alignment begins with the leadership team developing a scorecard. This vital tool, the cornerstone of the Six Sigma business improvement campaign, translates strategy into tactical operating terms. The scorecard also defines metrics an organization can use to determine success. Just as a scoreboard at a sporting event tells you who is winning, the scorecard tells leadership how well the company is meeting its goals.

2. Mobilize—The leadership should enable teams to take action by providing clear direction, feasible scope, a definition of success, and rigorous reviews.

   Mobilizing sets clear boundaries, lets people go to work, and trains them as required.

The key to mobilizing is focus; lack of focused action was one of the pitfalls of previous business improvement efforts. True focus means the project is correctly aligned with the organization's scorecard. Mobilized teams have a valid reason for engaging in improvement efforts; they can see benefits for the customer. The project has strategic importance, and they know it. They know exactly what must be done and the criteria they can use to determine success.

3. Accelerate—The leadership should drive a project to rapid results through tight clock management, training as needed, and shorter deadlines.

More than 70% of all improvement initiatives fail to achieve the desired results in time to make a difference. For projects to make an impact, they must achieve results quickly, and that is what acceleration is all about.

The Accelerate Leadership Principle involves three main components:

a. Action learning
b. Clock management
c. Effective planning

Accelerate employs an "action learning" methodology to quickly bridge the gap from "learning" to "doing." Action learning mixes traditional training with direct application. Training is received while working on a real-world project, allowing plenty of opportunity to apply new knowledge. The instructor is not simply a trainer, but a coach as well, helping work with real-world projects. Action learning accelerates improvement over traditional learning methods. One can receive training and also complete a worthwhile project at the same time. In addition to the four-to-six-month time frame, Accelerate requires teams to set deadlines that are reinforced through rigorous reviews.

4. Govern—The leadership must visibly sponsor projects and conduct regular and rigorous reviews to make critical midcourse corrections.

The fourth Leadership Principle is Govern. Once the leadership selects an improvement opportunity, their work is not done. They must remain ultimately responsible for the success of that project. Govern requires leaders to drive for results.

While governing a Six Sigma project, you need

- A regular communications plan and a clear review process
- Active sponsorship of teams and their projects
- Encouragement of proactive dialogue and knowledge sharing in the team and throughout the organization

2.10.2.2 Six Sigma Team

Teamwork is absolutely vital for complex Six Sigma projects. For teams to be effective, they must be engaged: involved, focused, and committed to meeting their goals. Engaged teams must have leadership support. There are four types of teams:

1. Black Belts
2. Green Belts
3. Breakthrough
4. Blitz

A brief description of these teams follows.

2.10.2.2.1 Black Belt

Black Belt teams are led by a Black Belt, and may have Green Belts and functional experts assigned to complex, high-impact process improvement projects or designing new products, services, or complex processes. Black Belts are internal Six Sigma practitioners, skilled in the application of rigorous statistical methodologies, and they are crucial to the success of Six Sigma. Their additional training and experience provide them with the skills they need to tackle difficult problems. Black Belts have many responsibilities:

- Function as a Team Leader on Black Belt projects
- Integrate their functional discipline with statistical, project, and interpersonal skills
- Serve as internal consultants
- Tackle complex, high-impact improvement opportunities
- Mentor and train Green Belts

2.10.2.2.2 Green Belt

Led by a Green Belt and comprising nonexperts, Green Belt teams tackle less complex, high-impact process improvement projects. Green Belt teams are often coached by Black Belts or Master Black Belts.

Green Belts are also essential to the success of Six Sigma. They perform many of the same functions as Black Belts, but their work requires less complex analysis. Green Belts are trained in basic problem-solving skills and the statistical tools needed to work effectively as members of process improvement teams.

Green Belt responsibilities include:

- Acting as Team Leader on business improvements requiring less complex analysis
- Adding their unique skills and experiences to the team

- Working with the team to come up with inventive solutions
- Performing basic statistical analysis
- Conferring with a Black Belt as questions arise

2.10.2.2.3 Breakthrough

While creating simple processes, sophisticated statistical tools may not be needed. Breakthrough teams are typically used to define low-complexity new processes.

2.10.2.2.4 Blitz

Blitz teams are put in place to quickly execute improvements produced by other projects. These teams can also implement Digitization for Efficiency using a new analytic tool set.

For typical Six Sigma projects, four critical roles exist:

1. Sponsor
2. Champion
3. Team leader
4. Team member

A Sponsor typically

- Remains ultimately accountable for a project's impact
- Provides project resources
- Reviews monthly and quarterly achievements, obstacles, and key actions
- Supports the project Champion by removing barriers as necessary

A Champion typically

- Reviews weekly achievements, obstacles, and key actions
- Meets with the team weekly to discuss progress
- Reacts to changes in critical performance measures as needed
- Supports the Team Leader, removing barriers as necessary
- Helps ensure project alignment

A Team Leader typically

- Leads improvement projects through an assigned, disciplined methodology
- Works with the Champion to develop the Team Charter, review project progress, obtain necessary resources, and remove obstacles

- Identifies and develops key milestones, timelines, and metrics for improvement projects
- Establishes weekly, monthly, and quarterly review plans to monitor team progress
- Supports the work of team members as necessary

Team members typically

- Assist the Team Leader
- Follow a disciplined methodology
- Ensure that the Team Charter and timeline are being met
- Accept and execute assignments
- Add their views, opinions, and ideas

2.10.3 Analytic Tool Sets

The following are the analytic tools used in Six Sigma projects.

2.10.3.1 Ford Global 8D Tool

What problem needs solving? →

Who should help solve problem? →

How do we quantify symptoms? →

How do we contain it? →

What is the root cause? →

What is the permanent corrective action? →

How do we implement? →

How can we prevent this in future? →

Whom should we reward? →

Ford Global 8D Tool is primarily used to bring performance back to a previous level.

2.10.3.2 DMADV Tool Set Phases

Define → What is important?

Measure → What is needed?

Analyze → How will we fulfill?

Design → How do we build it?

Verify → How do we know it will work?

TABLE 2.3

Fundamental Objectives of Six Sigma DMADV Tool

DMADV	Phase	Fundamental Objective
1	Define—What is important?	Define the project goals and customer deliverables (internal and external)
2	Measure—What is needed?	Measure and determine customer needs and specifications
3	Analyze—How is it fulfilled?	Analyze process options and prioritize based on capabilities to satisfy customer requirements
4	Design—How do we build it?	Design detailed processes capable of satisfying customer requirements
5	Verify—How do we know it will work?	Verify design performance capability

The DMADV tool is used primarily for the invention and innovation of modified or new products, services, or processes. Using this toolset, Black Belts optimize performance before production begins. DMADV is proactive, solving problems before they start. This tool is also called DFSS (Design for Six Sigma).

Table 2.3 lists fundamental objectives of DMADV.

2.10.3.3 DMAIC Tool

Define → What is important?

Measure → How are we doing?

Analyze → What is wrong?

Improve → What needs to be done?

Control → How do we guarantee performance?

The DMAIC tool refers to a data-driven quality strategy and is used primarily for improvement of an existing product, service, or process.

Table 2.4 lists the fundamental objectives of DMAIC.

2.10.3.4 DMADDD Tool

Define → Where must we be leaner?

Measure → What's our baseline?

Analyze → Where can we free capacity and improve yields?

Design → How should we implement?

Digitize → How do we execute?

Draw Down → How do we eliminate parallel paths?

TABLE 2.4

Fundamental Objectives of Six Sigma DMAIC Tool

DMAIC	Phase	Fundamental Objective
1	Define—What is important?	Define the project goals and customer deliverables (internal and external)
2	Measure—How are we doing?	Measure the process to determine current performance
3	Analyze—What is wrong?	Analyze and determine the root cause(es) of the defects
4	Improve—What needs to be done?	Improve the process by permanently removing the defects
5	Control—How do we guarantee performance?	Control the improved process's performance to ensure sustainable results

TABLE 2.5

Fundamental Objectives of Six Sigma DMADDD Tool

DMADDD	Phase	Fundamental Objective
1	Define—Where must we be leaner?	Identify potential improvements
2	Measure—What's our baseline?	Analog touch points
3	Analyze—Where can we free capacity and improve yields?	Task elimination and consolidated operations Value added/non-value-added tasks Free capacity and yield
4	Design—How should we implement?	Future state vision Define specific projects Define draw down timing Define commercialization plans
5	Digitize—How do we execute?	Execute project
6	Draw Down—How do we eliminate parallel paths?	Commercialize new process Eliminate parallel path

The DMADDD tool is primarily used to drive the cost out of a process by incorporating digitization improvements. These improvements can drive efficiency by identifying non-value-added tasks and using simple web-enabled tools to automate certain tasks and improve efficiency. In doing so, employees can be freed up to work on more value-added tasks.

Table 2.5 lists the fundamental objectives of DMADDD.

DMADV PROCESS

Define Phase: What is important?

(Define the project goals and customer deliverables.)

The key deliverables of this phase are

- Establish the goal.
- Identify the benefits.

- Select the project team.
- Develop a project plan.
- Project charter.

Measure Phase: What is needed?
(Measure and determine customer needs and specifications.)
The key deliverable in this phase is

- Identify specification requirements.

Analyze Phase: How do we fulfill it?
(Analyze process options and prioritize based on capability to satisfy customer requirements.)
The key deliverables in this phase are

- Design generation (data collection)
- Design analysis
- Risk analysis
- Design model (prioritization of data under major variables)

Design Phase: How do we build it?
(Design detailed processes capable of satisfying customer requirements.)
The key deliverables in this phase are

- Constructing a detail design
- Converting CTQs (Critical to Quality) into CTPs (Critical to Process) elements
- Estimating the capabilities of the CTPs in the design
- Preparing a verification plan

Verify Phase: How do we know it works?
(Verify design performance capability.)
The key deliverable in this phase is

- Designing a control and transition plan

THE DMAIC PROCESS
The majority of the time, Black and Green Belts approach their projects with the DMAIC analytic tool set, driving process performance to never-before-seen levels.

DMAIC has the following five fundamental objectives:

1. Define Phase: Define the project and customer deliverables
2. Measure Phase: Measure the process performance and determine current performance

3. Analyze: Collect, analyze and determine the root cause(s) of variation and process performance
4. Improve: Improve the process by diminishing defects with alternative remedial
5. Control: Control improved process performance

The DMAIC process contains five distinct steps that provide a disciplined approach to improving existing processes and products through the effective integration of project management, problem solving, and statistical tools. Each step has fundamental objectives and a set of key deliverables, so the team member will always know what is expected of him or her and his or her team.

DMAIC stands for the following:

- Define opportunities
- Measure performance
- Analyze opportunity
- Improve performance
- Control performance

Define Opportunities (What is important?)

The objective of this phase is to identify and validate the improvement opportunities that will help achieve the organization's goals and provide the largest payoff, develop the business process, define critical customer requirements, and prepare to function as an effective project team.

Key deliverables in this phase include the following:

- Team charter
- Action plan
- Process map
- Quick win opportunities
- Critical customer requirements
- Prepared team

Measure Performance (How are we doing?)

The objectives of this phase are

- To identify critical measures that are necessary to evaluate success or failure, meet critical customer requirements, and begin developing a methodology to effectively collect data to measure process performance

- To understand the elements of the Six Sigma calculation and establish baseline sigma for the processes the team is analyzing

Key deliverables in this phase include

- Input, process, and output indicators
- Operational definitions
- Data collection format and plans
- Baseline performance
- Productive team atmosphere

Analyze Opportunity (What is wrong?)
The objectives of this phase are

- To stratify and analyze the opportunity to identify a specific problem and define an easily understood problem statement
- To identify and validate the root causes and thus the problem the team is focused on
- To determine true sources of variation and potential failure modes that lead to customer dissatisfaction

Key deliverables in this phase include

- Data analysis
- Validated root causes
- Sources of variation
- Failure modes and effects analysis (FMEA)
- Problem statement
- Potential solutions

Improve Performance (What needs to be done?)
The objectives of this phase are

- To identify, evaluate, and select the right improvement solutions
- To develop a change management approach to assist the organization in adapting to the changes introduced through solution implementation

Key deliverables in this phase include

- Solutions
- Process maps and documentation
- Pilot results

- Implementation milestones
- Improvement impacts and benefits
- Storyboard
- Change plans

Control Performance (How do we guarantee performance?)
The objectives of this phase are

- To understand the importance of planning and executing against the plan and determine the approach to be taken to ensure achievement of the targeted results
- To understand how to disseminate lessons learned, identify replication and standardization opportunities/processes, and develop related plans

Key deliverables in this phase include

- Process control systems
- Standards and procedures
- Training
- Team evaluation
- Change implementation plans
- Potential problem analysis
- Solution results
- Success stories
- Trained associates
- Replication opportunities
- Standardization opportunities

Six Sigma methodology is not so commonly used in construction projects; however, the DMAIC tool can be applied at various stages in construction projects:

1. Detailed Design Stage—To enhance the coordination method in order to reduce repetitive work
2. Construction Stage—Preparation of builder's workshop drawings and composite drawings, as this requires a lot of coordination among different trades
3. Construction Stage—Preparation of the contractor's Construction Schedule
4. Execution of works

Impact of Six Sigma Strategy

The Six Sigma strategy affects five fundamental areas of business:

1. Process improvement
2. Product and service improvement
3. Customer satisfaction
4. Design methodology
5. Supplier improvement

2.11 TRIZ

TRIZ is short for teirija rezhenijia izobretalenksh zadach (theory of inventive problem solving), developed by the Russian scientist Genrish Altshuller. TRIZ provides systematic methods and tools for analysis and innovative problem solving to support the decision-making process.

Continuous and effective quality improvement is critical for an organization's growth, sustainability, and competitiveness. The cost of quality is associated with both chronic and sporadic problems. Engineers are required to identify, analyze the causes, and solve these problems by applying various quality improvement tools. Any of these quality tools taken individually does not allow a quality practitioner to carry out the complete problem solving cycle. These tools are useful for solving a particular phase of the problem, and need combinations of various tools and methods to find a solution. TRIZ is an approach that starts at a point where fresh thinking is needed to redesign a process or develop a new process. It focuses on a method for developing ideas to improve a process, get something done, design a new approach, or redesign an existing approach. TRIZ offers a more systematic, although still universal, approach to problem solving. It has advantages over other problem-solving approaches in terms of time efficiency and has a low-cost quality improvement solution. The pillar of TRIZ is the realization that contradictions can be methodically resolved through the application of innovative solutions. Altshuller defined an inventive problem as one containing a contradiction. He defined contradiction as a situation where an attempt to improve one feature of system detracts from another feature.

2.11.1 TRIZ Methodology

Traditional processes for increasing creativity have a major flaw in that their usefulness decreases as the complexity of the problem increases. At times, a trial-and-error method is used in every process, and the number

TABLE 2.6

Level of Inventiveness

Level	Degree of Inventiveness	% of Solutions	Source of Solution
1	Obvious solution	32%	Personal skill
2	Minor improvement	45%	Knowledge within existing systems
3	Major improvement	18%	Knowledge within the industry
4	New concept	4%	Knowledge outside industry and is found in science, not in technology
5	Discovery	1%	Outside the confines of scientific knowledge

of trials increases with the complexity of the problem. In 1946, Altshuller was determined to improve the inventive process by developing the "science" of creativity, which led to the creation of TRIZ. TRIZ was developed by Altshuller as a result of analysis of many thousands of patents. He reviewed over 200,000 patents, looking for problems and how they are solved. He selected 40,000 as representative of inventive solutions; the rest were direct improvements easily recognized within the system. Altshuller recognized a pattern in which some fundamental problems were solved with solutions that were repeatedly used from one patent to another, although the patent subject, applications, and timings varied significantly. Altshuller categorized these patterns into five levels of inventiveness. Table 2.6 summarizes Altshuller's findings.

He noted that with each succeeding level, the source of the solution required broader knowledge and more solutions to consider before an ideal solution could be found.

TRIZ is a creative thinking process that provides a highly structured approach for generating innovative ideas and solutions for problem solving. It provides tools and methods for use in problem formulation, system analysis, failure analysis, and pattern of system evolution. TRIZ, in contrast to techniques such as brainstorming, aims to create an algorithmic approach to the invention of new systems and refinement of old systems. Using TRIZ requires some training and a good deal of practice.

The TRIZ Body of Knowledge contains 40 creative principles drawn from analysis of how complex problems have been solved:

- The laws of systems solution
- The algorithm of inventive problem solving
- Substance-field analysis
- Seventy-six standard solutions

2.11.2 Application of TRIZ

Engineers can apply TRIZ for solving the following problems in construction projects:

- Nonavailability of specified material
- Regulatory changes in the use of a certain type of material
- Failure of a dewatering system
- Casting of a lower grade of concrete to that of a specified higher grade
- Collapse of a trench during excavation
- Collapse of formwork
- Collapse of a roof slab while casting is in progress
- Chiller failure during the peak hours in the summer
- Modifying the method statement
- The quality auditor can use TRIZ to develop corrective actions to the audit findings during auditing

2.11.3 TRIZ Process

Altshuller has recommended four steps to invent a new solution to a problem:

Step 1—Identify the problem.
Step 2—Formulate the problem.
Step 3—Search for a precisely well-solved problem.
Step 4—Generate multiple ideas and adapt a solution.

The aforementioned methods are primarily used for low-level problems. To solve more difficult problems, more precise tools are used:

1. ARIZ (Algorithm for Inventive Problem Solving)
2. Separation Principles
3. Substance-Field Analysis
4. Anticipator Failure Determination
5. Direct Product Evaluation

A quality function deployment (QFD) matrix is also used to identify new functions and performance levels to achieve a truly exciting level of quality by eliminating technical bottlenecks at the conceptual stage. QFD may be used to feed data into TRIZ, especially using the "rooftop" to help develop contradictions.

The different schools for TRIZ and individual practitioners have continued to improve and add to the methodology.

3

Quality Tools for Construction Projects

3.1 Introduction

The tools discussed in Chapter 2 have a variety of applications in construction projects. From the perspective of managing construction projects, these tools can be grouped into the following ten areas:

1. Project development
2. Project planning
3. Project monitoring and control
4. Quality management
5. Procurement/contract management
6. Risk management
7. Safety management
8. Quality assessment/measurement
9. Training and development
10. Customer satisfaction

3.2 Project Development

Project development is a process beginning with the project initiation and ending with the closeout and finalizing of project records after the construction project. The project development process is initiated in response to an identified need. It covers a range of time-framed activities extending from identification of a project need to a finished set of contract documents to the actual construction.

Construction project development has three major elements:

1. Study
2. Design
3. Construction

As the project develops, more information and specifications are developed.

Figure 3.1 illustrates the major elements of the construction project development process. Figure 3.2 illustrates the construction project development stages.

3.3 Project Planning

Project planning is the key to a successful project. It determines how the project will be successfully completed. It is the process of identifying all the activities needed to successfully complete the project.

Table 3.1 lists the basic reasons for planning the project.

Project planning is concerned with completing a project within a certain time frame, usually with defined stages and designated resources. Planning describes what needs to be done, when, by whom, and to what standards. The following are the steps to prepare a project plan:

1. The first step of project planning is to clearly define the problem to be solved by the project.
2. Once the problem is clearly defined, the next step is to define the project objectives or goals. Establishing properly defined objectives and goals is the most fundamental element of project planning. Therefore, the goals/objectives must be
   - Specific
   - Measurable
   - Agreed upon/Achievable
   - Realistic
   - Time (cost) limited
3. The next step in project planning is to identify the project deliverables.
4. The project deliverables are subdivided into smaller activities to enable preparation time and cost estimates.
5. The next step is to estimate activity resources, activity duration, and develop a schedule.
6. The next step is to estimate costs and develop a budget based on project deliverables and schedule.
7. Simultaneously, supporting plans such as Quality Management, Human Resources, Communication, Risk Management, and Contract Management are developed.
8. The next step is to compare the plan for compliance with the original project objectives.

Sr. No.	Stages	Elements	Description
1	Study	Problem Statement/ Need Identification	Define project needs, goals, objectives.
		Need Assessment	Identification of needs.
			Prioritization of needs.
			Leveling of needs.
			Deciding what needs to be addressed.
		Need Analysis	Perform project need analysis/study to outline the scope of issues to be considered in the planning phase.
		Feasibility Study	Technical studies, economics assessment, financial assessment, market demand, environmental and social assessment.
		Establish Project Objectives/Goals	Scope, time, cost, quality.
		Identify Project Delivery System	Establish how the participants, owner, designer (A/E), and contractor will be involved in the construction of the project/facility. (Design/Build/Bid, Design/Build, Guaranteed Maximum Price, CM type, PM type, BOT, Turnkey, etc.)
		Identify Project Team	Select design (A/E) firm if Design/Bid/ Build type of contract system is selected. Select other team members based on project delivery system requirements.
		Identify Alternatives (Conceptual Alternatives)	Identify alternatives based on a predetermined set of performance measures.
		Preliminary Resources	Estimate resources.
		Preliminary Schedule	Estimate the duration for completion of project/facility.
		Preliminary Financial Implications	Preliminary budget estimates of total project cost (life-cycle cost) on the basis of any known research and development requirements. This will help arrange the finances (funding agency).
		Authorities Clearance	Identify issues, sustainability, impacts, project constraints and potential approvals (environmental authorities, permits) required so that subsequent design and authority approval processes.
		Select Alternatives	Select preferred alternative.
		Project Initiation	Project charter.

FIGURE 3.1
Major elements of construction project development process.

2	Design	Develop Concept Design	Report, drawings, models, presentation.
		Project Planning	Prepare project plan.
		Schematic Design	Preliminary design, value engineering.
		Design development	Detailed design.
		Construction Documents	Construction contract documents.
3	Construction	Bidding/Tendering	Bidding package and request for proposal.
		Contractor Selection	Most competitive bidder.
		Construction	Contractor to execute contracted works.
		Monitoring & Control	Monitor and control scope, schedule, cost, quality, risk, procurement of the project.
		Commissioning and Handover	Testing, commissioning, and handover.
		Project Closeout	Close the project.

FIGURE 3.1 ***(Continued)***

9. The plan is updated, if required, to meet the original objectives/goals keeping in mind three principal aspects of construction projects: scope, time, and cost. In order to complete a project successfully, these three aspects must be balanced.
10. The project is launched based on this plan.

Figure 3.3 illustrates the construction project planning steps.

3.4 Project Monitoring and Control

Monitoring is collecting, recording, and reporting information concerning project performance that project managers and others wish to know.

Controlling is using the actual data collected through monitoring and comparing the same to the planned performance to bring actual performance in line with planned performance by correcting the variances or implementing approved changes.

Monitoring and controlling of construction projects is normally done by collecting and recording the progress and status of various activities and compiling them in the form of progress reports, comparing actual performance to the agreed-upon objectives (scope, schedule, budget, quality), and identifying the variances and taking appropriate action if they warrant any change request.

Monitoring and control of construction projects starts from the inception of the project and continues until the project closing. Project monitoring and controlling is a continuous improvement process. The principle of the PDCA

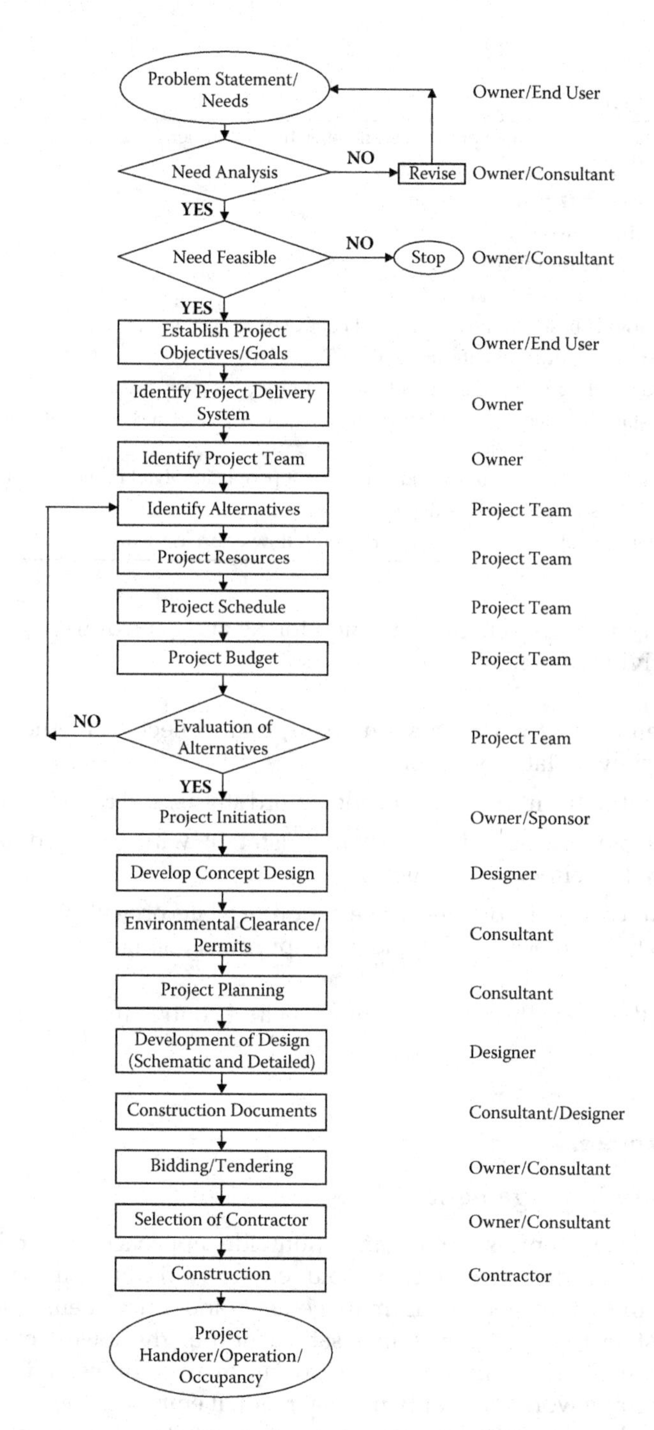

FIGURE 3.2
Construction project development stages.

TABLE 3.1

Reasons for Planning

1	To execute work in an organized and structured manner
2	To eliminate or reduce uncertainty
3	To reduce risk to the minimum
4	To reduce rework
5	To improve the efficiency of the process
6	To establish quality standards
7	To provide basis for monitoring and control of project work
8	To know the duration of each activity
9	To know the cost associated with each activity
10	To establish a benchmark for tracking the quantity, cost, and timing of work required to complete the project
11	To know the responsibility and authority of people involved in the project
12	To establish a timely reporting system
13	Integration of project activities for smooth flow of project work

cycle model can be used to describe monitoring and controlling construction process activities as follows:

Plan: Determine the information needs, data collection methods, and frequency of data collection.

Do: Collect status information needs, record status, and report progress.

Check: Compare actual performance (status) with planned performance (baseline) and analyze issues.

Act: Take corrective or preventive action, update project plan, update project documents, and implement approved changes.

Figure 3.4 illustrates the project monitoring and controlling process cycle.

3.5 Quality Management

Quality management is an organizationwide approach to understanding customer needs and delivering solutions to fulfill and satisfy the customer. Quality management is management and implementation of the quality system to achieve customer satisfaction at the lowest overall cost to the organization while continuing to improve the process. The quality system is a framework for quality management. It embraces the organization structure, policies, procedures, and processes needed to implement the quality management system.

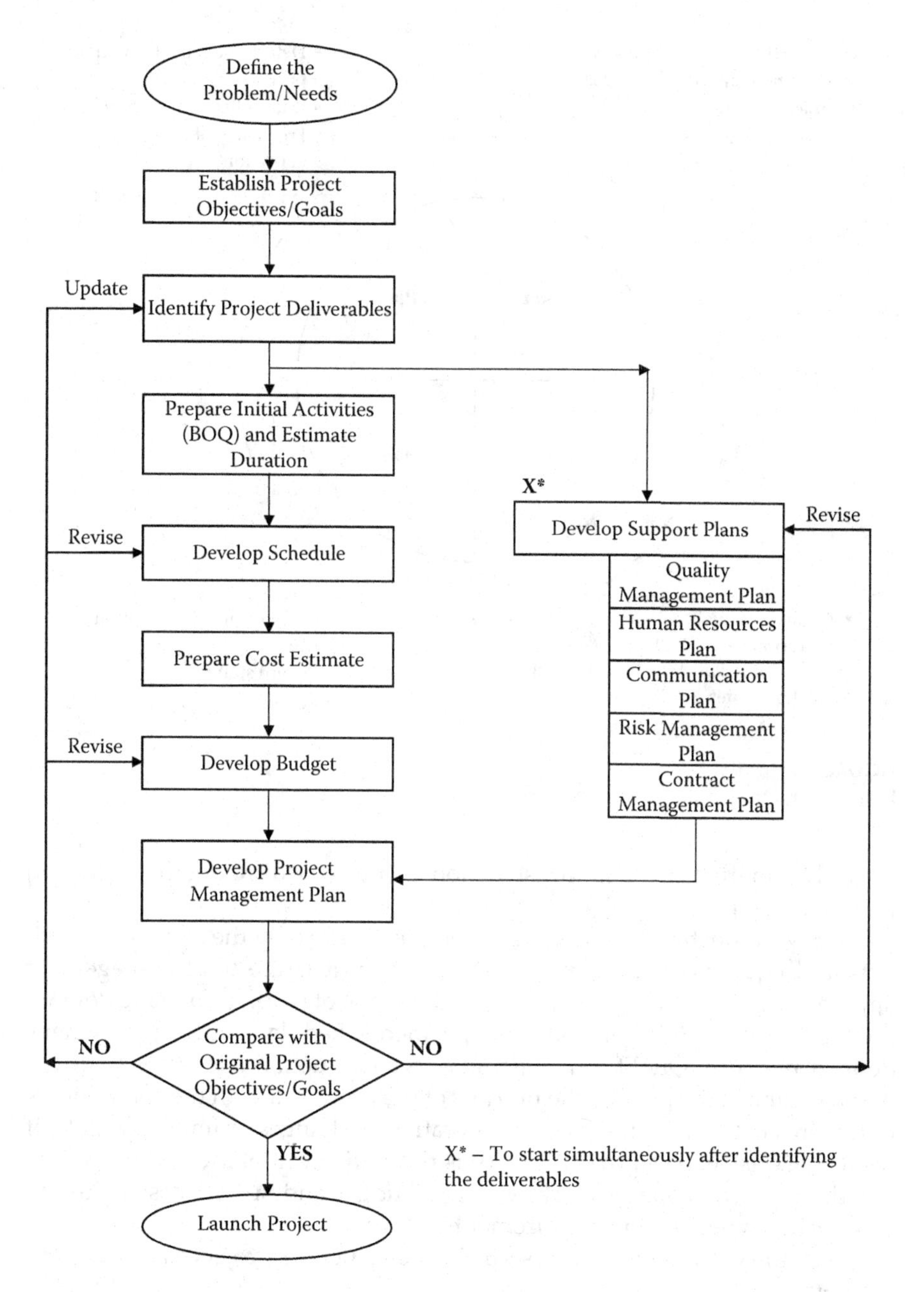

FIGURE 3.3
Construction project planning steps.

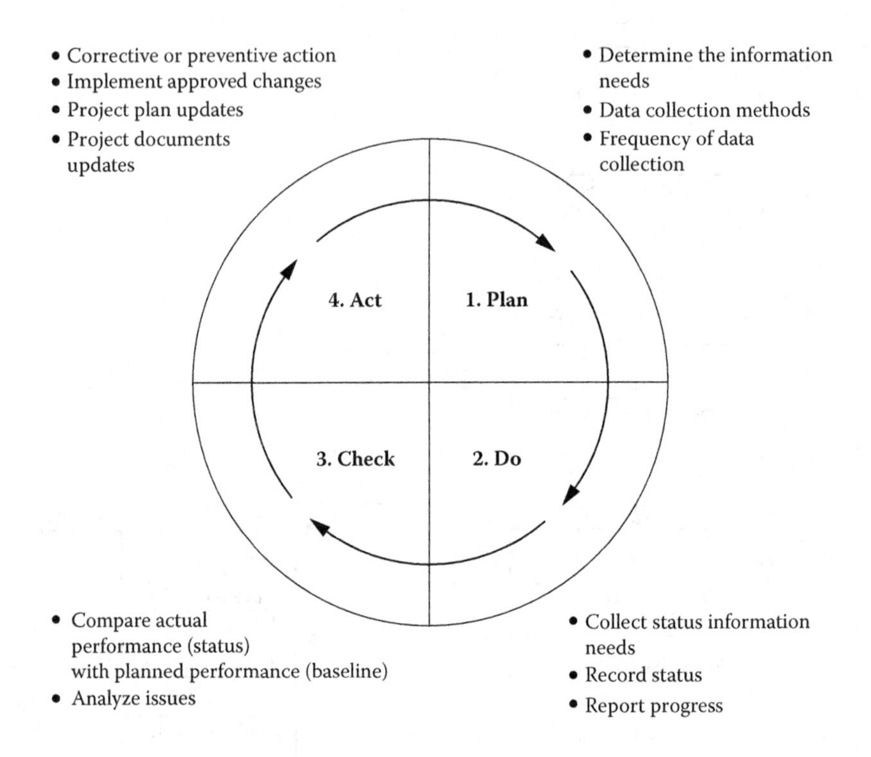

FIGURE 3.4
Project monitoring and controlling process cycle.

Quality management in construction projects is different from that in manufacturing.

Quality in construction projects encompasses not only the quality of products and equipment used in the construction, it is the total management approach to completing the facility per the scope of work to customer/owner satisfaction within the budget and specified schedule to meet the owner's defined purpose. Quality management in construction addresses both the management of the project, the product of the project, and all the components of the product. It also involves incorporation of changes or improvements, if needed. Construction project quality is the fulfillment of the owner's needs per the defined scope of work within a budget and specified schedule to satisfy the owner's/user's requirements.

The quality management system in construction projects mainly consists of

- Planning quality
- Performing quality assurance
- Performing quality control

Planning Quality:

The quality plan for construction projects is part of the overall project documentation consisting of

- Well-defined specification for all the materials, products, components, and equipment to be used to construct the facility
- Detailed construction drawings
- Detailed work procedure
- Details of the quality standards and codes to be compiled
- Cost of the project
- Manpower and other resources to be used for the project
- Project completion schedule

Performing Quality Assurance:

Quality assurance in construction projects covers all activities performed by the design team, contractor, and quality controller/auditor (supervision staff) to meet the owner's objectives as specified and to ensure and guarantee that the project/facility is fully functional to the satisfaction of the owner/end user. Auditing is part of the quality assurance function.

Performing Quality Control:

Quality control in construction projects is performed at every stage through use of various control charts, diagrams, checklists, etc., and can be defined as

- Checking of executed/installed works to confirm that they have been performed/executed as specified, using specified/approved materials, installation methods, and specified references, codes, and standards to meet the intended use
- Controlling budget
- Planning, monitoring, and controlling the project schedule

3.5.1 Project Quality Management Plan

Construction projects involve the owner, designer (consultant), and contractor. In order to achieve project objectives, both the designer as well as the contractor have to develop a project quality management plan. The designer's quality management plan should be based on the owner's project objectives, whereas the contractor's plan should take into consideration the requirements in the contract documents.

Figure 3.5 illustrates the elements of a quality management plan.

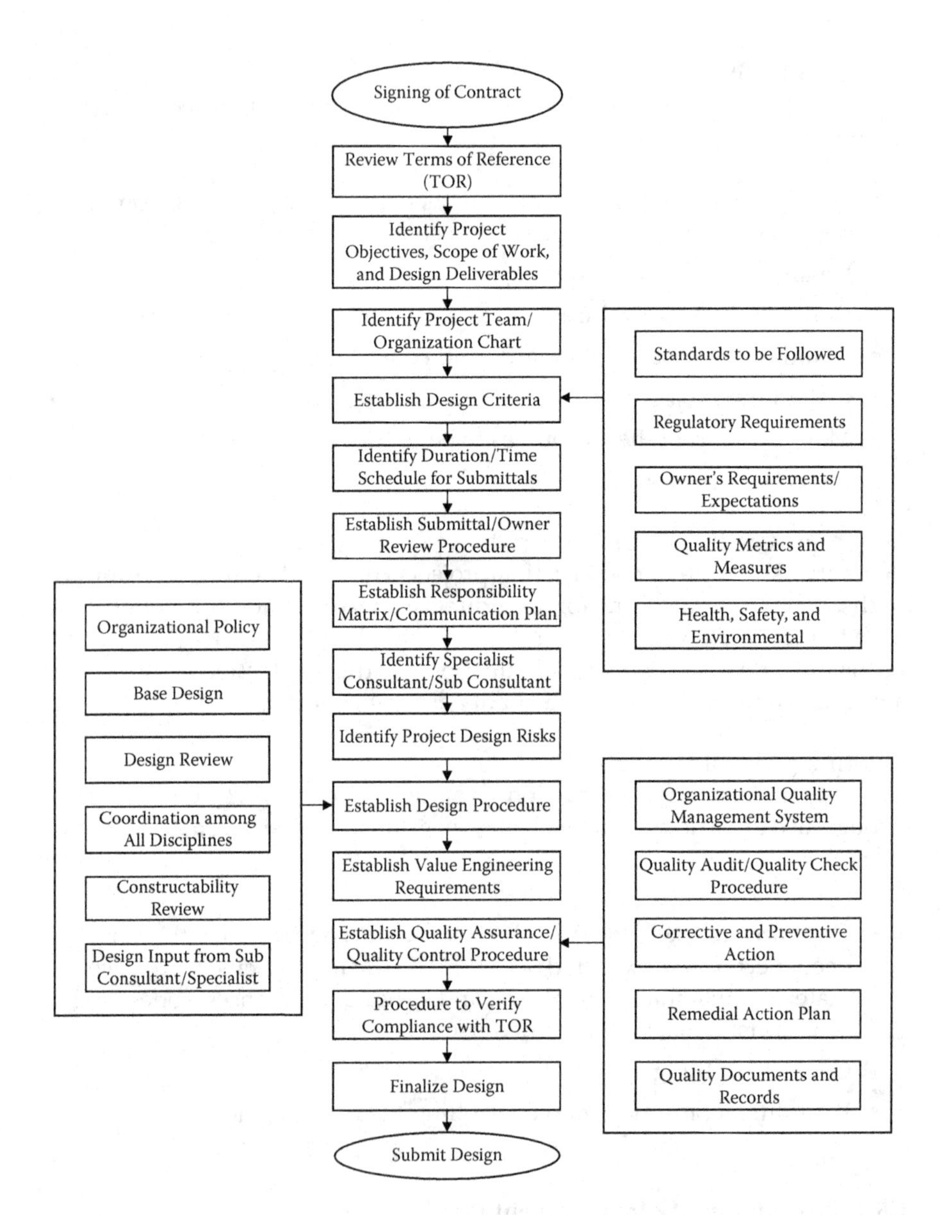

FIGURE 3.5
Major elements of quality management plan.

3.6 Risk Management

Risk management is the process of identifying, assessing, and prioritizing different kinds of risks; planning risk mitigation; implementing a mitigation plan; and controlling the risks. It is a process of thinking systematically about the possible risks, problems, or disasters before they happen, and setting up a procedure that will avoid the risk, or minimize or cope with the impact. The objectives of project risk management are to increase the probability and impacts of positive events and decrease the probability and impacts of events adverse to the project objectives. Risk is the probability that the occurrence of an event may turn into an undesirable outcome (loss, disaster). It is virtually anything that threatens or limits the ability of an organization to achieve its objectives. It can be unexpected and unpredictable events that have the potential to damage the functioning of the organization in terms of money, or in the worst-case scenario, it may cause the business to close. Risk management process consists of the following seven steps:

1. Identify the potential sources of risk in the project.
2. Analyze their impact on the project.
3. Select those with a significant impact on the project.
4. Determine how the impact of risk can be reduced.
5. Select the best alternative.
6. Develop and implement a mitigation plan.
7. Monitor and control the risks by implementing a risk response plan, tracking identified risks, identifying new risks, and evaluating the risk impact.

Figure 3.6 illustrates the risk management cycle.

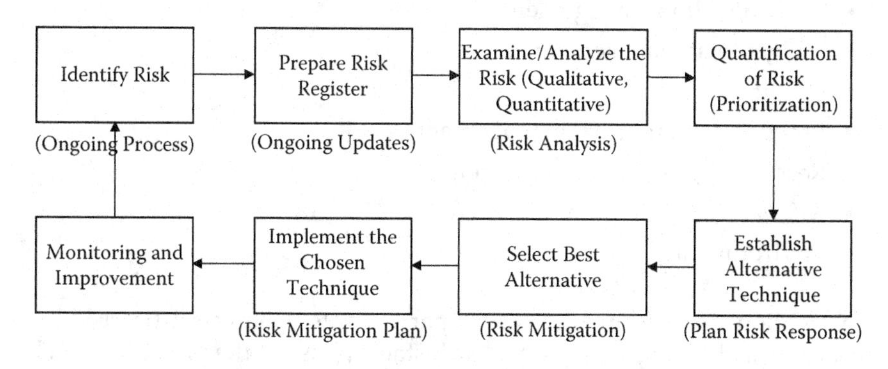

FIGURE 3.6
Risk management cycle.

Risk can be managed, mitigated, shared, transferred, or accepted. It cannot be ignored. The process of risk management is designed to reduce or eliminate the risk of certain kinds of events happening or having an impact on the business.

Construction projects are characterized as very complex projects, where uncertainty comes from various sources. Construction projects involve a cross section of many different participants. They have varying project expectations and both influence and depend on each other in addition to the other players involved in the construction process. The relationships and the contractual groupings of those who are involved are also more complex and contractually varied. Construction projects often require large amounts of materials and physical tools to move or modify these materials. Most items used in construction projects are normally produced by other construction-related industries/manufacturers. Therefore, risk in construction projects is multifaceted. Construction projects inherently contain a high degree of risk in their projections of cost and time as each is unique. No construction project is without any risk. The risk starts from the inception of the project and is tracked until completion of the project and beyond.

Risk management in construction projects is mainly focused on delivering the project with

- What was originally accepted (per the defined scope)
- Agreed-upon time (per the schedule, without any delay)
- Agreed-upon budget (no overruns)

Quality, safety, and environmental management are frequently considered subsets of risk management.

Construction project risks mainly relate to the following:

- Scope and change management
- Schedule/time management
- Budget/cost management
- Quality
- Procurement/contract management
- Resources (manpower) management
- Safety
- Environment

Construction projects involve many participants. Characteristic risks are associated with each participant at each stage/phase of the project life cycle. Therefore, it is necessary to

- Identify and record the risks that affect the project at each stage.
- Assess the likelihood of each risk occurring in the future and the impact on the project should this happen.
- Plan cost-effective management actions with clearly identified responsibilities to avoid, eliminate, or reduce any significant risk identified.
- Monitor and report the status of these risks and the effectiveness of planned risk management action.

3.6.1 Construction Project Risks

Construction projects have an abundance of risks affecting business owners, developers, design firms (consultants), contractors, and financial institutions funding the project. Each of these parties are involved in certain portions of the overall construction project risk; however, the owner has a greater share of risks, as the owner is involved from the inception until completion of the project and beyond. The owner must take initiatives to develop risk consciousness and awareness among all the parties, emphasizing the importance of explicit consideration of risk at each stage of the project. Traditionally:

- The owner/client is responsible for the investment/finance risk.
- The designer (consultant) is responsible for design risk.
- Contractors and subcontractors are responsible for the construction risk.

Risk management is an ongoing process. In order to reduce the overall risk in construction projects, the risk assessment (identification, analysis, and evaluation) process must start as early as possible to maximize project benefits. Early risk identification can lead to better estimation of the costs in the project budget, whether through contingencies, contractual plans, or insurance. Risk identification is the most important function in construction projects. Table 3.2 lists major risk factors that may affect the owner at different stages of the project, and Table 3.3 illustrates the major risks affecting the designer (consultant).

Risk factors in construction projects can be categorized in a number of ways according to the level of detail or selected viewpoints. These are categorized based on various risks factors and sources of risk. The contractor has to identify related risks affecting the construction, analyze them, evaluate the effects on the contract, and evolve a strategy to counter these risks before bidding for a construction contract. Figure 3.7 illustrates typical categories of risks in construction projects.

TABLE 3.2

Major Risk Factors Affecting the Owner

Serial Number		Risk Factor
1	*Pre-Design*	
	1.1	Long procedure for authority approval and permits
	1.2	Incomplete/improper feasibility study
	1.3	Delay in raising the funds to proceed with the work
	1.4	Change in law
	1.5	Late internal approval process from the owner team
2	*Design*	
	2.1	Incompetent consultant
	2.2	Incomplete design scope
	2.3	Inadequate or ambiguous specification
	2.4	Delays in authority approval
	2.5	Delay in completion of design
3	*Tendering*	
	3.1	Inadequate tendering price if estimate is wrong
4	*Construction*	
	4.1	Contractor lacks knowledge of and experience in similar type of construction
	4.2	Delay in site transfer
	4.3	Impractical planning and scheduling
	4.4	Omissions in design and specifications
	4.5	Inflation
	4.6	Fluctuation in exchange rate
	4.7	Increase in material cost
	4.8	Coordination problems
	4.9	Ineffectiveness and lack of supervision by consultant
	4.10	Delays in approval of transmittals
	4.11	Contracting system
	4.12	Lack of coordination between project participants
	4.13	Unpredicted contingencies, force majeure
	4.14	Political disturbances

3.6.1.1 Risk Register

A risk register is a document recording the details of all the identified risks at the beginning of the project and during the life cycle of the project in a format that consists of comprehensive lists of significant risks along with the actions and costs estimated for the identified risks. The risk register is updated every time a new risk is identified or relevant actions are taken. Table 3.4 illustrates the contents of the risk register, and Table 3.5 is an example risk assessment plan for material delivery and handling of material at the site.

TABLE 3.3

Major Risk Factors Affecting Designer (Consultant)

Serial Number		Risk Factor
1	*Concept Design*	
	1.1	Selected alternatives not suitable for final solution
	1.2	Lack of information or information transfer with wrongly estimated objectives of the proposal
	1.3	Environmental considerations
2	*Detail Design*	
	2.1	Project performance specifications
	2.2	Technical specifications
	2.3	Follow-up of codes and standards
	2.4	Difficulties in dealing with specifications and standards concerning existing conditions and client's requirements
	2.5	Lack of knowledge about the technical conditions
	2.6	Incomplete design
	2.7	Prediction of possible changes in design during construction phase
	2.8	Design completion schedule
	2.9	Incorrect estimation
	2.10	Environmental considerations
	2.11	Regulatory requirements
	2.12	Constructability

3.6.2 Reduction in Construction Project Risks

Risk response is a process of developing options and actions to enhance opportunities and to reduce threats to project activities. The following are the four main risk response strategies in construction projects:

1. Avoid
2. Transfer
3. Mitigate (reduce)
4. Accept (retain)

Risk analysis tools and techniques such as Qualitative Risk Analysis can be used to choose the most appropriate responses/actions.

The feasibility study plays an important role in deciding about the future of a construction project. It is the basis for making investment decisions and to justify the selection of the owner's objectives. Table 3.6 lists the major risk variables to be considered, using a probability matrix to conduct a feasibility study to achieve the best results.

Sr. No.	Category	Types
1	Technical	Incomplete design
		Incomplete scope of work
		Design changes
		Design mistakes
		Errors & omissions
		Incomplete specifications
		Inappropriate schedule/plan
		Inappropriate construction method
		Conflict with different trades
		Improper coordination with regulatory authorities
		Inadequate site investigation data
		New technology
2	Construction	Delay in transfer of site
		Different site conditions from the information provided
		Delay in mobilization
		Changes in scope of work
		Resource (labor) productivity issues
		Equipment/plant productivity issues
		Insufficiently skilled workforce
		Union and labor unrest
		Failure/delay of machinery and equipment
		Poor quality of material
		Failure/delay of material delivery
		Delay in approval
		Extensive subcontracting
		Subcontractor's subcontractor
		Information flow breaks
		Failure of project team members to perform as expected
3	Financial	Inflation
		Recession
		Fluctuations in exchange rate
		Availability of foreign exchange (certain countries)
		Availability of funds
		Delays in payment
		Local taxes
4	Economical	Variation of construction material price
		Sanctions
5	Commercial	Import restrictions
		Custom duties

FIGURE 3.7
Typical categories of risks in construction.

6	Logistic	Resources availability
		Spare parts availability
		Inconsistent fuel supply
		Transportation facility
		Access to worksite
		Unfamiliarity with local conditions
7	Physical	Damage to equipment
		Structure collapse
		Damage to stored material
		Leakage of hazardous material
		Injuries
		Theft
		Fire
8	Political	Change in laws and regulations
		Constraints on employment of expatriate workforce
		Use of local agent and firms
		Civil unrest
		War
9	Legal	Permits and licenses
		Professional liability
		Litigation
		Environmental protection rules
		Pollution
		Disposal of waste
		Health and safety rules
10	Natural	Flood
		Earthquake
		Cyclone
		Sandstorm
		Heavy rains
		High humidity
		Landslide
11	Contract	Dispute resolution technique
		Delay in change order negotiations

FIGURE 3.7 *(Continued)*

The feasibility study establishes the objectives of the project, and the client knows what kind of products/projects are required. It is at this stage that the client can help reduce the project risk by clearly defining the project objectives to the designer during the briefing stage. The owner always expects a high-performance/high-quality project. In order to achieve the higher expected performance/quality, the owner should define the performance/ quality standards of the project based on rational research of their own and/

TABLE 3.4

Components of Risk Register

Serial Number	Components
1	Risk identification number (Risk ID)
2	Description of risk
3	The owner of risk
4	Estimated likelihood of occurrence
5	Impact on quality, schedule, budget, safety, environment, and performance
6	Estimated severity (seriousness) of impact
7	Prioritization of risk compared to others
8	List of activities influenced (Work Breakdown Structure)
9	Leading indicators for the risk and when they must be evaluated
10	Risk mitigation and reduction plan currently in operation
11	Risk mitigation and reduction contingency plan on leading indicators
12	Timeline for mitigation actions
13	Tracking of the leading indicators and prioritization over earlier ones
14	Date of review/update
15	Forecasting risk happenings in future
16	Action to be taken in future

or the market need. The client has to ensure that the project scope is clearly defined and well documented by the designer (consultant).

Inadequate construction program scheduling is often observed in construction projects with a tight schedule when some programs need to be reduced to meet the project time line. A tight project schedule is one of the most significant risks, and it is therefore necessary to prepare a practical schedule allowing sufficient but not redundant time to accommodate all the design and construction activities. Table 3.7 lists the typical factors to be considered when reviewing the project schedule to avoid/mitigate future risk.

As time and cost are always closely correlated, a lengthy schedule will undoubtedly wreck the project cost benefit. Therefore, the project cost has to be estimated with great accuracy to minimize risk occurrence. An incomplete and inaccurate cost estimate is directly related to the designer's/consultant's knowledge and attitude toward work. The selection of an experienced designer by the owner can help minimize the difference between the proposed and practical program scheduling and estimated cost. Similarly, the designer has to clearly understand the client's needs in a technically competent way to reduce the risk in the project. Table 3.8 lists typical factors to review in cost estimation for a project to minimize the occurrence of risks.

Table 3.9 illustrates the recommendations to minimize the occurrence of risks in construction projects.

TABLE 3.5
Risk Plan

RISK ASSESSMENT OWNER NAME PROJECT NAME										
ACTIVITY = MATERIAL HANDLING										
			Initial Risk Rating				**FINAL RISK RATING**			
SL. NO.	**Hazard**	**Potential Accident or Identified Risk**	**Probability**	**× Severity**	**Risk Factor**	**CONTROL MEASURES IN PLACE/ MITIGATION ACTIONS**	**Probability**	**× Severity**	**Residual Risk Factor**	**Risk Acceptable**
1	Fall of Materials	Serious Injury or Permanent Disability	3	3	9	a. Secure materials stored in tiers by stacking, racking, blocking, or interlocking to prevent them from falling. b. Stack the materials properly in marked location. c. When lifting objects, lift with your legs, keep your back straight, do not twist, and use handling aids. d. For loads with sharp or rough edges, wear gloves or other hand and forearm protection. e. Dock boards must have handholds or other effective means for safe handling.	2	1	2	Low

(*Continued*)

TABLE 3.5 (*Continued*)

Risk Plan

RISK ASSESSMENT OWNER NAME PROJECT NAME										
ACTIVITY = MATERIAL HANDLING										
			Initial Risk Rating				**FINAL RISK RATING**			
SL. NO.	**Hazard**	**Potential Accident or Identified Risk**	**Probability**	**× Severity**	**Risk Factor**	**CONTROL MEASURES IN PLACE/ MITIGATION ACTIONS**	**Probability**	**× Severity**	**Residual Risk Factor**	**Risk Acceptable**
2	Fall of Personnel	Injury to Personnel	3	3	9	a. Store materials safely to avoid being struck by/crushed by falling material. b. Do not stand in corners where emergency movement is restricted. c. When lifting objects, lift with your legs, keep your back straight, do not twist, and use handling aids. d. Wear PPE. e. Use taglines to control the load movement. f. Avoid lifting above the shoulder level.	3	1	3	Low

RPN-Risk Priority Number

8—— 16 High Risk

4——6 Medium Risk

1—— 3 Low Risk

TABLE 3.6

Risk Variables for Feasibility Study

Serial Number	Risk Variables
1	Marketing aspect
2	Investment risk in next 10 years
3	Financial aspect (payback)
4	Economic aspect
5	Technical and technological risk
6	Environmental and spatial plan aspect
7	Regulations and policy aspect
8	Political
9	Social and cultural

TABLE 3.7

Factors for Schedule Review

Serial Number	Item to be Reviewed
1	Scope of work
2	All the activities identified and listed
3	Activity relationship is established
4	Work Breakdown Structure developed
5	Activity duration estimated
6	Project schedule network diagram prepared
7	Logical relationship between activities established
8	Critical activities established and are reasonable
9	Project critical path established
10	Construction schedule prepared

TABLE 3.8

Cost Estimate Review

Serial Number	Item to be Reviewed
1	Scope of work
2	All the activities identified and listed
3	All quantities are accurately worked out
4	Assessment of productivity and availability of resources (labor)
5	Availability of material and its cost
6	Cost escalation assumptions
7	All project-related costs, direct and indirect
8	Bidding practices
9	Identification of contingency
10	Reserves

TABLE 3.9

Recommendations to Minimize Risk Occurrence

Serial Number	Recommendation
1	Owner should know what kind of project is needed and clearly define the objectives.
2	Designer should clearly understand the owner's need and what kind of project is expected by the client.
3	Project quality and performance measures to be reasonable.
4	Project delivery system to be properly selected.
5	Project contracting/procurement system to be properly selected.
6	Selection of experienced design team.
7	Designer to carry out in-depth investigation of site conditions and design technically competent project within the limitations of the client.
8	Selection of contractor to be on performance and quality basis having full knowledge of similar types of projects.
9	Allocation of risks to the proper party.
10	Constructability of innovative design/technology to be examined.
11	Efficient communication system among all the project participants.
12	Cooperative team members with professional competency and expertise.
13	Project schedule to be practical.
14	Cost estimate to be reasonably accurate.
15	Regulatory requirements.
16	Environmental aspects.

3.6.3 Monitoring and Controlling Project Risks

A project risk action plan is a systematic process of tracking and evaluating the effectiveness of implementation of actions against established levels of risk in the areas of quality, time, and cost.

Figure 3.8 is an example project risk action plan.

3.7 Procurement/Contract Management

Project procurement/contract management comprises organizational methods, processes, and procedures to obtain the required construction products. It includes the processes to acquire construction-related products/materials, equipment, and services from outside companies.

Conventional notions of the procurement/purchasing cycle that apply in batch production, mass production, or in merchandising are less appropriate to construction projects. The procurement for construction projects involves commissioning professional services and creating a specific solution. The process is complex, involving interaction between the owner/client, designer/consultant, contractor(s), suppliers, and various regulatory bodies.

Project Risk Register
Likelihood: 5 = Certain; 4 = Likely; 3 = Possible; 2 = Unlikely; 1 = Rare
Impact: 5 = Critical; 4 = Major; 3 = Moderate; 2 = Minor; 1 = Low
Risk Level: 15–25 = High; 6–12 = Medium; 1–5 = Low

				INITIAL ASSESSMENT							MITIGATED RISK			
Serial No.	**Risk ID#**	**Risk**	**Owner of Risk**	**Likelihood**	**Impact**	**Risk Score**	**Risk Level**	**MITIGATION/ CONTROLS – TO BE ACTIONED**	**Action to be taken by**	**Date of Review/ Update**	**Likelihood**	**Impact**	**Risk Score**	**Risk Level**
1	1	Any of the many projectwide risks actually occurring and delaying project delivery.		5	4	**20**	**High**	Hold monthly risk management meetings with whole project team, continually update risk register and take necessary action.					**0**	
2	2	Resident engineer's (RE) lack of procedural and management ability creating liabilities to the employer.		4	4	**16**	**High**	Employer to replace resident engineer or order them to implement proper management systems with competent staff immediately.					**0**	

FIGURE 3.8
Project risk action plan.

3	3	Contractor delays due to non-performance to an approved FIDIC Clause 14 program.		4	4	**16**	**High**	Pursue FIDIC Clause 14 program and enforce the agreed program. Ensure timely issuance of information requested in Request for Information (RFI). Track information release against program and RFIs.					**0**	
4	4	Unagreed main works contract resulting in latent dispute and delay to project.		3	5	**15**	**High**	Expedite finalization of Bill of Quantity (BOQ) (Incl. approx quant's for provisional items). Discuss revised terms of contract with target as soon as is practicable.					**0**	
5	5	Split liability for design of major structural elements and supervision thereof, resulting in denial of responsibilities and frustration regarding project delivery.		5	3	**15**	**High**	Award major design elements and supervision responsibility to one single lead consultant who may be responsible for subconsultants, including the supervision consultant.					**0**	

FIGURE 3.8 (*Continued*)

6	6	Latent defects due to earlier failure of the dewatering system under the enabling works contractor, resulting in claims from the main contractor or end users.		3	4	**12**	**Medium**	Consultant RE for the enabling works is required to prepare a risk analysis of the past and current situation with a record of events and all notices given to the enabling works contractor. Liabilities must be clarified and the employer's rights protected. Site conditions at handover to the main contractor to be verified and confirmation given that current dewatering under the main contractor is now satisfactory.					**0**	
7	7	Delay to the works due to late design information or instructions.		3	4	**12**	**Medium**	Enforce FIDIC Clause 6.3. Ensure that contractor's RFIs include reasonable dates for information required and the potential impact of late information.					**0**	

FIGURE 3.8 ***(Continued)***

8	8	Delay and cost uncertainty due to lack of design information in contract for significant provisional items, e.g., bathroom fittings; lighting; bedroom furniture, etc., constituting large contingency items that may not be properly programmed for with uncertain installation costs, OH&P.		4	3	**12**	**Medium**	Designers to provide approximate quantities for typical items to the required level of specification to establish installation, Overhead and profit (OH&P) rates based upon supply, and installation of prime cost items.					**0**	
9	9	Limited project management resource on site resulting in lack of control of project team and processes.		3	4	**12**	**Medium**	Ensure that consultancy agreements cover all project deliverables and clear lines of responsibility and communication are established.					**0**	

FIGURE 3.8 (*Continued*)

10	10	Failure of the dewatering system provided through enabling works contractor, resulting in consequential damage and claims from main contractor.		3	3	**9**	**Medium**	Transfer responsibility of dewatering from enabling works contractor to main contractor as soon as practicable. Ensure main contractor then provides a suitable fully functioning system.					**0**	
11	11	Uncertainty of final cost.		3	3	**9**	**Medium**	Enforce change management procedures and early warnings through clause 6.3/RFIs.					**0**	
12	12	Damage to surrounding property resulting in third party claims.		3	3	**9**	**Medium**	Ensure that contractor carries out an initial property condition survey with photographic evidence and reviews on regular basis. Implement all necessary risk assessments and brief staff and operatives accordingly.					**0**	

FIGURE 3.8 ***(Continued)***

13	13	Lack of private dedicated project management offices, resulting in inability to carry out tasks in private, with the risk of misunderstandings from those observing or hearing information not agreed upon or formally disseminated resulting in disputes.		3	3	**9**	**Medium**	Employer to instruct main contractor to build dedicated project management/ consultants' office as soon as practicable.					**0**	
14	14	Latent defects to structural elements resulting in major losses and claims from end users.		2	4	**8**	**Medium**	Provision of Inherent Defects Insurance (IDI). IDI insurance may not be commercially obtainable at this stage of the project. Ensure competent engineering and construction. Ensure that consultants and contractors have robust P.I. insurance cover.					**0**	

FIGURE 3.8 ***(Continued)***

15	15	Nuisance to neighbors.		4	2	**8**	**Medium**	Contractor to implement risk analysis and mitigate noise, dust, traffic, vibration, etc., as necessary.					**0**	
16	16	Third party claims for damage while working on restricted site with heavy plant & tower crane, etc.		2	3	**6**	**Medium**	Ensure that risk assessments are carried out and all reasonable mitigation is implemented. Ensure contractor has robust insurance.					**0**	

FIGURE 3.8 (*Continued*)

Generally, a construction project comprises building materials (civil), electromechanical items, finishing items, and equipment. Construction involves installation and integration of various types of materials/products, equipment, systems, or other components to complete the project/facility.

In construction projects, the involvement of outside companies/parties starts at the early stage of the project development process. The owner/client has to decide which work is to be procured or constructed by others. Figure 3.9 illustrates the stages at which the procurement management process takes place, and Figure 3.10 illustrates the procurement management process life cycle.

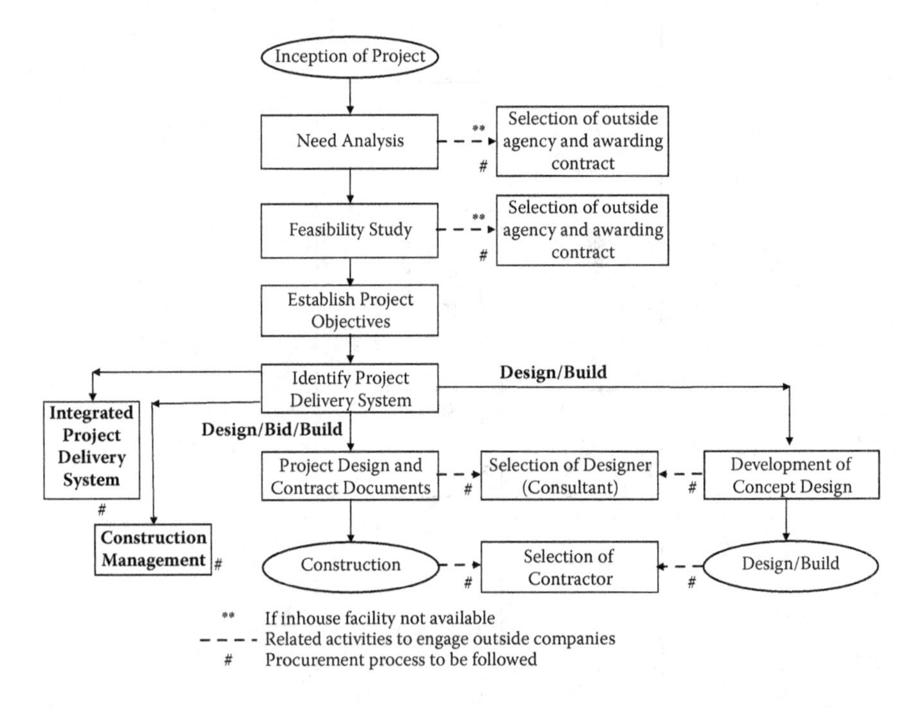

FIGURE 3.9
Procurement management process.

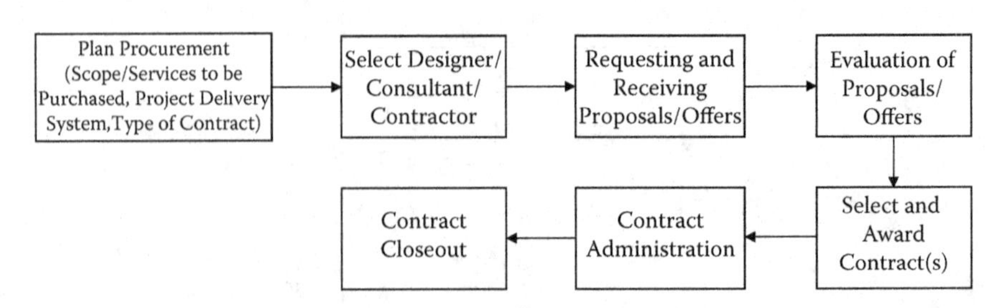

FIGURE 3.10
Procurement process life cycle.

3.7.1 Need Analysis

The procurement of construction projects is vast in scope because it involves the gathering and organizing of a myriad of separate individuals, firms, and companies to design, manage, and build numerous types of construction projects for specific clients or customers.

The procurement process in construction projects begins at an early stage, immediately after inception of the project, with identification of the need, want, or desire for one or more new facilities/projects or to improve the facility. The need should be based on real (perceived) requirements or deficiency.

The owner/client may engage a specialist consultant to perform a need analysis if there is no in-house capability to perform the study.

3.7.2 Feasibility Study

A feasibility study is defined as an evaluation or analysis of the potential impact of the identified need of the proposed project. It takes its starting point from the output of the project identification need. Once the owner's need is identified, the traditional approach is pursued through a feasibility study or an economical appraisal of owner needs for benefits. A feasibility study is performed to more clearly define the viability and form of the project that will produce the best or most profitable results. It assists decision makers (investors/owners/clients) in determining whether or not to implement the project. It may be conducted in-house if the capability exists. However, the services of a specialist or architect/engineer (A/E) are usually commissioned by the owner/client to perform such a study. Since the feasibility study stage is very crucial, in which all kinds of professionals and specialists are required to combine kinds of knowledge and experience into a broad-ranging evaluation of feasibility, it is required to engage a firm having expertise in the related fields. The feasibility study establishes the broad objectives for the project and so exerts an influence throughout the subsequent stages. The successful completion of the feasibility study marks the first of several transition milestones and is therefore most important to determine whether or not to implement a particular project or program. The feasibility study decides the possible design approaches that can be pursued to meet the need. Table 3.10 illustrates the required qualifications to engage a specialist firm to perform a feasibility study.

After completion of the feasibility study, it is possible to define the project objectives. The project objective definition will usually include the following information:

- Project scope and project deliverables
- Preliminary project schedule
- Preliminary project budget
- Specific quality criteria the deliverables must meet

TABLE 3.10

Consultant's Qualification for Feasibility Study

Serial Number	Description
1	Experience in conducting feasibility studies
2	Experience in conducting feasibility studies for similar types of projects
3	Fair and neutral with no prior opinion about what decision should be made
4	Experience in strategic and analytical analysis
5	Knowledge of analytical approach and background
6	Ability to collect a large number of important and necessary data via work sessions, interviews, surveys, and other methods
7	Market knowledge
8	Ability to review and analyze market information
9	Knowledge of market trends in similar types of projects/facilities
10	Multidisciplinary experienced team having proven record: a. Financial analyst b. Engineering/technical experts c. Policy experts
11	Experience in review of demographic and economic data

- Type of contract to be employed
- Design requirements
- Regulatory requirements
- Potential project risks
- Environmental considerations
- Logistic requirements

The completion of the feasibility study and submission of project objectives are the starting point for the project by the owner/client, if the project/objective is approved. Thus, the project objectives are established.

3.7.3 Project Delivery Systems

After establishment of the project objectives, the owner/client develops the project procurement strategy by selecting a particular type of project delivery system. The type selected varies from project to project, taking into consideration the objectives of the project.

A project delivery system is defined as an organizational arrangement among various participants comprising the owner, designer, contractor, and many other professionals involved in the design and construction of a project/facility to translate/transform the owner's needs/goals/objectives into a finished facility/project. The project delivery system establishes responsibility for how the project is delivered to the owner. The project delivery system defines the responsibility/obligations each of the participants is expected to perform during the various phase of the construction project life cycle (concept design, schematic design, detailed design, construction, testing, commissioning, and handover). The project delivery system is the approach by which the project is delivered to the owner, but is separate and distinct from the contractual arrangements for financial compensation. Figure 3.11 illustrates the most common project delivery system followed in construction projects.

Figure 3.12 shows details about the Design/Bid/Build type of project delivery system, and Figure 3.13 depicts the contractual relationship.

Figure 3.14 shows details about the Design/Build type of project delivery system, and Figure 3.15 depicts the contractual relationship.

Serial Number	Category	Classification	Subclassification
1	Traditional System (Separated & Cooperative)	Design/Bid/Build	Design/Bid/Build
		Variant of Traditional System	Sequential Method
			Accelerated Method
2	Integrated System	Design/Build	Design/Build
		Design/Build	Joint Venture (Architect and Contractor)
		Variant of Design/ Build System	Package Deal
		Variant of Design/ Build System	Turnkey Method: EPC (Engineering, Procurement, Construction)
		Variant of Design/ Build System (Turnkey)	Build/Operate/Transfer (BOT)
			Build/Own/Operate/Transfer (BOOT)
			Build/Transfer/Operate (BTO)
			Design/Build/Operate/Maintain (DBOM)
		Variant of Design/ Build System (Funding Option)	Lease/Develop/Operate (LDO)
			Wraparound (Public–Private Partnership)
		Variant of Design/ Build System	Build/Own/Operate (BOT)
			Buy/Build/Operate (BBO)
3	Management-Oriented System	Management Contracting	Project Manager (Program Management)
		Construction Management	Agency Construction Manager
			Construction Manager-At-Risk
4	Integrated Project Delivery System	Integrated Form of Contract	

FIGURE 3.11
Categories of project delivery systems.

Project Delivery System	Main Aspects	Advantages	Disadvantages
Traditional System (Design/Bid/Build)	In this method, the owner contracts design professionals to prepare detailed design and contract documents. These are used to receive competitive bids from the contractors. A design build/bid/build contract has well-defined scope of work. This method involves three steps: 1. Preparation of complete detailed design and contract documents for tendering. 2. Receiving bids from pre-qualified contractors. 3. Award of contract to successful bidder. In this method, two separate contracts are awarded, one to the designer/consultant and one to the contractor. In this type of contract structure, design responsibility is primarily that of the architect or engineer employed by the client and the contractor is primarily responsible for construction only.	• Traditional, well-known delivery system. • Suitable for all types of client. • Client has control over design process and quality. • Project scope is defined. • Project schedule is known. • Cost certainty at the time of awarding the contract. • Client has direct contractual control over designer (consultant) and contractor. • Low risk for client. • Contractual roles and responsibilities of all parties are clearly defined and known and understood by each participant. • Well-defined relationship among all the parties. • Higher level of competition resulting in low bid cost.	• Long project life cycle • Client has no control over selection of contractor/subcontractor. • Client must have the resources and expertise to administer the contracts of designer and contractor. • Higher level of inspection by supervision agency. • Lack of input from constructors may result in design and constructability issues. • Project cost may be higher than estimated by the designer. • If contract documents are unclear, it raises the unexpected costs drastically. • Project likely to be delayed as it may not be possible to complete as per estimated schedule. • Exposure to risk where unreasonable time is set to complete the design phase. • No incentives for the contractor.

FIGURE 3.12
Design/bid/build delivery system.

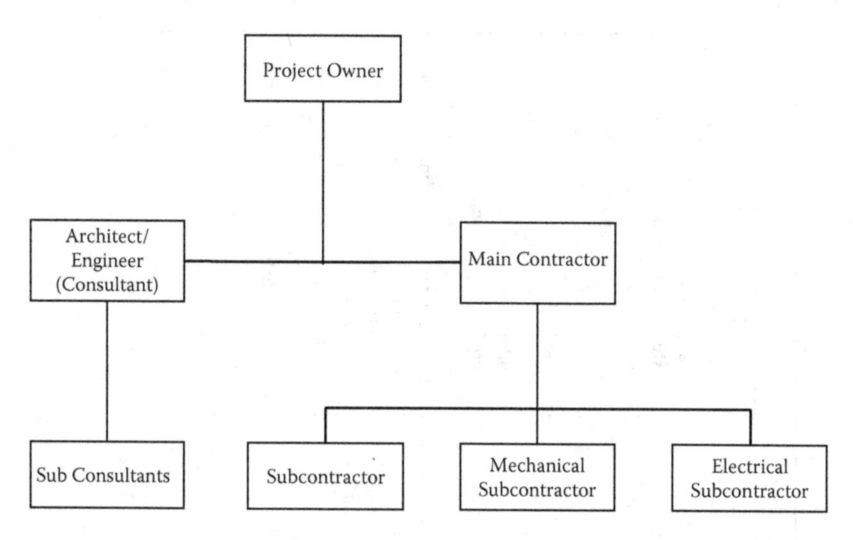

FIGURE 3.13
Design/bid/build contractual relationship.

Figure 3.16 shows the typical logic flow diagram for the Design/Bid/Build type of construction project, and Figure 3.17 depicts the typical logic flow diagram for the Design/Build type of contracting system.

Figure 3.18 shows details about the Project Manager type of project delivery system, and Figure 3.19 depicts the contractual relationship.

Figure 3.20 gives details about construction management in the risk type of project delivery system, and Figure 3.21 depicts the contractual relationship.

Figure 3.22 gives details about the Integrated Project Delivery type of project delivery system.

3.7.4 Types of Contracts/Pricing

Regardless of the type of project delivery system selected, the contractual arrangement by which the parties are compensated must also be established. In construction projects, determining how to procure the product is as important as determining what to procure and when. In most procurement activities, there are several options available for purchase, or subcontracts. The basis of compensation type relates to the financial arrangement among the parties as to how the designer or contractor is to be compensated for their services. The following are the most common types of contract/compensation methods:

1. Firm fixed price or lump sum contract
    a. Firm fixed price
    b. Fixed price incentive fee
    c. Fixed price with economic adjustment price

Project Delivery System	Main Aspects	Advantages	Disadvantages
Design/Build	In a design/build contract, the owner contracts a firm contractor singularly responsible for design and construction of the project. In this type of contracting system, the contractor is appointed based on an outline design or design brief to understand the owner's intent of the project. The owner has to prepare a comprehensive scope of work and has to clearly define his or her needs and performance requirements prior to signing of the contract. It is a must that project definition is understood by the contractor to avoid any conflict in future as the contractor is responsible for detailed design and construction of the project. Owner has to be involved in taking decisions during the selection of design alternatives and the monitoring of costs and schedules during construction, and therefore the owner has to maintain/hire a team of qualified professionals to perform these activities.	• Reduces overall project time because construction begins before completion of design. • Singular responsibility; contractor takes care of the schedule, design, quality, methods, technology. • For the owner/client, the risk is transferred to design/build contractor. • Project cost defined in early stage and has certainty. • Early involvement of contractor assists constructability. • Suitable for straightforward projects where significant changes or risks are not anticipated and the owner is able to precisely specify the objectives/requirements. • Risk management is better than Design/Bid/Build.	• Not suitable for complex projects. • Owner has reduced control over design quality. • Extensive involvement of owner during entire life cycle of project. • Real price for a contract cannot be estimated by the owner/client in the beginning. • Changes by owner can be expensive and may result in heavy cost penalties to the owner. • Poor identification of owner need and wrong understanding of project brief/concept can cause serious problems during the project realization. • For the contractor, there are more risks in this type of contract. • Not suitable for renovation projects.

FIGURE 3.14
Design/build delivery system.

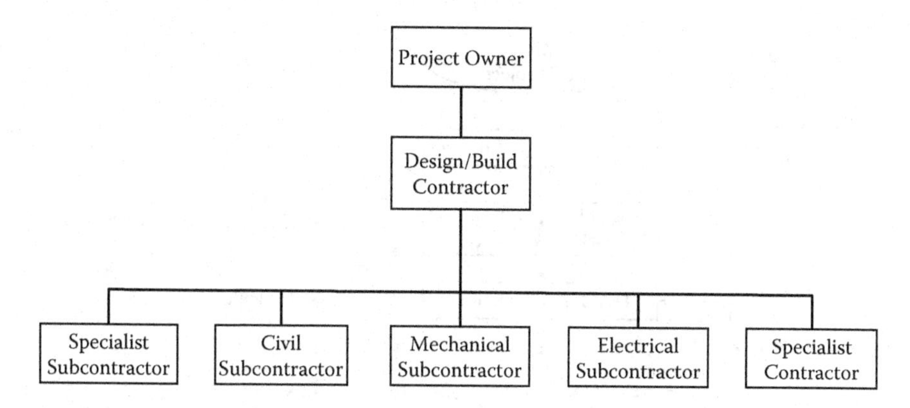

FIGURE 3.15
Design/build contractual relationship.

2. Cost reimbursement contract (cost plus)
    a. Cost plus percentage fee
    b. Cost plus fixed fee
    c. Cost plus incentive fee
    d. Cost plus award fee
3. Remeasurement contract
    a. Bill of quantities
    b. Schedule of rates
    c. Bill of materials
4. Target price contract
5. Time and material contract
6. Guaranteed maximum price (GMP)
    a. Cost plus fixed fee GMP contract
    b. Cost plus fixed fee GMP and bonus contract
    c. Cost plus fixed fee GMP with an arrangement for sharing any cost saving type contract

Figure 3.23 illustrates the various type of contracting systems.

3.7.5 Selection of Project Delivery Teams (Designer/Consultant, Contractor)

Construction projects involve three main parties. These are

1. Owner
2. Designer/consultant
3. Contractor

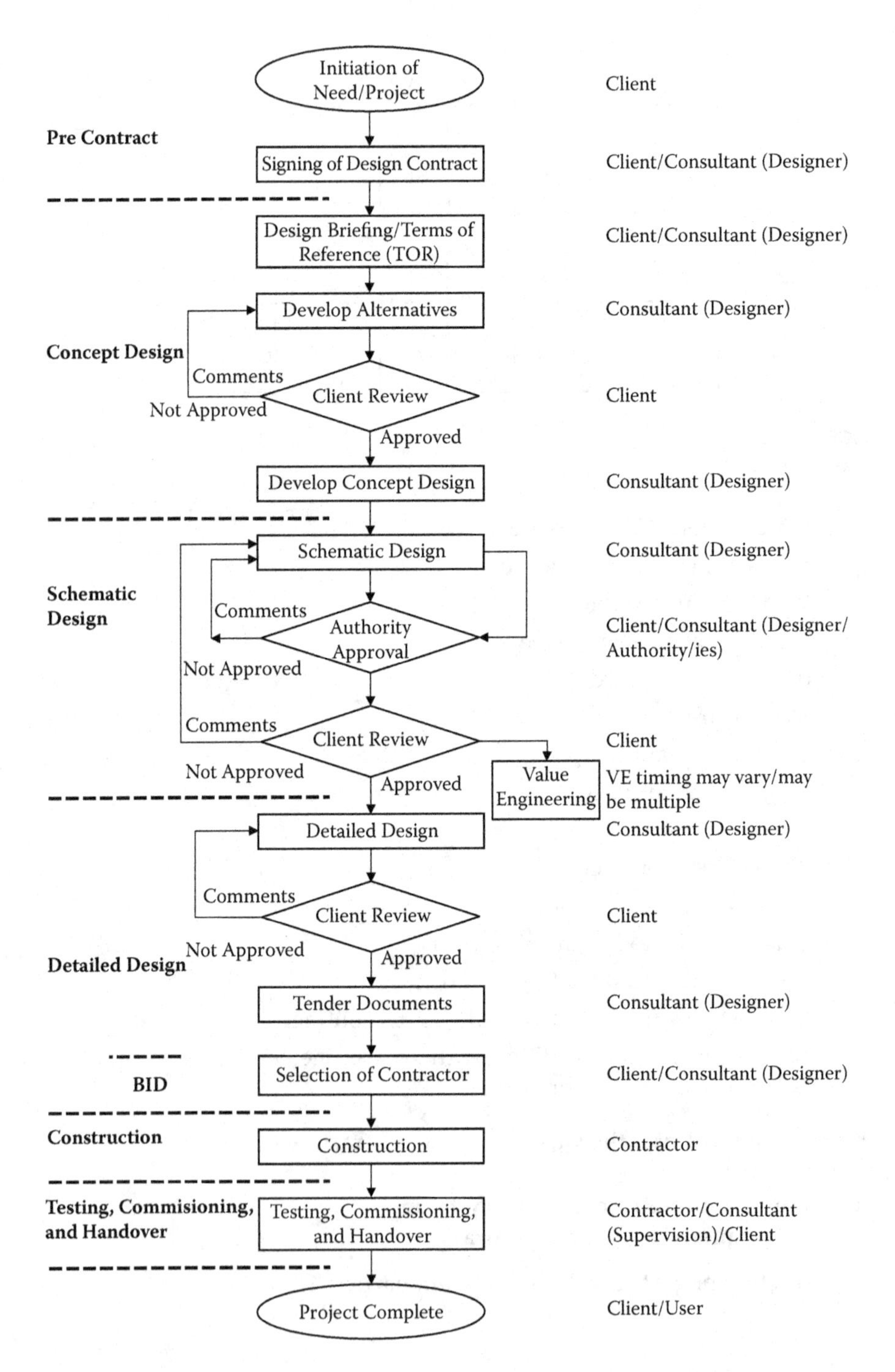

FIGURE 3.16
Logic flow diagram for design/bid/build system.

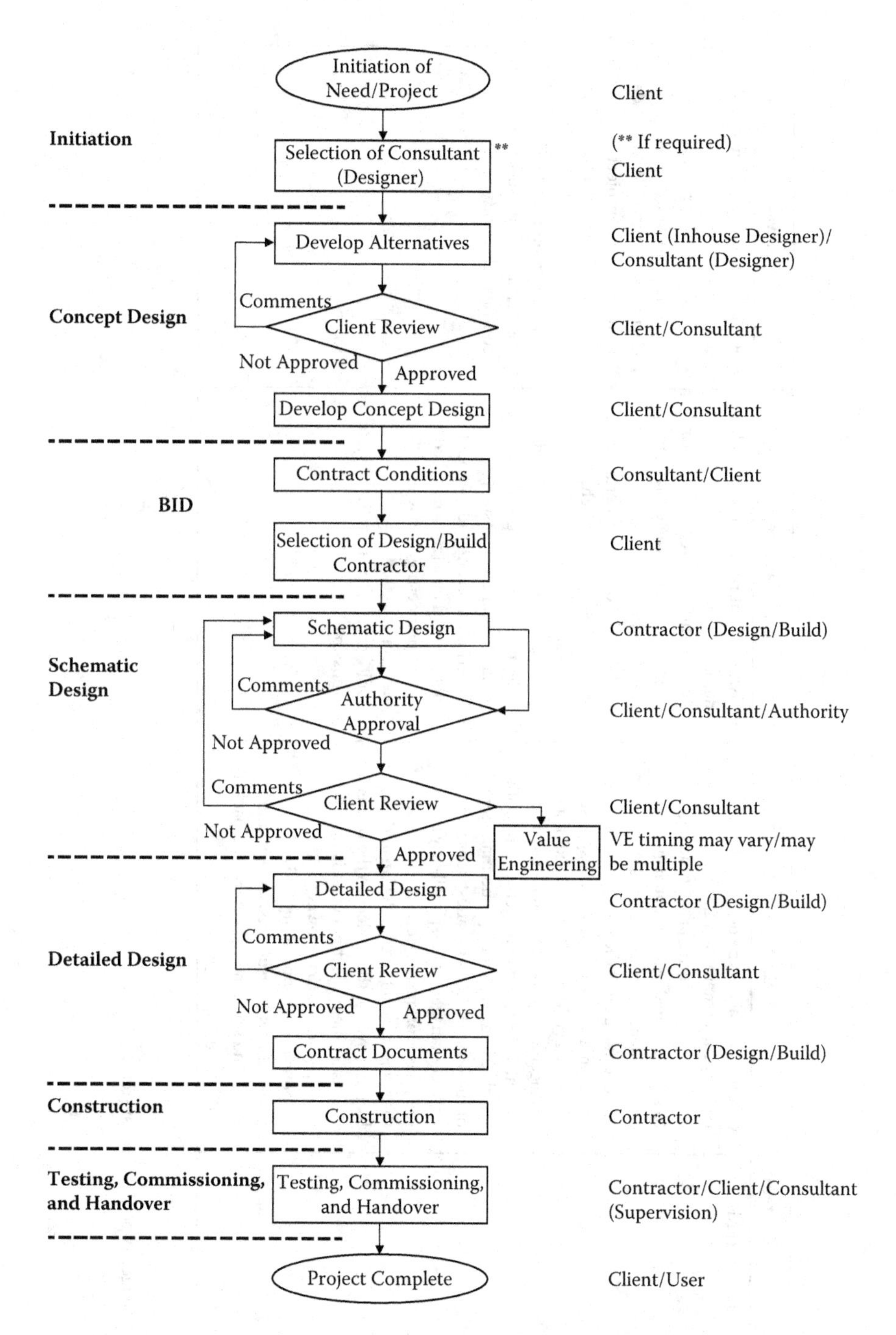

FIGURE 3.17
Logic flow diagram for design/build system.

Project Delivery System	Main Aspects	Advantages	Disadvantages
Project Manager (PM)	A project manager type delivery system is used by the owner when the owner decides to turn over the entire project management to a professional project manager. In a project manager type of contract, the project manager is the owner's representative, and is directly responsible to the owner. The project manager acts as a management consultant on behalf of the owner. The project manager is responsible for planning, monitoring, and management of the project. In its broadest sense, the project manager has responsibility for all the phases of the project from inception of the project until the completion and handing over of the project to the owner/end user. The project manager is involved in giving advice to the owner and is responsible for appointing design professionals, consultants, supervision firms, and selecting the contractor/package contractor for the project.	• Owner/client retains full control of design. • Provide the opportunity for "Fast Track" or overlapping design and construction phases. • Reduces the owner's general management and oversight responsibilities. • Changes can be accommodated in unlet packages as long as there is no impact on time and cost. • Expert opinion and independent view toward constructability, cost, value engineering, and team member selection. • Multi-prime type of delivery system is possible.	• Added project management cost to the owner. • Owner must have resources and expertise to deal with PM. • Owner cedes much of day-to-day control over the project to the PM. • Project manager is not at risk to the cost. • Owner continues to hold construction contracts and retains contractual liability.

FIGURE 3.18
Project manager type delivery system.

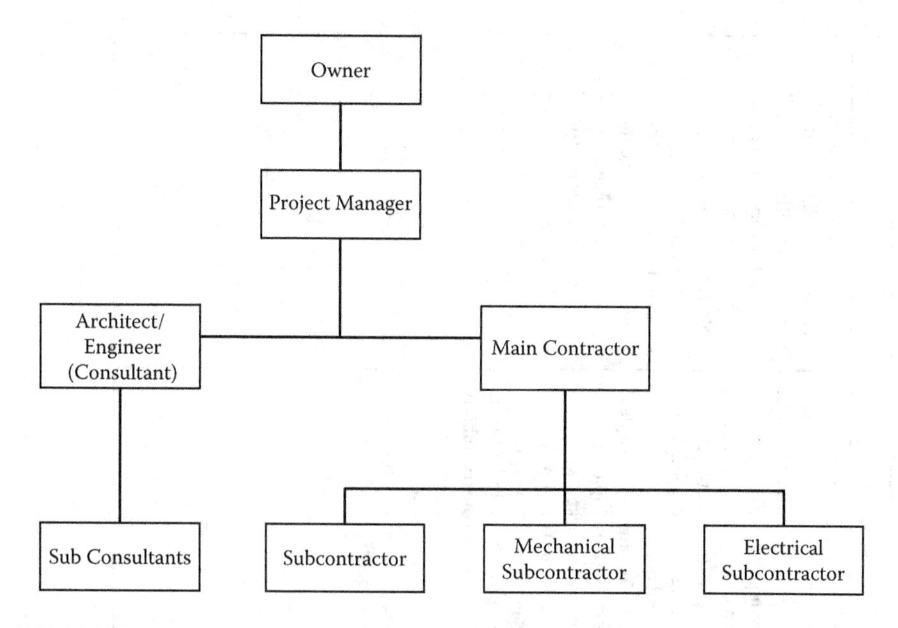

FIGURE 3.19
Project manager contractual relationship.

The participation and involvement of all three parties at different construction phases is required to develop a successful facility/project. These parties are involved at different levels and their relationship and responsibilities depend on the type of project delivery system and contracting system. The following are the three most common procurement methods for selection of project teams:

1. Low bid
2. Best value
   a. Total cost
   b. Fees
3. Qualification-based selection (QBS)

Figure 3.24 illustrates the main aspects of the selection methods and selection criteria for the project delivery system team.

In construction projects, the selection process can range from simply deciding to directly award a contract, to a multistage process that involves information gathering about the seller through a request for information (RFI), prequalification questionnaire (Request for Qualification or RFQ), and soliciting activities. Normally, every owner maintains a list of prospective and previously qualified sellers. The existing list of potential sellers (designer/consultant, contractor) can be expanded by placing advertisements in general publications such as newspapers, trade publications, or professional journals. An Internet search can also help generate a list of prospective sellers.

Project Delivery System	Main Aspects	Advantages	Disadvantages
Construction Management	In this method, the owner contracts a construction management firm to coordinate the project for the owner and provide construction management services. The construction management type of contract system is a four-party arrangement involving owner, designer, construction management firm, and the contractor. The construction manager provides advice to the owner regarding cost, time, safety, and the quality of materials/products/systems to be used on the project. The architect/engineer or supervisor is responsible for maintaining the construction quality and supervising the construction process. The basic concept of construction manager type of contract is that the firm is knowledgeable and capable of coordinating all aspects of the project to meet the intended use of the project by the owner. There are two general types of construction manager type contracts. These are 1. Agency construction manager 2. Construction manager at-risk Agency construction manager acts as an advisor to the owner/client, whereas construction manager at-risk is responsible for on-site performance and actually performs some of the project work. The agency CM firm performs no design or construction, but assists the owner in selecting design firms and contractors to build the project.	• Client retains full control of project. • Agency CM helps the owner by providing advice during the design phase, bid evaluation, overseeing construction, managing project schedule, cost, and quality. • Suitable for large and complex projects with multiple phases and contract packages. • Since the construction manager is an expert entity, he or she provides an independent view regarding constructability, cost, and value engineering. • Shorter project schedule • Changes can be accommodated in unlet packages.	• Construction manager takes control of packages and interaction with contractors/subcontractors, but has no contractual role. • No cost certainty until final package is let. • Client retains all the risk. • Client has to manage the contractual agreements with each package contractor. • Client has to carry majority of risk.

FIGURE 3.20
Construction management delivery system.

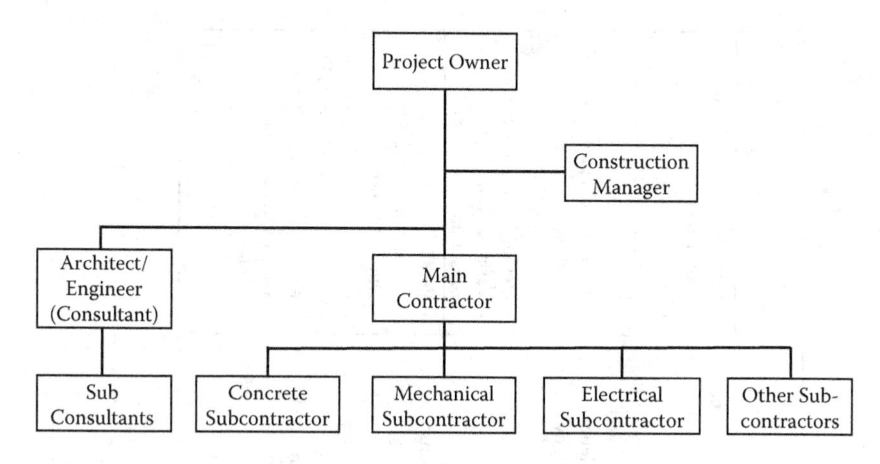

FIGURE 3.21
Construction manager contractual relationship.

Project Delivery System	Main Aspects	Advantages
Integrated Project Delivery System	Integrated project delivery system is an alternative project delivery method distinguished by a contractual agreement between a minimum of owner, design professional, and contractor where the risks and rewards are shared and stakeholders' success is dependent on the project. It is a relational contracting arrangement that sets performance and results expectations. It creates an integrated team that results in optimization of the project.	• Integrated form of agreement between owner, architect/engineer, and contractor. • Owner's business plan serves as the basic design criteria. • Design and cost of the project stay in balance. • Common understanding of project parameters, objectives, and constraints. • Better relationship among project teams. • Better collaboration. • More certainty of outcome within the agreed-upon target cost, schedule, and program.

FIGURE 3.22
Integrated project delivery system.

3.7.5.1 Request for Information (RFI)

An RFI is a procurement procedure where potential suppliers are provided with a general or preliminary description of the problem or need and are requested to provide information or advice about how to better define the problem or need or alternative solutions. It may be used to assist in preparing a solicitation document.

Serial Number	Project Contracting Type	Main Aspects	Advantages	Disadvantages
1	Fixed price or lump sum	With this type of contract, the contractor agrees to perform the specified work/services at a fixed price (it is also called lump sum) for the specified and contracted work. Any extra work is executed only upon receipt of instruction from the owner. Fixed price contracts are generally inappropriate for work involving major uncertainties, such as work involving new technologies.	• Low financial risk to the owner. • Total cost of project known before construction. • Suitable for projects, which can be completely designed and whose quantities are definable. • Minimal control by the owner over project cost.	• High financial risk to the contractor. • Owner bears the risk of poor quality from the contractor trying to maximize the profit within fixed cost. • Contractor's price may include high contingency. • Variation can be time consuming and costly.
2	Cost reimbursement contract (Cost plus)	It is a type of contract in which the contractor agrees to do the work for the cost related to the project plus an agreed-upon fee that covers profits and nonreimbursable overhead costs. The following are the different types of cost plus contracts: 1. Cost-plus% fee contract 2. Cost-plus fixed fee contract 3. Cost-plus incentive fee contract	• Suitable for starting construction concurrently while design is in progress. • Suitable for renovation type of projects. • Suitable for projects in which major changes are expected. • Suitable for projects with possible introduction of latest technology.	• Possibility of overspending by the contractor. • Difficult to predict final cost. • Project quality is likely to be affected as the fee will be same no matter how low the costs are.

FIGURE 3.23
Contracting types.

3	Remeasurement contract	It is a contract in which the contractor is paid as per the actual quantities of the work done. In this type of contract the contractor agrees to do the work based on one of the following criteria: 1. Bill of quantities 2. Schedule of rates 3. Bill of material In this type of contract, payment is linked to measured work completion.	• Suitable for competitive bidding. • Fair competition. • For contractor, low risk. • Suitable for projects where quantities of work cannot be determined in advance of construction. • Fair basis for competition.	• Requires adequate breakdown and design definition of work unit. • Final cost not known with certainty until the project is complete. • Additional requirements of administrative staff to measure, control, and report.
4	Target price contract	A target cost contract is based on the concept of a top-down approach, which provides a fixed price for an agreed range of out-turn costs around the target. In this type of contract overrun or underspend are shared by the owner and the contractor at predetermined agreed-upon percentages.	• Final cost is known. • Contractor may share the savings per agreed-upon percentages.	• Tight cost control is required. • Difficult to adjust major variations or cost inflation.
5	Time and material contract	A time and material contract has elements of unit price and cost plus type of contract. This type of contract is mainly used for maintenance contracts and small projects.	• The owner pays the contractor based on actual cost for material and time per pre-agreed rates. • Suitable for design services where it is difficult to determine total expected efforts in advance.	• Contractor must be accurate in estimating the price, which normally includes indirect and overhead cost.

FIGURE 3.23 (*Continued*)

6	Cost plus Guaranteed maximum price contract	With this type of contract, the owner and the contractor agree to a project cost guaranteed by the contractor as maximum. In this method, the contractor is compensated for actual costs incurred, in connection with design and construction of the project, plus a fixed fee, all subject, however, to a ceiling above which the client is not obligated to pay. This type of contract needs adequate cost pricing information for establishing a reasonable firm target price at the time of initial contract negotiation.	• Maximum contract price is certain. • Construction can start at an early stage.	• The contractor shares the maximum risk. • High administration cost from the owner side to monitor what the contractor is actually spending to get the benefit of underspending. • The contractor's tendering/bidding department should have adequate information to provide necessary data to support negotiation of final cost and incentive price revision.

FIGURE 3.23 (*Continued*)

Serial Number	Selection Type	Description	Selection Criteria			
			D/B/B	D/B	CM at Risk	IPD
1	Low Bid	Total construction cost including the cost of work is the sole criteria for final selection.	Yes	Yes	Not Typical	Not Typical
2	a. Best Value; Total cost	Both total cost and other factors are criteria in the final selection.	Yes	Yes	Not Typical	Not Typical
	b. Best Value; Fees	Both fees and qualifications are factors in the final selection.	Not Applicable	Yes	Yes	Yes
3	Qualification-Based Selection (QBS)	Total costs of work are *not* a factor in the final selection. Qualification is the sole factor used in the final selection.	Not Applicable	Yes	Yes	Yes

FIGURE 3.24
Procurement selection type and selection criteria for project delivery system.

3.7.5.2 Request for Qualification (RFQ)/Prequalification Questionnaire (PQQ)

QFQ/PQQ is a process that enables prequalification of proponents for a particular requirement and avoids having to struggle with a large number of lengthy proposals. It is a solicitation document requesting proponents to submit the qualifications and special expertise in response to the required scope of services. This process is used to shortlist qualified proponents for the procurement process. Figure 3.25 illustrates a Request for Qualification or the prequalification questionnaire to select a contractor.

3.7.5.3 Request for Proposal (RFP)

An RFP is a project-based process involving solution, qualifications, and price as the main criteria that define a winning proponent. It is a solicitation document inviting prospects to submit proposals in response to the required scope of services. The document does not specify in detail how to accomplish or perform the required services. A Request for Proposal can range from a single-step process for straightforward procurement to a multistage process for complex and significant procurement. Table 3.11 illustrates the requirements to be included in the Request for Proposal (RFP) document for a construction project designer/consultant.

3.7.5.4 Request for Quotation (RFQ)

RFQ is a priced-based bidding process that is used when complete documents consisting of defined project deliverables, solution, specifications,

Serial Number	Element	Description
1	General Information	a. Company name
		b. Full address
		c. Registration details/Business permit
		d. Management details
		e. Nature of company such as partnership, share holding
		f. Stock exchange listing, if any
		g. Affiliated/group of companies, if any
		h. Membership of professional trade associations, if any
		i. Award winning project, if any
		j. Quality management certification
2	Financial Information	a. Yearly turnover
		b. Current workload
		c. Audited financial report
		d. Tax clearance details
		e. Bank overdraft/letter of credit capacity
		f. Performance bonding capacity
		g. Insurance Limit
3	a. Organization Details (General)	a. Core area of business
		b. Number of years in the same field of operation
		c. Quality control/assurance organization
		d. Grade/Classification, if any
	b. Organization Details (Experience)	a. Number of years in the same business
		b. Technical capability i. Engineering ii. Shop drawing production
		c. List of previous contracts i. Name of project ii. Value of each contract iii. Contract period of each contract iv. Contract completion delay
		d. List of failed/uncompleted contracts
		e. Overall tender success rate
		f. Claims/Dispute/Litigation

FIGURE 3.25
Request for qualification (prequalification of contractor).

4	Resources	a. Human resources i. Management ii. Engineering staff iii. CAD Technicians iv. List of key project personnel v. Skilled labor: permanent/temporary iv. Unskilled labor: permanent/contracted
		b. List of equipment, machinery, plant
		c. Human resource development plan
5	Health, Safety, and Environmental	a. Medical facility
		b. Safety record
		c. Environmental awareness
6	Project reference	
7	Bank reference	

FIGURE 3.25 (*Continued*)

performance standards, and schedules are known. Potential bidders are provided with all the related information (documents)—except price—and are requested to submit the price. The evaluation of the bids is only on *price,* subject to their submitting all the required information. Most construction contractors are selected on a price basis.

Normally, design professionals and consultants are hired on the basis of *qualifications.* Qualification-based selection can be considered as meeting one of the 14 points of Deming's principles of transformation, which states: "End the practice of awarding business on the basis of price alone." The selection is solely based on demonstrated competence, professional qualification, and experience for the type of services required. In quality-based selection (QBS), the contract price is negotiated after selection of the best-qualified firm.

In construction projects, the seller's (designer/consultant, contractor) bids and proposals depend on the type of delivery system and contracting methods. Normally, the RFP for design/consultancy services is sent to shortlisted/prequalified firms. Figure 3.26 is an example evaluation criterion for shortlisting of contractors and Table 3.12 lists the criteria for selection of a construction project designer/consultant on a QBS basis.

Similarly, the selection of project manager/construction manager is done using a QBS procedure. Table 3.13 illustrates the services performed by the project manager.

3.7.5.4.1 Designer's Responsibilities

Figure 3.27 illustrates the procedure for submission of a design services proposal, and Figure 3.28 illustrates the overall scope of work normally performed by the designer (consultant).

TABLE 3.11

Contents of Request for Proposal (RFP) for Designer/Consultant

Serial Number	Content
Project Details (Project Objectives)	
1	Introduction
2	Project description
3	Project delivery system
4	Designer's/consultant's scope of work
5	Preliminary project schedule
6	Preliminary cost of project
7	Type of project delivery system
Sample Questions (Information for Evaluation)	
1	Consultant name
2	Address
3	Quality management system certification
4	Organization details
5	Type of firm such as partnership or limited company
6	Whether the firm listed in the stock exchange
7	List of awards, if any
8	Design production capacity
9	Current workload
10	Insurance and bonding

11	Experience and expertise
12	Project control system
13	Design submission procedure
14	Design review system
15	Design management plan
16	Design methodology
17	Submission of alternative concept
18	Quality management during design phase
19	Design firm's organization chart a. Responsibility matrix b. CVs of design team members
20	Designer's experience with green building standards or highly sustainable projects
21	Conducting value engineering
22	Authorities approval
23	Data collection during design phase
24	Design responsibility/professional indemnity
25	Designer's relationship during construction
26	Preparation of tender documents/contract documents
27	Review of tender documents
28	Evaluation process and criteria
29	Any pending litigation
30	Price schedule

Evaluation Criteria	Weightage	Key Points for Consideration	Review Result
1. General Information			
1 Company information		Company's current position-a MUST information	**Yes or No**
2. Financial	**25%**		
1 Total Turnover (last 5 years)	25%	Sum of the turnover for the last five uears	
2 Values of Current work-in-hand	25%	Project value/Value of current work-in-hand	
3 Audit Financial Reports	10%	To confirm the ratio given in point three	
4 Financial Standing	30%		
33% Assets		Current Assets/Current Liabilities	
34% Liabilities		Total Liabilities - Total Equity/Total assets	
33% Profit/Loss		Net Profit before Tax/Total Equity	
5 Bonding and Insurance Limit	10%		
60% a.Performace & Bonding Capacity		Provided or Not provided	
40% b. Insurance		Provided or Not provided	
3. Organization Details			
3a. Business	**20%**		
1 Company's Core Area of Business	30%	Degree of satisfactory answer	
2 Experience of years in business	30%	No. of Years	
3 ISO Certification	15%	Yes or No	
4 Registration/Classification Status	15%	Grade or Classification	
5 Organizational Chart	10%	Key staff indicated (Name/Title), Balanced Resources, Departmental (Specialization) Diversity, Lines of Communication	
6 Dispute/Claims		Degree of satisfactory answer	
3b. Experiene	**30%**		
50% a. Projects' value		No. of projects with comparable value	
50% b. Projects' type (similar type and complexity)		No. of projects with similar complexity	
4. Resources	**20%**		
1 Personnel	60%		
30% Management		No. of managerial staff	
30% Engineers		No. of engineers and project staff	
30% Technicaians		No. of CAD technicians and foreman	
10% Staff Development		% turnover spent on training	
2 Technology	10%	% of turnover spent on acquiring latest construction technology	
3 Plant & Equipment	30%	List of plant and equipment	
5. General	**5%**		
1 Bank References	30%	Provided or Not provided	
2 Project References	30%	Provided or Not provided	
2 Health, Safety and Environmant narration	40%	Degree of Satisfactory answer	

FIGURE 3.26
Contractor selection criteria.

TABLE 3.12

Qualification-Based Selection of Architect/Engineer (Consultant)

Serial Number	Items to be Evaluated
1	Organization's registration and license
2	Management plan and technical capability
3	Quality certification and quality management system
4	LEED or similar certification
5	Number of awards
6	Design capacity to perform the work
7	Financial strength and bonding capacity
8	Professional indemnity
9	Current load
10	Experience and past performance in similar type of work
11	Experience and past performance of proposed individuals in similar projects
12	Professional certification of proposed individuals
13	Experience and past performance with desired project delivery system
14	Design of similar-value projects in the past
15	List of successfully completed projects
16	Proposed design approach in terms of
	a. Performance
	b. Effectiveness
	c. Maintenance
	d. Logistic support
	e. Environment
	f. Green building
17	Design team composition
18	Record of professionals in timely completion
19	Safety consideration in design
20	Litigation
21	Price schedule

Upon signing of the contract, the designer (consultant) has to prepare a schedule showing the milestones for completion of each phase for the project life-cycle design process. Figure 3.29 illustrates an example schedule for design services.

3.7.6 Contract Management

Contract management is a process that describes the procedure to be followed by both parties to a contract to meet their contractual obligations to ensure the delivery of the product/facility/project required per the contract. The aim of the contract management process is to ensure that the project is continuing to the satisfaction of both the parties without any claims, disputes, friction, or surprises. Contract management is about resolving or reducing

TABLE 3.13

Project Management Services

Serial Number	Content	
1	*During Design Phase Management*	
	1.1	Assist owner to select design team/consultant
	1.2	Design phase coordination
	1.3	Constructability review
	1.4	Scheduling services
	1.5	Cost estimation
	1.6	Value engineering coordination and review
	1.7	Authority approval coordination
	1.8	Project risk management
	1.9	Bid and contract award phase
2	*During Construction*	
	2.1	Assist owner to select contractor's core staff
	2.2	Establishment of an IT platform to exchange real-time information between owner, project manager, designer (consultant), and contractor.
	2.3	Establish document control solution
	2.4	Technical correspondence between contractor
	2.5	Develop project management plan
	2.6	Conduct progress meetings
	2.7	Review change requests
	2.8	Review progress payment
	2.9	Advise owner on budget status
	2.10	Advise owner about the schedule
	2.11	Advise owner about construction quality
	2.12	Inform owner about possible cost increase
	2.13	Inform owner about possible project delay
	2.14	Inform owner about major safety violation/accident
	2.15	Coordinate on-site and off-site inspection
	2.16	Testing and commissioning of services
	2.17	Review project closure documents and reports

such friction and achieving the completion of the project as envisaged. The following factors are essential for good contract management:

- Properly defined scope of work
- Type of contract

Changes to the contract, procedure, or works may have an effect on completion, influence performance, and may affect the financial viability of the contract. Table 3.14 shows the major reasons for claims.

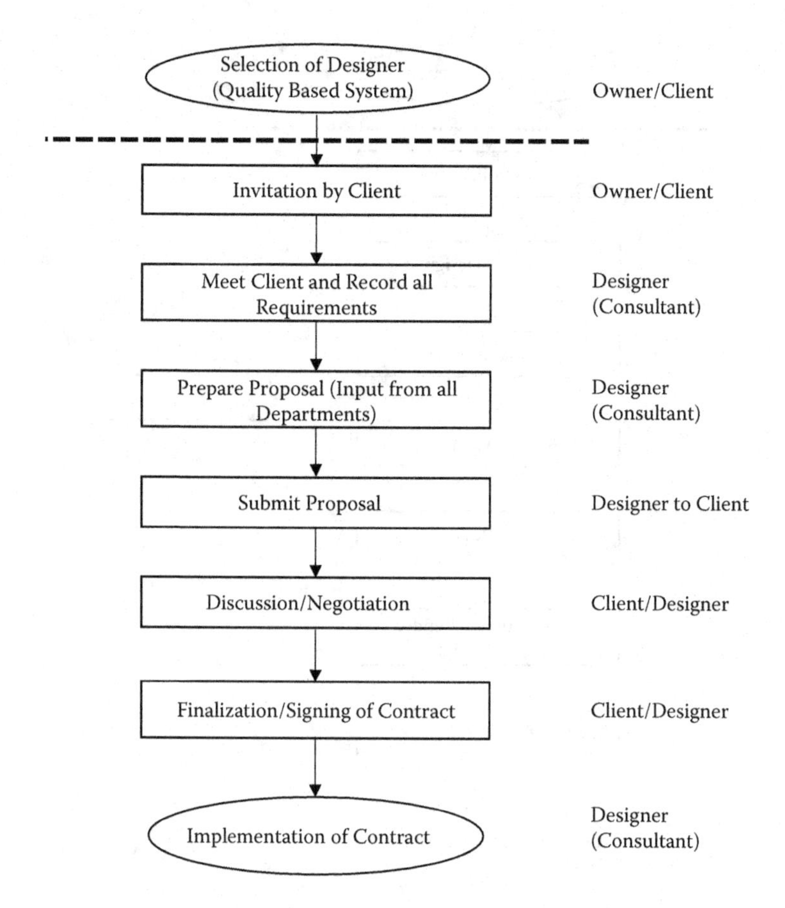

FIGURE 3.27
Contract proposal procedure for designer.

A contract management plan normally consists of three broader areas. These are

1. Service delivery management
2. Relationship management
3. Contract administration

3.7.6.1 Service Delivery Management

Service delivery management is concerned with ensuring the service is being fully delivered, as agreed in the contract, to the specified level of performance. It includes clearly defined performance targets and measures combined with a regular reporting system.

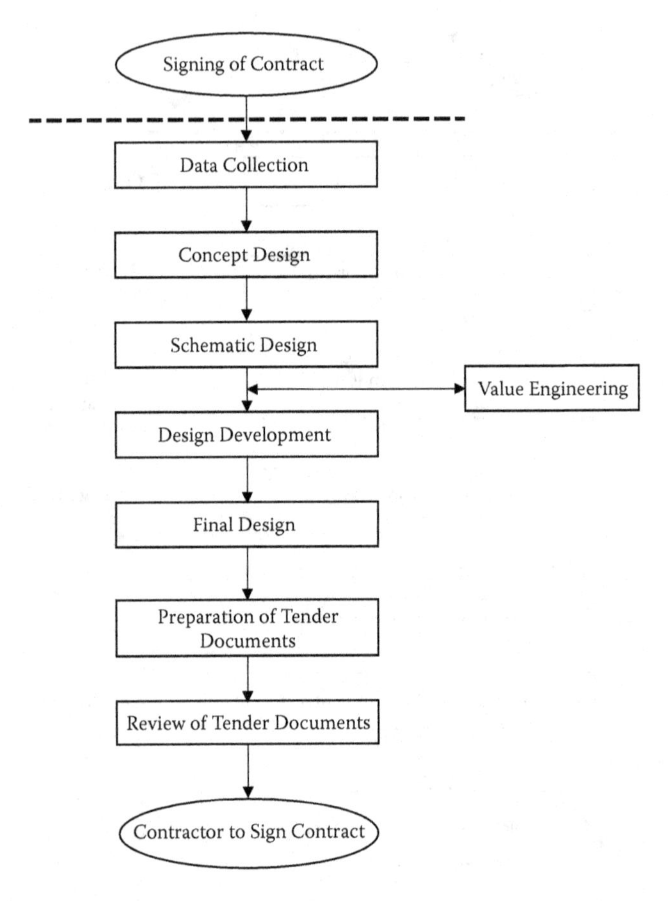

FIGURE 3.28
Overall scope of work of designer.

3.7.6.2 Relationship Management

Relationship management focuses on keeping the relationship between two parties open and constructive, resolving or easing tensions, and identifying problems early. It is important that both parties

- Understand each other's business, obligations, and responsibilities
- Communicate openly by timely sharing of plans
- Develop confidence and trust
- Recognize mutual interest

3.7.6.3 Contract Administration

Contract administration is the process of formal governance of a contract and how changes are made to the contract document. It is concerned with

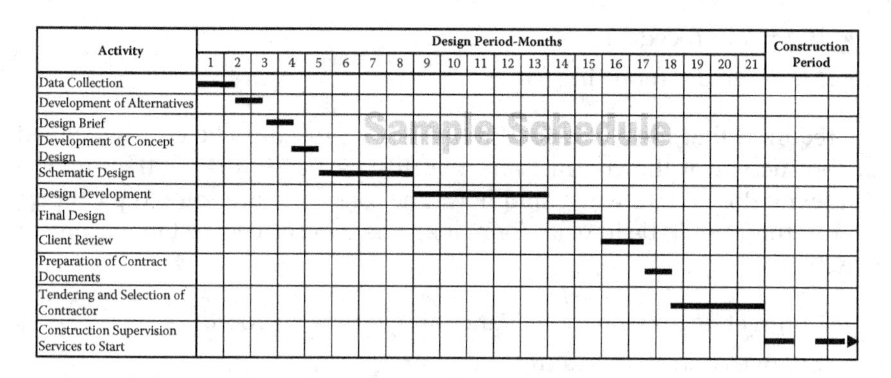

FIGURE 3.29
Design schedule.

TABLE 3.14

Major Reasons for Claims

Serial Number	Reasons
1	Delay in making the site available on time
2	Inadequate specifications a. Omission b. Errors
3	Conflict between contract documents
4	Lack of coordination among trades and services
5	Additional requirements by the owner/client
6	Owner's change of schedule
7	Payment delays
8	Not meeting regulatory requirements
9	Introduction of latest technology
10	Process/method improvement
11	Design changes/modifications
12	Lack of discipline
13	Unforeseen circumstances
14	Charges payable to outside party due to cancellation of certain item/product
15	Delay in approval

managing the contractual relationship between various participants to successfully complete the facility and meet the owner's objectives. It includes tasks such as

- Monitoring contract performance (scope, cost, schedule, quality, risk)
- Making changes to the contract documents by taking corrective action as needed
- Variation order process

- Payment procedures
- Management reporting

It is required that the contract administration procedure be clearly defined for the success of the contract and that the parties to the contract understand who does what, when, and how. The following are some typical procedures that should be in place for management of the contract management activities:

- Contract document maintenance and variation procedure
- Change control procedure
- Variation order procedure
- Payment procedure
- Resource management and planning procedure
- Performance review system
- Management reporting procedure

Table 3.15 lists the contents of a contract management plan.

3.7.7 Contract Closeout

The contract closeout is the process of completing the contract, acceptance of all the deliverables, and issuance of notice of completion. Normally, the contract documents stipulate all the requirements be completed to close the contract. Table 3.16 lists all the activities required to be recorded to close the contract.

3.8 Safety Management

The safety management system is a systematic and coordinated approach to managing safety, including the necessary organizational structures, procedures, and accountabilities to optimally manage safety.

Safety management in construction projects refers to a set of measures concerning safety and health during construction to be implemented and operated by the contractor on a continuous basis. This system is conducted in conjunction with other management systems. The safety management system is established per OHSAS 18001 requirements. Construction projects are complex and nonstandardized. They are exposed to dynamic site conditions and involve coordination among various parties and participants.

The safety management process for construction projects is required to ensure that the construction project is executed with appropriate care to

TABLE 3.15

Contents of Contract Management Plan

Serial Number	Topics
1	Contract summary, deliverables, and scope of work
2	Type of contract
3	Contract schedule
4	Contract cost
5	Project team members with roles and responsibilities
6	Core staff approval procedure
7	Contract communication matrix/management reporting
8	Coordination process
9	Liaison with regulatory authorities
10	Material/product/system review/approval process
11	Shop drawing review/approval process
12	Project monitoring and control process
13	Risk identification and management
14	Review of variation/change requests
15	Contract change control process
	a. Scope
	b. Material
	c. Method
	d. Schedule
	e. Cost
16	Quality of performance
17	Inspection and acceptance criteria
18	Progress payment process
19	Claims, disputes, conflict, and litigation resolution
20	Project holdup areas
21	Contract documents and records
22	Post contract liabilities
23	Contract closeout and independent audit

prevent accidents that could cause personal injury or property damage, and may endanger the safety and health of neighbors to the construction site and the general public who may be exposed to the project. Safety management also includes human resource programs, such as drug and alcohol treatment programs, which contribute to reducing accidents on construction job sites.

3.8.1 Causes of Accidents

Safety statistics for construction indicate high fatality and injury rates. There are significantly more injuries and lost workdays due to injuries or illness in construction than in virtually any other industry. The causes

TABLE 3.16

Contract Closeout Checklist

Sr. No.	Description	Yes/No
Project Execution		
1	Contracted works completed	
2	Site work instructions completed	
3	Job site instructions completed	
4	Remedial notes completed	
5	Noncompliance reports completed	
6	All services connected	
7	All the contracted works inspected and approved	
8	Testing and commissioning carried out and approved	
9	Any snags?	
10	Is the project fully functional?	
11	All other deliverable completed	
12	Spare parts delivered	
13	Is waste material disposed of?	
14	Whether safety measures for use of hazardous material have been established	
15	Whether the project is safe for use/occupation	
16	Whether all the deliverables have been accepted	
Project Documentation		
17	Authorities' approval obtained	
18	Record drawings submitted	
19	Record documents submitted	
20	As-built drawings submitted	
21	Technical manuals submitted	
22	Operation and maintenance manuals submitted	
23	Equipment/material warrantees/guarantees submitted	
24	Test results/test certificates submitted	
Training		
25	Training to owner/end user's personnel imparted	
Negotiated Settlements		
26	Whether all the claims and disputes have been negotiated and settled	
Payments		
27	All payments to subcontractors/specialist suppliers released	
28	Bank guarantees received	
29	Final payment released to main contractor	

TABLE 3.16 (*Continued*)

Contract Closeout Checklist

Sr. No.	Description	Yes/No
Handing over/Taking over		
30	Project handed over/taken over	
31	Operation/maintenance team taken over	
32	Excess project material handed over/taken over	
33	Facility manager in action	

Source: Abdul Razzak Rumane (2010), *Quality Management in Construction Projects*. CRC Press, Boca Raton, FL. Reprinted with permission from Taylor & Francis Group.

TABLE 3.17

Major Causes of Construction Accidents

Serial Number	Cause of Accident
1	Poor attitude toward safety
2	Lack of safety awareness
3	Lack of proper training
4	Deficient enforcement of training
5	Unsafe site conditions
6	Unsafe access
7	Poor lighting
8	Safe equipment not provided
9	Personal protective equipment/safety gear not used
10	Unsafe methods or sequencing
11	Hoisting/lifting machinery not checked for safe operating conditions
12	Unsafe loading/unloading techniques
13	Unskilled operator
14	Barriers or guards not provided
15	Scaffolding not properly secured
16	Warning signs not displayed
17	Escape route-site map not displayed and not explained
18	Flammable and combustible material not stored in safe area
19	Fire
20	Isolated, sudden deviation from prescribed behavior

of injuries in construction are numerous. Table 3.17 lists the major causes of accidents on construction sites, and Table 3.18 shows accident prevention measures.

Traditionally, a site-specific health and safety plan is developed, implemented, monitored, and updated throughout the construction stage by the main contractor until the project is complete and all trades have left the site. In order to achieve an accident-free atmosphere at the construction site,

TABLE 3.18

Preventive Actions to Avoid Construction Accidents

Serial Number	Cause of Accident	Preventive Action
1	Poor attitude toward safety	Interact with everyone and explain the importance of safety and using safety equipment. Evaluate the response and ensure the employee understands.
2	Lack of safety awareness	Arrange training and awareness programs.
3	Lack of proper training	Engage trainer/facilitator having expertise in the task, who can interview, test, or observe employee, and access prior training records. Arrange safety drills.
4	Deficient enforcement of training	Monitor the safety program on regular basis. Enforce the plan and penalize noncompliance.
5	Unsafe site conditions	Know proper site conditions. Study soil conditions and ground reality.
6	Unsafe access	Provide proper ladders and access routes. Keep site clean with proper housekeeping.
7	Poor lighting	Provide temporary lighting on all routes frequently used by the employees which have poor visibility.
8	Safe equipment not provided	Check the equipment before use. Check the license validity if applicable.
9	Personal protective equipment/safety gear not used	Ensure that everyone is using personal protective equipment. Issue warning letters for noncompliance.
10	Unsafe methods or sequencing	Worker to know standard methods and sequence of the task he or she is performing.
11	Hoisting/lifting machinery not checked for safe operating conditions	Inspect and check for maintenance requirement before use. Only authorized employees to operate.
12	Unsafe loading/unloading techniques	Provide proper deck for unloading. Use proper safety gear while lifting the material. Barricade the loading/unloading area. Display signs.
13	Unskilled operator	Engage trained operator.
14	Barriers or guards not provided	Use barriers and guards as necessary. Use warning signs.
15	Scaffolding not properly secured	Tighten scaffolding and check the form work and scaffolding is erected in accordance with applicable standards.
16	Warning signs not displayed	Display signs to keep employees away from hazardous/unsafe areas.
17	Escape route-site map not displayed and not explained	Provide escape route map at various locations. Give training and inform about the escape route and gathering points.

TABLE 3.18 (*Continued*)

Preventive Actions to Avoid Construction Accidents

Serial Number	Cause of Accident	Preventive Action
18	Flammable and combustible material not stored in safe area	Designate specific area for storage of flammable and combustible material. Display notice as "Hazardous Area."
19	Fire	Provide temporary firefighting system per local code. In case of fire, inform local fire brigade immediately.
20	Isolated, sudden deviation from prescribed behavior	Cannot predict or prevent, unless employee's emotional or physical condition contributed and this condition was obvious to others.

everyone at the construction project site (client, designer/consultant, main contractor, subcontractors, and employees) should be a party to implementing and monitoring a safety management system. Particularly, the architect/engineer (A/E) should play a greater role in safety during construction. A/E team members must be given proper training in construction safety to

- Identify unsafe conditions
- Evaluate unsafe conditions
- Identify and prevent some hidden unsafe conditions
- Acquire expertise to identify both visible and unsafe conditions

3.8.2 Total Safety Management

The concept of a safety management system has to be considered to be like TQM. Safety has to be considered as the responsibility of everyone involved at the project site. All the main participants (client, designer/consultant, and contractor) should be made responsible for safety at the project site to achieve zero accidents, analogous to the quality principle Zero Defect. Thus, like TQM, safety can be termed Total Safety Management (TSM).

Figure 3.30 illustrates the application of TQM principles to TSM.

There are three core aspects of a safety management system:

1. Systematic—Safety management activities are in accordance with predetermined safety policies, objectives, and responsibilities to prevent accidents, which cause, or have potential to cause, personal injury, fatalities, or property damage.
2. Proactive—An approach that emphasizes safety hazard identification, investigation and analysis, and risk control and mitigation before events that affect safety occur.
3. Explicit—All safety management activities are documented, communicated, and implemented.

Serial Number	Basic Requirements for Total Quality Management	Total Safety Management Approach
1	Set organization vision	Safety policy
2	Identify customers and their needs	Everyone on site is vulnerable to safety risks and is to be protected
3	Develop team approach by involving everyone in the organization	Safety is responsibility of all individuals involved
4	Define critical processes and measures	Hazard identification, risk assessment and control
5	Develop strategic plan	Documenting and detailing safety requirements and regulatory requirements
6	Review planning	Review site communication plan
7	Education and training	System education and training
8	Defect prevention not detection	Accident investigation to determine corrective and preventive actions
9	Improvement of process	System feedback and continuous improvement
10	Identify major products and process	Identify measures for emergency situations
11	Systematic approach to TQM	Systematic and coordinated approach to managing safety

FIGURE 3.30
TQM approach to TSM.

3.8.3 Safety Management Plan

In construction projects, the requirements for a contractor to prepare a Safety Management Plan (SMP) are specified in the contract documents. The contractor has to submit the plan for review and approval by the supervisor/consultant during the mobilization stage of the construction phase. The following are the items to be specified by the designer and to be considered by the contractor to establish a safety management system for construction project sites to raise the level of safety and health on construction sites and prevent accidents:

1. Project scope describing the project and detailing safety requirements
2. Safety policy statement documenting the contractor's/subcontractor's commitment and emphasis on safety
3. Regulatory requirements on safety
4. Roles and responsibilities of all individuals involved
5. Site communication plan detailing how safety information will be shared

6. Hazard identification, risk assessment, and control
7. Accident investigation to document root causes and determine corrective and preventive actions
8. Accident reporting system
9. Plant, equipment maintenance, and licensing
10. System education and training
11. Measures for emergency situations
12. Routine inspections
13. Continuous monitoring and regular assessment
14. System feedback and continuous improvement
15. Safety assurance measures
16. Evaluation of subcontractor's safety capabilities
17. Health surveillance
18. Site neighborhood characteristics and constraints
19. Safety audit
20. Documentation
21. Records
22. System update
23. Emergency evacuation plan

Table 3.19 illustrates an example safety management plan for a construction project detailing hazardous identification, risk assessment, and control for scaffolding work. However, the contractor has to take into consideration all the requirements listed under the contract documents, depending on the complexity and nature of the project.

3.9 Quality Assessment/Measurement

A quality assessment system in construction projects is a system or method to measure and evaluate the quality of workmanship, the finish, and performance of the constructed works, based on approved standards and quality requirements, to the satisfaction of the client/end user. In order to achieve a competitive advantage and growth, it is essential that the organization have a well-established quality system and an effective quality improvement procedure. Quality assessment/measurement needs to be continuous, ongoing, and performed in a timely manner. In construction projects, assessment/measurement is carried out by using checklists and a project monitoring/control

TABLE 3.19

Safety and Health Plan

PROJECT NAME

RISK ASSESSMENT IMPACT MATRIX

RISK FACTOR MATRIX

Safety Plan			**Probability**				**RISK PRIORITY**		**Methodology**
PROBABILITY	S		4	3	2	1	8 to 16	High Risk	1. Identify hazards
4 Probable: Likely to occur immediately or shortly	e	4	16	12	8	4	4 to 6	Medium Risk	2. List hazard event
3 Reasonably probable: Probably will occur in time	v	3	12	9	6	3	1 to 3	Low Risk	3. Calculate risk factor with initial risk rating i.e., risk rating is probability × severity
2 Remote: May occur in time	e	2	8	6	4	2			
1 Extremely remote: Unlikely to occur	r	1	4	3	2	1			4. Detail controls
	i								5. Calculate residual risk factor with final risk rating
	t								6. Enter risk priority
	y								7. Evaluate acceptability of risk

Severity

4 *Catastrophic:* Imminent danger exists, hazard capable of causing death and illness on a wide scale.

3 *Critical*: Hazard can result in serious illness, severe injury, property and equipment damage.

2 *Marginal:* Hazard can cause illness, injury, or equipment damage, but the results would not be expected to be serious.

1 *Negligible:* Hazard will not result in serious injury or illness, remote possibility of damage beyond minor first-aid cases.

system to ensure meeting the owner's needs. It is mainly performed during the following three phases:

1. Design phase
2. Construction phase
3. Testing, commissioning, and handover phase.

A quality assessment/measurement system in construction projects can be classified as follows:

- Internal quality assessment (internal audit)
- External quality assessment

3.9.1 Internal Quality System

An internal quality system is a process in which a quality audit is performed by the organization itself.

3.9.1.1 Quality Assessment by Designer

Quality assessment during the design phase is normally carried out by the design team per the quality management system before it is sent for review by the owner (project manager). The designer (consultant) also submits the design to regulatory authorities to obtain their approval for compliance with regulatory requirements prior to submission of drawings (documents) to the owner/client.

3.9.1.2 Quality Assessment by Contractor

The contractor has to check the executed works prior to the submission of the checklist to the supervision engineer (consultant).

3.9.2 External Quality System

External assessment in construction projects can be divided into the following:

1. Quality inspection/checking by owner-appointed supervision engineer (consultant)
2. Independent testing agency appointed by the owner (consultant per contract requirements) to inspect/test mainly the following items/works:
    a. Concrete strength testing
    b. Reinforcement testing

c. Elevator works
d. Chiller testing
e. Diesel generator testing
f. Electrical switchboards

The following are the four main steps needed to establish a quality assessment/measurement process in construction projects:

1. Identify the component/items/products/systems to be reviewed and checked in each of the trades (architectural, structural, mechanical, HVAC, electrical, low voltage, landscape, external etc.)
2. Identify the frequency of checking (timing)
3. Identify the persons/agency authorized to check and approve
4. Identify the indicators/criteria for performance monitoring

Table 3.20 lists items to be verified and checked internally by the designer before submission to regulatory authorities and subsequently submission to the owner (project manager).

Table 3.21 lists major activities to be assessed/checked/inspected by the contractor and supervision engineer to ensure compliance of executed works with contract documents.

TABLE 3.20

Checklist for Design Drawings

Serial Number	Items to be Checked
1	Whether the design meets owner requirements (TOR)
2	Whether designs were prepared using authenticated and approved software
3	Whether design calculation sheets are included in the set of documents
4	Whether the design is fully coordinated for conflict between different trades
5	Whether the design has taken into consideration relevant collected data requirements
6	Whether the reviewer's comments were responded to
7	Whether regulatory approval was obtained and comments, if any, incorporated and all review comments responded to
8	Whether design has environmental compatibility
9	Whether energy efficiency measures have been considered
10	Whether the design constructability has been considered
11	Whether the design matches the property limits
12	Whether legends match the layout
13	Whether design drawings are properly numbered
14	Whether design drawings have owner logo/designer logo per standard format
15	Whether the design formats of different trades are uniform
16	Whether the project name and contract reference is shown on the drawing

TABLE 3.21

Checklists for Executed Works

Serial Number	Description of Works	Approved	Not Approved	Comments
Shoring				
1	Shoring			
2	Dewatering			
Earth Works				
1	Excavation			
2	Backfilling			
Concrete Substructure				
1	Blinding concrete			
2	Termite control			
3	Reinforcement of steel			
4	Concrete casting			
5	Shuttering for beams and columns			
Concrete Superstructure				
1	Reinforcement of steel for beams and columns			
2	Shuttering for columns and beams			
3	Concrete casting of beams and columns.			
4	Formwork for slab			
5	Reinforcement for slab			
6	Concrete casting of slab			
7	Precast panels			
Masonry				
1	Block work			
2	Concrete unit masonry			
Partitioning				
1	Installation of frames			
2	Installation of panels			
Metal Work				
1	Structural steel work			
2	Installation of cat ladders			
3	Installation of balustrade			
4	Installation of space frame			
5	Installation of handrails and railings			

(*Continued*)

TABLE 3.21 (*Continued*)

Checklists for Executed Works

Serial Number	Description of Works	Approved	Not Approved	Comments
Thermal and Moisture Protection				
1	Application of waterproofing membrane			
2	Installation of insulation			
Doors and Windows				
1	Aluminum windows and doors			
3	Glazing			
4	Steel doors			
6	Wooden frames			
7	Wooden doors			
8	Curtain wall			
9	Hatch doors			
11	Louvers			
Internal Finishes				
1	Plastering			
2	Painting			
3	Cladding-ceramic			
4	Cladding-marble			
5	Ceramic tiling			
6	Stone flooring			
7	Acoustic ceiling			
8	Demountable partitions			
Furnishing				
1	Carpeting			
2	Blinds			
3	Furniture			
External Finishes				
1	Painting			
2	Brickwork			
3	Stone			
4	Cladding-aluminum			
5	Cladding-granite			
6	Curtain wall			
7	Glazing			
Equipment				
1	Installation of maintenance equipment			
2	Installation of kitchen equipment			
3	Installation of parking control			

TABLE 3.21 (*Continued*)

Checklists for Executed Works

Serial Number	Description of Works	Approved	Not Approved	Comments
Roof				
1	Parapet wall			
2	Thermal insulation			
3	Waterproofing			
4	Roof tiles			
5	Installation of drains			
6	Installation of gutters			
Elevator System				
1	Installation of rails			
2	Installation of door frames			
3	Installation of cabin			
4	Cabin finishes			
5	Installation of wire rope			
6	Installation of drive machine			
7	Installation of controller			
HVAC Works				
1	Installation of piping			
2	Installation of ducting			
3	Installation of insulation			
4	Installation of dampers, grills, and diffusers			
5	Installation of cladding			
6	Installation of fans			
7	Installation of fan coil units (FCUs)			
8	Installation of air handling units (AHUs)			
9	Installation of pumps			
10	Installation of chillers			
11	installation of cooling towers			
12	Installation of thermostat and controls			
13	Installation of starters			
14	Building management system (BMS)			
Mechanical Works				
A. Water Supply				
1	Installation of piping system			
2	Installation of pumps			

(*Continued*)

TABLE 3.21 (*Continued*)

Checklists for Executed Works

Serial Number	Description of Works	Approved	Not Approved	Comments
3	Filter units			
4	Toilet accessories			
5	Installation of insulation			
6	Water heaters			
7	Hand dryers			
8	Water tank			
B. Drainage System				
1	Installation of pipes below grade			
2	Installation of pipes above grade			
3	Installation of manholes			
4	Installation of clean out, floor drains			
5	Installation of sump pumps			
6	Installation of gratings			
C. Irrigation System				
1	Installation of piping system			
2	Installation of pumps			
3	Installation of controls			
D. Firefighting System				
1	Installation of piping system			
2	Installation of sprinklers			
3	Installation of foam system			
4	Installation of pumps			
5	Installation of hose reels/cabinet			
6	Installation of fire hydrant			
7	External fire hose cabinet			
Electrical Works				
1	Conduiting—raceways			
2	Cable tray—trunking			
3	Floor boxes			
4	Wiring			
5	Cabling			
6	Installation of bus duct			
7	Installation of wiring devices/ accessories			
8	Installation of light fittings			
9	Grounding			
10	Distribution switchboards			

TABLE 3.21 (*Continued*)

Checklists for Executed Works

Serial Number	Description of Works	Approved	Not Approved	Comments
Fire Alarm System				
4	Installation of detectors, bells, pull stations, interface modules			
5	Installation of repeater panel			
6	Installation of mimic panel			
7	Installation of fire alarm panel			
Telephone/Communication System				
1	Installation of racks			
2	Installation of switches			
Public Address System				
5	Installation of speakers			
6	Installation of racks			
7	Installation of equipment			
Audiovisual System				
5	Installation of speakers			
6	Installation of monitors/screens			
7	Installation of racks			
8	Installation of equipment			
CCTV/Security System				
1	Installation of cameras			
2	Installation of panels			
3	Installation monitors/screens			
Access Control				
1	Installation of RFID proximity readers, fingerprint readers			
2	Installation of magnetic locks, release buttons, door contacts			
3	Installation of panel			
4	Installation of server			
Systems Integration				
1	Installation of switches			
2	Installation of servers			
External Works				
1	Site works			
2	Asphalt work			

(*Continued*)

TABLE 3.21 (*Continued*)

Checklists for Executed Works

Serial Number	Description of Works	Approved	Not Approved	Comments
3	Pavement works			
4	Piping works			
5	Electrical works			
6	Manholes			
7	Road marking			

Source: Abdul Razzak Rumane (2010), *Quality Management in Construction Projects.* CRC Press, Boca Raton, FL. Reprinted with permission from Taylor & Francis Group.

Apart from project quality check/inspection, it is also required to assess the following six activities on a regular basis to track project progress and its compliance with contract documents:

1. Schedule
2. Budget
3. Resources
4. Project team performance
5. Risk assessment
6. Safety

Monitoring and control of the schedule, budget, and resources (project progress) is performed on a regular basis by project monitoring and control team members. Also, team members from the supervision team as well as from the contractor side are assigned to carry out risk assessment and safety assessment. Performance evaluation is done by the line manager in coordination with the corporate office requirements.

3.10 Training and Development

Training and development is a set of systematic processes designed to meet learning objectives related to an organization's strategic plan. It is a subsystem of an organization. The main objective of training and development is to ensure the availability of required skilled professionals to an organization.

Training is a means of acquiring proficiency in skills or sets of skills, new knowledge, and changing attitudes to achieve the desired objectives in an efficient and economical manner. It is the process of identifying performance requirements and the gap between what is required and what exists.

It is a process that includes a variety of methods to help the employee attain a specific level of knowledge, skills, and abilities (KSA) through professional development.

Development includes the acquisition of behavioral skills including communication, interpersonal relations, and conflict resolution. Development may also include processes aimed at the acquisition and development by employees of knowledge, understanding, behaviors, or attitudes specifically required to perform their professional duties.

Training and development helps in optimizing the utilization of human resources, which further helps the employee achieve the organizational goals as well as their individual goals. It also helps to provide opportunities and a broad structure for the development of human resources' technical and behavioral skill in an organization. The training and development process cycle mainly has five steps or phases:

1. Needs assessment
2. Development
3. Training implementation
4. Assessment (evaluation)
5. Follow-up

Figure 3.31 illustrates the training process cycle.

Construction projects are constantly increasing in technological complexity. Further, the competitive nature of business, comprehensive management system, accelerated changes in construction processes, and use of the latest equipment, machinery, products, and methods require more and more that the project team members and other stakeholders have technical training in addition to formal education in their respective fields. Also, as the technological changes are occurring at an ever-increasing rate, the need for continual individual development is crucial.

There is a need for improved quality-based performance of construction projects and therefore effective management services. Training in construction project management is considered an essential issue. Training can play an important role in improving the ability and efficiency of construction professionals and construction workers, whether the training is a part of an ongoing process of professional skill development or simply about learning a specific skill. In both cases, it can improve people's skills and knowledge and help them perform their tasks more effectively.

Construction projects are facing challenges of ever-changing construction technology, knowledge ideology, management techniques, project delivery systems, contracting practices, and the dynamic nature of site works. To meet these challenges, construction professionals require training on a regular basis to keep themselves abreast of these changes. Construction projects involve mainly three parties: owner, designer (consultant), and contractor.

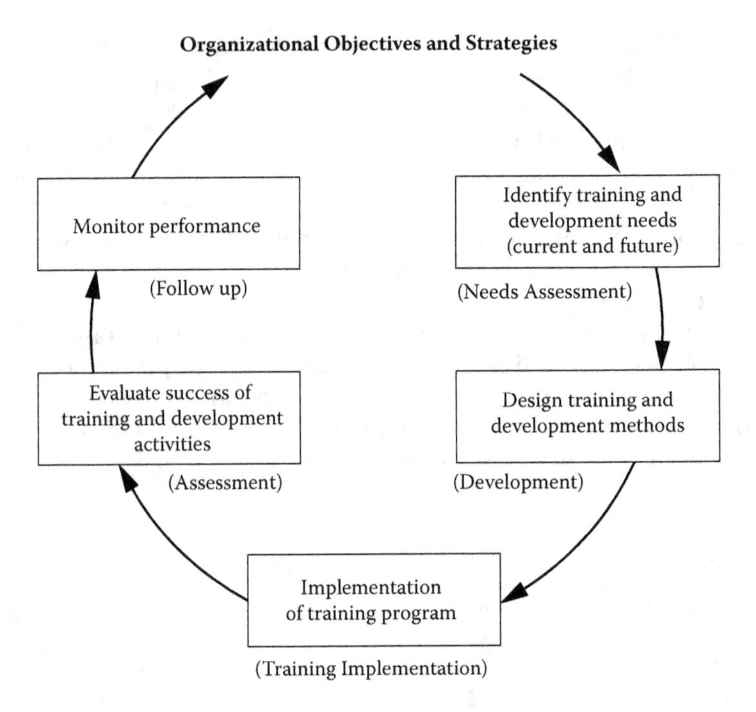

FIGURE 3.31
Training process cycle.

There are many other participants such as subcontractors, suppliers, and testing and auditing agencies who are directly or indirectly involved in construction projects. Professionals working with each of these parties need to have the skills and working knowledge to satisfy the organizational requirements for competitive advantage of the organization as well as individual growth. The area of training can depend on the type of business and the field of expertise in a particular management field. In general, there is a need to follow these management areas in the construction sector:

1. Project inception/development/feasibility
2. Project cost estimation and cost management
3. Project planning and scheduling
4. Quality management
5. Contract conditions
6. Procurement and tendering procedures
7. Risk management
8. Health and safety management
9. Environmental management
10. Value management

3.10.1 Needs Assessment

Training is directed toward agreed upon standards and objectives. The person being trained participates with the trainer or facilitator in the training activity. Training is a means of communicating new knowledge and skills to the trainee and changing his or her attitude toward work performance. It can raise awareness and provide the opportunity to the people to explore their existing knowledge and skills. But to be effective, training should be based on the needs of the person(s) who is (are) being trained. The training need is the gap between what somebody already knows and what they need to know to do their job or fulfill their role effectively.

The first step in the training and development process is to conduct a needs assessment. The assessment begins with a need that can be identified in several ways but is generally described as a gap between what is currently in place and what is needed, now and in future. In order to establish need, it is necessary to assess the current level of knowledge and skills and what level of knowledge and skills is needed to do their job or task. In general, there are four objectives/levels of training and development:

1. Organizational objectives (organization's goals)
2. Occupational objectives (jobs and related tasks that need to be learned)
3. Individual objectives (competencies and skills that are needed to perform the job)
4. Societal objectives

However, the first three are more commonly used in construction projects to conduct a needs assessment by organizations. Societal objectives are mainly taken care of by governmental agencies.

In order to establish the needs, following are to be considered:

1. Is the need mandated by a regulatory body (safety, fire, handling of hazardous material from environmental perspective, health care, food quality)?
2. Is the need mandated by management's strategic requirements to achieve the following?
   a. Organizational objectives
   b. Occupational objectives
   c. Individual objectives

Once the need is established, it is required to set objectives to

1. Identify
   - Who needs training (Individuals who are to be trained)
   - What training is needed

- Establish how the data to be collected
- Analyze the data
- Summarize the findings

3.10.1.1 Organizational Objectives

The first level of training assessment is organizational objectives. Organizational analysis looks into the effectiveness of the organization's existing capability and determines where training is needed and under what conditions it will be conducted. The following are the major items to be identified and analyzed to establish organizational training objectives:

- Organizational goals
- Strategic plans, objectives
- Political impact
- Economical impact
- Environmental impact
- Global/world market
- Future market trend
- Technological changes
- Competitor's approach to training and development
- Organizational structure
- Significance and importance of training in the organization
- Codes, standards, and regulations
- Willingness by the employees to participate in training

A proper assessment of needs will help the organization to prioritize availability of resources to address the current/future requirements.

3.10.1.2 Occupational Objectives

The assessment for occupational training is to determine where a change or new way of working about a job or group of jobs is needed and to establish the knowledge, skills, attitude, and abilities needed to achieve optimum performance.

3.10.1.3 Individual Objectives

The assessment for individual training analyzes how an individual employee is performing the job and determines which employees need training and what training is needed. Individual training requirements are based on annual performance (appraisal) reports with the employees and their line manager. Individual training helps in enhancing the skills of the employee to improve performance on the current job, or to deal with forthcoming changes or developmental needs that will enable individuals to program their careers.

All three levels of needs analysis discussed above are interrelated; therefore, the needs analysis at these three levels should be considered in conjunction with one another. The data collected from each level is critical for a thorough and effective needs assessment.

3.10.1.4 Data Collection

Data collection is a critical part of needs assessment. The following ten types of data are normally collected for needs assessment:

1. Business plans, objectives
2. Interviews
3. Observations
4. Work sample
5. Questionnaires
6. Performance evaluation
7. Performance problems
8. Best practices followed by other companies/competitors
9. Analysis of operating problems (work time)
10. Survey

3.10.2 Development

Once the needs assessment is completed and the training objectives are clearly identified, the development phase (design of training and development methods) is initiated. This phase/step consists of

1. Selection of trainer/facilitator (internal or external)
2. Design of course material
3. Techniques or methods to be used to facilitate training
4. Selection of an appropriate method (on the job or any other methods)
5. Duration of training

The type of training methods depends on the training objectives or expected learning outcomes. It depends on how the professional needs to be trained and what the needs of the participant (trainee) are. While planning for the type of training, the following points are to be considered:

- Needs and abilities of the participants
- Communication method
- Regulatory requirements, if any

3.10.3 Training Implementation

The following are the seven most common training methods:

1. Seminars
2. Workshops
3. Lectures
4. Self-study tutorials
5. Coaching
6. On-the-job training
7. Field visits

The following are the most common materials used for training:

1. Projectors
2. Slides
3. Audiovisual aids
4. Videos
5. Computer-based instruction techniques
6. Flip charts
7. Wall charts
8. Flash cards
9. Workbooks
10. Handouts
11. User manuals
12. Case studies
13. Job aids

3.10.4 Assessment

Evaluation of training is crucial for ascertaining the effectiveness of a training program and plan for future training. Evaluation is to be done for both the trainer as well as the trainee. Normally, evaluation forms are distributed to all the participants at the end of the training session to evaluate understanding of the subject of training by the trainee, and also to know their opinion of the trainer.

3.10.5 Follow-up

The training program is only a part of the knowledge and skills acquisition process. Putting the acquired knowledge into practice is the most important step. The training will be considered successful only if the desired positive changes have taken place and an improvement in the organizational and individual performance is recorded.

In order to verify the effectiveness of training, a short session to discuss the training process and its on-the-job implementation is required.

Construction projects involve the owner, designer (consultant), and contractor. The roles and responsibilities of personnel vary according to their affiliation to a particular working group and also may vary from project to project. For example, an engineer working with a designer must have knowledge and skills related to design, whereas an engineer working with a contractor must know how to execute/implement the project. The knowledge and skills of the team members should be closely linked to the project requirements. Deployment of project staff is done in accordance with the specific qualification listed in the contract documents. Therefore, while establishing the needs assessment for project personnel, contract requirements should be taken into consideration.

Normally, the construction professionals assigned to a particular project are, apart from client/owner representative(s), from the designer's (consultant's) or contractor's regular staff, supplemented by additional hiring, if needed, to fulfill contract requirements. It is assumed that these personnel are fully capable of taking up the assigned project responsibilities. Sometimes it is required to hire local labor to comply with regulatory requirements. In such cases, it is possible that these personnel may not posses the requisite knowledge and skills to perform the required task. Therefore, special training should be arranged for these personnel to maintain the organization's reputation.

The training needs of team members assigned to, or about to be assigned to, a construction project are different from those of normal employees working with the organization. The assigned team members should have knowledge of the site conditions, safety, environmental rules, quality, and climatic conditions. Therefore, special training is required by establishing a special needs assessment taking into consideration project-specific roles and responsibilities. Project personnel hired for a specific project should be given induction training to become fully oriented in their respective jobs. Further, they should be made fully aware of the company's quality management system.

Table 3.22 lists the training needs for a contractor's project manager to make him or her fully conversant with project management.

TABLE 3.22

Training Needs for Contractor's Project Manager

Serial Number	Areas of Training
1	Human relations
2	Construction methods
3	Project management
4	Communication skills
5	Resource management
6	Quality analysis/quality control
7	Cost management
8	Project planning and scheduling
9	Regulatory requirements about labor
10	Safety, health and environmental requirements
11	Basic design principles (all the trades)
12	Human resources management
13	Management of subcontractors
14	Valuation of work in progress
15	Contract conditions
16	Building contract laws
17	Interpersonal skills
18	Logistical/demographic conditions
19	Organization planning

3.11 Customer Satisfaction

Quality is meeting customer requirements and ensuring customer satisfaction. Customer satisfaction in construction is delivering a project or facility per the defined scope within the agreed-upon budget and specified schedule. It is delivering a project or facility that satisfies the owner/end user needs and requirements. Customer satisfaction is an important factor in the development of the construction process and customer relationship.

Customer satisfaction can be used for evaluation of quality and ultimately for assessment of an organization's quality improvement program, organizational focus, and project team performance. Most organizations solicit comprehensive feedback regularly from their customers to share openly with the entire organization. Meaningful data and feedback concerning the organization's operating environment, and the organization's performance in that arena, are essential for effective response and organizational success. However, to achieve competitive advantage, the meaningful data and feedback must be a continuous process. It needs to be as objective as possible to help identify areas needing improvement.

Construction is an increasingly competitive industry where organizational focus on exceeding expectations takes on an added significance. It is important for improved performance and essential to establish a competitive advantage. To achieve competitive advantage, a contractor should emphasize a high-quality construction facility/project. Customer satisfaction and continuous improvement are fundamental goals for achieving competitive advantage. Customer satisfaction enables construction companies to differentiate themselves from their competitors and create sustainable advantage.

In construction, a customer may be defined either as the owner of the project or facility or the one who needs the construction facility (end user). Customer satisfaction in the construction industry is how well a contractor meets the customer's (owner's) expectations. Customer satisfaction in construction does not guarantee that future contracts will be contracted to the same contractor based on customer satisfaction criteria. Generally, a contractor's selection criteria are based on price but also on the contractor's technical and financial capabilities and previous experience of the contractor's competence; however, the contractor's relationship and the level of satisfaction with the owner/client has significant importance for securing future jobs.

Customer satisfaction is therefore reflective of the customer's experience of and confidence in the contractor's abilities and cooperation. The essential objective in improving customer satisfaction is to achieve client loyalty, which can ensure that the contractor is considered and prequalified for future contracts or a partnering guarantee.

Customer satisfaction in construction is a function of several factors such as

- The customer's past or direct experience with the contractor
- References about the contractor having successfully completed similar projects
- The customer's personal needs that the contractor can satisfy
- Contract deliverable system
- Bidding procedures and requirements

Construction projects are a temporary endeavor, unique in having a finite duration. Construction projects involve numerous stakeholders who are closely related and interact during a given project. The level of stakeholder satisfaction directly influences the current project and subsequent projects. For contractors, completing a project in accordance with the defined scope within budget and on time satisfies owner needs and generates profit. The relationship between the customer and contractor is periodic and depends on the duration of the project. In construction, mutual cooperation and maintaining a harmonious working relationship between the customer (owner) and contractor is strongly emphasized. Further, in construction, the project organizations and collaborative relationships are of a one-off nature. The customer satisfaction survey model is different from those in other industries, where the relationship between the client (customer) and supplier is normally a long-term relationship. In construction, each customer has unique needs and expectations, and therefore it is necessary to make a concerted effort to identify customer satisfaction through direct discussions with the owner and participation meetings with the owner and their team prior to the project start. Therefore, while creating customer satisfaction surveys for construction, the following primary objectives should be considered:

- Establishing client expectations
- Identifying the most effective methods used to solicit customer feedback
- Procedure to collect the feedback
- Investigate how customer feedback is utilized

An example model of a customer satisfaction survey in construction is shown in Figure 3.32.

Implementation of customer satisfaction surveys helps manage and improve customer relationships. In addition, measuring customer satisfaction has several benefits for organizations:

- Improvement in communication between the client and contractor/supplier
- Improvement in mutual understanding
- Identifying the areas for improvement
- Better understanding of the problems
- Evaluation of progress toward the objectives
- Monitoring accomplishments
- Monitoring changes

Customer satisfaction in the construction industry is different from that in manufacturing, process, or services industries, where evaluation models or indices exist for accessing customer satisfaction and collecting data to correct or prevent defects/mistakes. The function of the construction industry is to provide customers with facilities that meet their needs and expectations. The data collected through customer satisfaction questionnaires in the construction industry can only be used to improve the next product/project. Table 3.23 lists customer satisfaction questionnaires that can be used to get related information about completed building construction projects.

Sr. No.	Stages	Elements	Description
1	Pre-construction	Establish client expectation	Contractor to have one-on-one discussion with owner/client, project manager, consultant (construction supervisor) prior to start of construction to help 1. Establish customer expectations 2. Determine performance criteria 3. Define team expectations 4. Determine required team members
2	Construction	To know how the work is progressing	Conduct meetings at regular interval to find out whether: 1. Client's expectation are met 2. Work progress is satisfactory a. Schedule b. Material c. Work force 3. Specified quality is achieved 4. Cost variance 5. Any discrepancies with documents 6. Staff performance and cooperation 7. Subcontractor's performance
3	Post-construction	Prepare survey questionnaire and get feedback from owner/client, project manager, consultant	The survey questionnaire should include items related to following factors: 1. Quality assurance at work 1.1 Management and implementation of agreed-upon quality assurance procedures 1.2 Quality of material 2. Workmanship 2.1 Quality of work 3. Skill of personnel 4. Cooperation of site staff 4.1 Agreement about changes/variations 5. Behavior of site staff 6. Work supervision 7. Attendance to defects 8. Safety at work 8.1 Safety measures 8.2 Handling of accidents 8.3 Attendance to injuries/accidents 9. Environmental consideration 9.1 Cleanliness of site 9.2 Handling of hazardous material 9.3 Disposal of waste 10. Subcontractor's performance 10.1 Competency of subcontractors 11. Adherence to schedule 12. Overrun 13. Information flow 14. Handover procedure 15. Attending to snags 16. Overall satisfaction 17. Consideration for future projects

FIGURE 3.32
Customer satisfaction survey in construction.

TABLE 3.23

Customer Satisfaction Questionnaire

Sir/Madam

We are conducting a survey to improve our performance. We would appreciate your response to the following questions. Your valuable information will help us to satisfy your requirements in a better way. (The information provided by you shall be kept confidential)

Customer Satisfaction Questionnaire

Name of Organization:

Name of Completed Project:

Q1 Workmanship of the following installed items

Sr. No.	Work Description	Very Good	Good	Average	Poor	No Comment
1	Structural work					
2	Precast work					
3	Internal finish					
4	External finish					
5	Plumbing					
6	Drainage					
7	Firefighting					
8	HVAC					
9	Elevator					
10	Escalator					
11	Electrical work					
12	Landscape work					
13	Pavements					
14	Streets/roads					
15	Waterproofing					

Q2 How are the following systems functioning?

Sr. No.	Systems	Very Good	Good	Average	Poor	No Comment
1	HVAC					
2	Water system					
3	Lighting					
4	Power					
5	Fire alarm system					
	Telephone system					
5	Public address system					
7	Security system					
8	Kitchen equipment					
9	Window cleaning system					
10	Irrigation system					
11	Elevators					
12	Escalators					

Q3 How was the behavior of our site staff?

Sr. No.	Position	Cooperative	Polite	Impolite
1	Managers			
2	Engineers			
3	Foreman			
4	Laborers			
5	Office staff			

(*Continued*)

TABLE 3.23 (*Continued*)

Customer Satisfaction Questionnaire

Q4	**How was site safety?**			
	Good	Average		Poor
Q5	**Was the project completed on schedule?**			
	Yes	No		
Q5A	**If your reply to Q5 is No, then please let us know:**			
	Have you incurred any revenue loss/any other damages due to noncompletion in time?			
	Yes	No		
Q6	**How was the response to your on-site problems?**			
	Immediate	Within one week	Within two weeks	No response
Q7	**Were our variation claims per contract?**			
	Yes	No		
Q8	**Are you satisfied with variation settlement method?**			
	Satisfied	Not satisfied		Needs improvement

Q9 **Are you satisfied with our performance?**

Satisfied | Not satisfied | Needs improvement

Q10 **How was the response to your problems from our head office?**

Good | Poor | Needs improvement

Q11 **How do you compare our performance with other competitors?**

Better than others | Same as others | Worse than others

Q12 **Would you consider us when inviting bidders for your new tenders?**

Yes | Can't say

Q13 **Any Suggestion(s)?**

Source: Abdul Razzak Rumane (2010). *Quality Management in Construction Projects.* CRC Press, Boca Raton, FL. Reprinted with permission from Taylor & Francis Group.

4

Application of Construction Tools in Construction Projects

4.1 Construction Projects

Construction projects are mainly capital investment projects. They are customized and nonrepetitive in nature. Construction projects have become more complex and technical, and the relationships and the contractual groupings of those who are involved are also more complex and contractually varied.

4.2 Project Inception

The project inception stage is sometimes referred to as the feasibility stage or inception phase. It is a relatively short period at the start of the project and is about understanding the project scope and objectives and getting enough information to either confirm that the project should proceed or be convinced that it should not proceed and abort it. Successful completion of inception-stage activities and approval of the project by the stakeholder/sponsor marks the initiation of the project.

The primary objectives of the inception phase are

- Establish a justification or business case for the project.
- Establish project objectives and scope and boundary conditions.
- Estimate an overall cost and schedule for the project.
- Conduct a cost/benefit analysis.
- Identify potential risks associated with the project.
- Conduct a feasibility study to assess whether the project is technically feasible and economically viable.
- Outline key requirements that will drive design trade-offs and identify which requirements are critical.

- Identify the type of contact and project delivery system.
- Identify initial project team members.
- Obtain stakeholders' approval to proceed with the project or to stop the project.

4.2.1 Identification of Need

Most construction projects begin with recognition of a new facility. The owner of the facility could be an individual, a public/private sector company, or a governmental agency. The need of the project is created by the owner and is linked to the available financial resources to develop the facility. The owner's needs are quite simple and are based on the following:

- To have the best facility for the money, that is, to have maximum profit or services at a reasonable cost
- On-time completion, that is, to meet the owner's/user's schedule
- Completion within budget, that is, to meet the investment plan for the facility

The owner's need must be well defined, indicating the minimum requirements of quality and performance, an approved main budget, and a required completion date. Sometimes the project budget is fixed, and therefore the quality of building system, materials, and finishes of the project need to be balanced with the budget. Table 4.1 lists major points to be considered for a needs analysis of a construction project, and Table 4.2 illustrates the need statement.

4.2.2 Feasibility Study

Once the owner's need is identified, the traditional approach is pursued through a feasibility study or an economical appraisal of owner needs or benefits, taking into account the many relevant moral, social, environmental, and technical constraints. The feasibility study takes its starting point from the output of the project identification need. Depending on the circumstances the feasibility study may be short or lengthy, simple or complex. In any case, it is the principal requirement in project development, as it gives the owner an early assessment of the viability of the project and the degree of risk involved. The outcome of the feasibility study helps selection of a defined project that meets the stated project objectives, together with a broad plan of implementation.

A project feasibility study is usually performed by the owner through his own team or by engaging individuals/organizations involved in preparation of economical and financial studies. The objective of a feasibility study is to review the technical/financial viability of the project to give sufficient information to enable the client to proceed or abort the project. Normally, the feasibility study is conducted by a specialist consultant in this field. The following are the contents of the feasibility study report:

TABLE 4.1

Major Considerations for Needs Analysis of a Construction Project

Serial Number	Points to be Considered
1	Is the project in line with the organization's strategy/strategic plan and mandated by management in support of a specific objective?
2	Is the project a part of the mission statement of the organization?
3	Is the project a part of the vision statement of the organization?
4	Is the need mandated by a regulatory body?
5	Is the need to meet government regulations?
6	Is the need to fulfill the deficiency/gap of such type of projects in the market?
7	Is the need created to meet market demand?
8	Is the need to meet the research and development requirements?
9	Is the need for technical advances?
10	Is the need generated to construct a facility/project that is innovative?
11	Is the need to improve the existing facility?
12	Is the need a part of mandatory investment?
13	Is the need to develop infrastructure?
14	Is the need to serve the community and fulfill social responsibilities?
15	Is the need to resolve a specific problem?
16	Will the need have any effect on the environment?
17	Is there any time frame to implement the need?
18	Does the need have financial constraints?
19	Doles the need involve major risk?
20	Is the need within the capability of the owner/client, either alone or in cooperation with other organizations?
21	Can the need be managed and implemented?
22	Is the need realistic and genuine?
23	Is the need measurable?
24	Is the need beneficial?
25	Does the need comply with environmental protection agency requirements?
26	Does the need comply with the government's health and safety regulations?

1. Project history (project background information)
2. Description of proposed project
   a. Project location
   b. Plot area
   c. Interface with adjacent/neighboring area
   d. Expected project deliverables
   e. Key performance indicators
   f. Constraints
   g. Assumptions

TABLE 4.2

Need Statement

Serial Number	Points to be Considered
1	Project purpose and need a. Project description
2	What is the purpose of the project? a. Project justification
3	Why is the project needed now?
4	How was the need for the project determined? a. Supporting data
5	Is it important to have the needed project?
6	Is such a facility/project required?
7	What are the factors contributing to the need?
8	What is the impact of the need?
9	Will the need improve the existing situation and be beneficial?
10	What are the hurdles?
11	What is the time line for the project?
12	What are funding sources for the project?
13	What are the benefits of the projects?
14	What is the environmental impact?

3. Feasibility study details
4. Business case
    a. Project need
    b. Stakeholders
    c. Project benefits
    d. Financial benefits
    e. Estimated cost
    f. Estimated time
    g. Justification
5. Risk
6. Environmental impact (considerations)
7. Social impact (considerations)

4.2.3 Selection of Project Delivery System

Construction projects have involvement of three major groups or parties:

1. Owner
2. Designer (consultant)
3. Contractor

These parties have different types of organizational arrangements considering the project objectives and procurement strategy of the owner/client. The following are the three primary project delivery systems (methods) used in construction projects:

1. Design/Bid/Build (Traditional Type)
2. Design/Build
3. Construction Management Type

Most governmental agencies and clients prefer to use the design/bid/build type of project delivery method. In this traditional method, the client selects the designer and then bids out the project for construction.

Construction project life cycles begin with identification of need or want for a new facility or improvement to an existing facility and extend through various phases such as conceptual design, schematic design, design development, construction and testing, and commissioning and handover.

4.3 Conceptual Design

Conceptual design is the first phase of the construction project life cycle and includes the following:

- Identification of need by the owner, and establishment of main goals
- Feasibility study based on the owner's objectives
- Selection of a project delivery system
- Identification of project team by selection of other members and allocation of responsibilities
- Identification of alternatives
- Financial implications, resources, based on estimation of life-cycle cost of the favorable alternative
- Time schedule
- Development of concept design

Figure 4.1 illustrates logic flow diagram of the conceptual design phase.

4.3.1 Client Brief (Terms of Reference)

Once the project delivery system is finalized and the designer/consultant is selected and contracted by the owner to proceed with the project design, a client brief or Terms of Reference (TOR) is issued to the designer/consultant to prepare the design proposal and contract documents. A client brief (TOR) is prepared by the owner/client or by the project manager on behalf of the owner

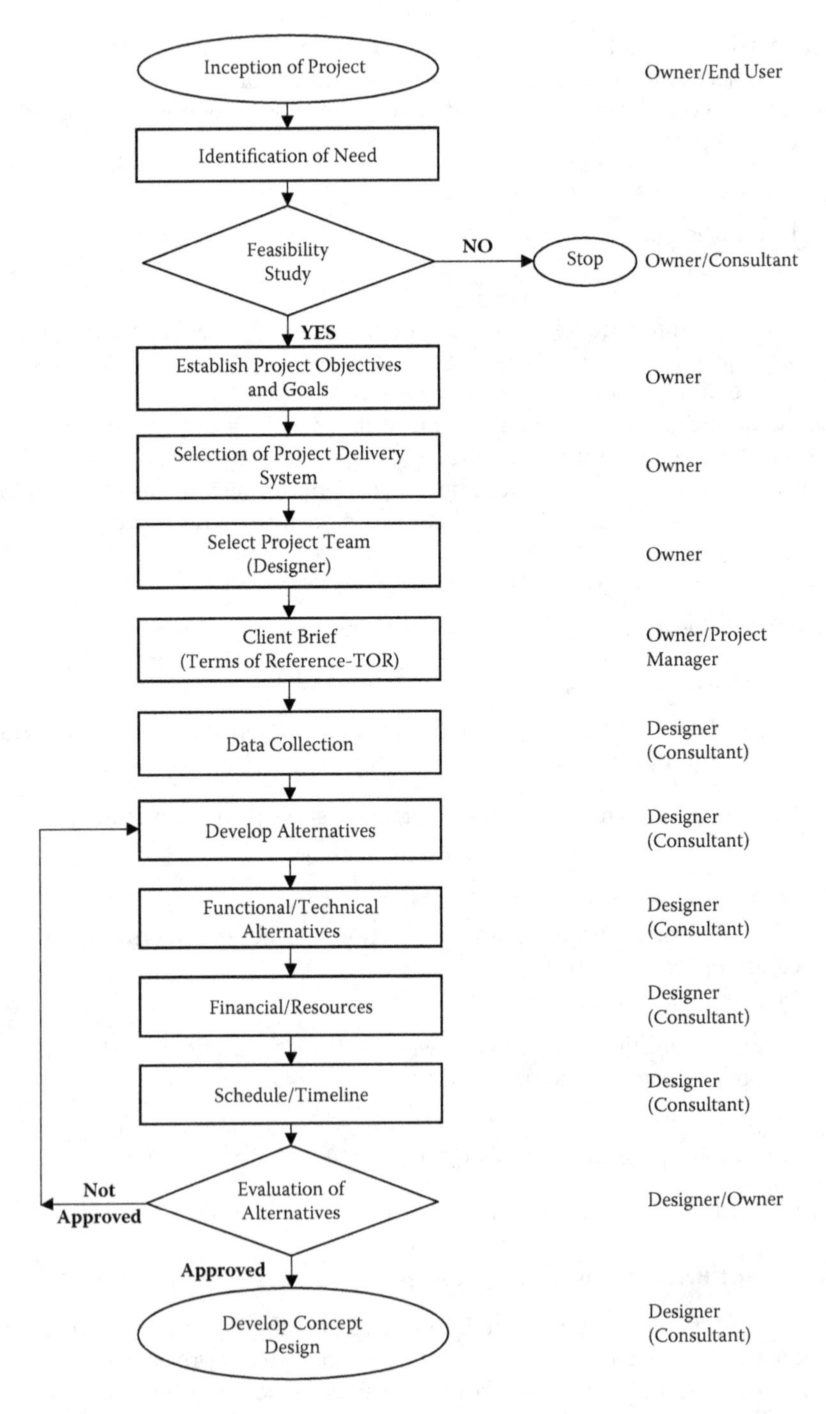

FIGURE 4.1
Logic flow of concept design phase.

describing the project objectives and requirements to develop the project. A well-prepared accurate and comprehensive TOR is essential to achieve a qualitative and competitive project. The TOR gives the project team (designer) a clear understanding for the development of project. Further, the TOR is used throughout the project as a reference to ensure that the established objectives are achieved. The client brief or TOR describes information such as

- The need or opportunity that has triggered the project
- Proposed location of the project
- Project/facility to be developed
- Procurement strategy
- Project constraints
- Estimated timescale
- Estimated cost

The TOR generally details the services to be performed by the designer (consultant), which include, but are not limited to, the following:

- Predevelopment studies, collection of required data and analysis of the same to prepare design drawings and documents for the project
- Development of alternatives
- Preparation of concept design
- Preparation of schematic design
- Preparation of detail design
- Obtaining authorities' approvals
- Compliance standards, codes, and practices
- Coordination of and participation in value engineering study
- Preparation of construction schedule
- Preparation of construction budget
- Preparation of contract documents for bidding purpose
- Prequalification/selection of contractor
- Evaluation of proposals
- Recommendation of contractor to the owner/client

The following are the requirements for a building construction project, normally mentioned in the TOR, to be prepared by the designer during the conceptual phase for submission to the owner:

1. Site Plan
   A. Civil
   B. Services

   C. Landscaping
   D. Irrigation
2. Architectural Design
3. Building and Engineering Systems
   A. Structural
   B. Mechanical (HVAC)
   C. Public Health
   D. Fire Suppression Systems
   E. Electrical
   F. Low Voltage Systems
   G. Others
4. Cost Estimates
5. Schedules

Table 4.3 shows an example TOR for a building construction project.

4.3.2 Identification of Project Team (Designer)

The owner is the primary member of the project team. His or her responsibilities and relationship with other team members depend upon the type of deliverable system the owner prefers.

For the design/bid/build type of contract system, the owner has to first select design professionals/consultants. Generally, the owner selects the designer/consultant on the basis of qualifications (qualifications-based system or QBS) and prefers to use one he or she has used before and with which he or she has had satisfactory results.

4.3.2.1 Project Team (Designer) Organization

Upon signing of the contract with the client to design the project and offer other services, the designer (consultant) assigns the project manager to execute the contract. The designer is responsible for managing development of the design and contract documents to meet the client's needs and objectives. The project manager coordinates with other departments and recruits design team members to develop a project design. A project team leader, along with design engineers, a quality engineer, and an AutoCAD technician from each trade are assigned to work for the project. The project team is briefed by the project manager about the project objectives and the roles, responsibilities, and authorities of each team member. A quality manager also joins the project team to ensure compliance with the organization's quality management system. Figure 4.2 illustrates the project design

TABLE 4.3

Contents of Terms of Reference (TOR) Documents

Serial Number	Topics
1	Project Objectives
	1.1 Background
	1.2 Project Information
	1.3 General Requirements
	1.4 Special Considerations
2	Project Requirements
	2.1 Scope of Work
	2.2 Work Program
	2.2.1 Study Phase
	2.2.2 Design Phase
	2.2.3 Tender Stage
	2.2.4 Construction Phase
	2.3 Reports and Presentations
	2.4 Schedule of Requirements
	2.5 Drawings
	2.6 Energy Considerations
	2.7 Cost Estimates
	2.8 Time Program
	2.9 Interior Finishes
	2.10 Aesthetics
	2.11 Lighting
	2.12 Heating and Cooling
3	Opportunities and Constraints
	3.1 Site Location
	3.2 Site Conditions
	3.3 Land Size and Access
	3.4 Climate
	3.5 Time
	3.6 Budget
4	Performance Target
	4.1 Financial Performance
	4.1.1 Performance Bond
	4.1.2 Insurance
	4.1.3 Delay Penalty
	4.2 Energy Performance Target
	4.2.1 Energy Conservation
	4.3 Work Program Schedule
5	Environmental Rating System
6	Design Approach
	6.1 Procurement Strategy
	6.2 Design Parameters
	6.2.1 Architectural Design
	6.2.2 Structural Design
	6.2.3 Mechanical Design
	6.2.4 Electrical Design
	6.2.5 Conveying System

(Continued)

TABLE 4.3 (*Continued*)

Contents of Terms of Reference (TOR) Documents

Serial Number	Topics
	6.2.6 Landscape
	6.2.7 External Works
	6.2.8 Parking
	6.3 Sustainable Architecture
	6.4 Engineering Systems
	6.5 Value Engineering Study
	6.6 Design Review by Client
7	Specifications and Contract Documents
8	Project Team Members
	8.1 Number of Project Personnel
	8.2 Staff Qualification
9	Visits
	9.1 Specialists
	9.2 Selection of Product/System

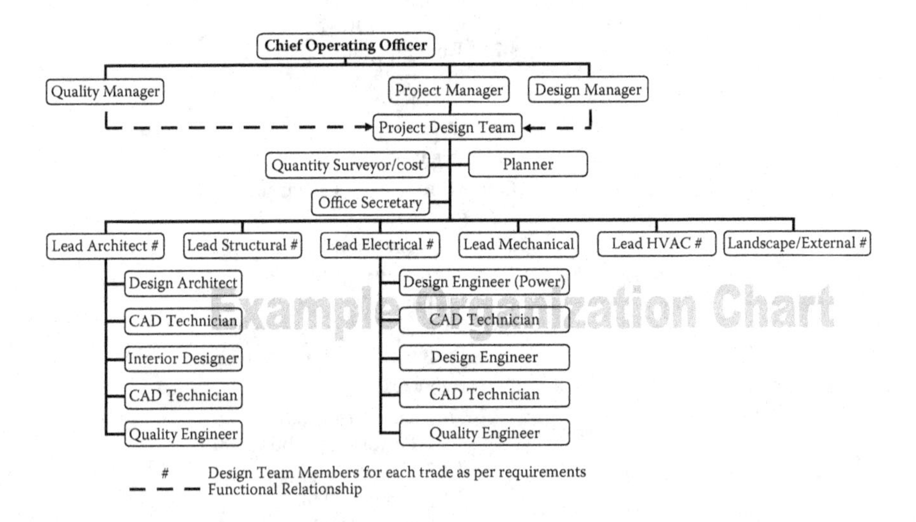

FIGURE 4.2
Project design organization chart.

organization structure, and Table 4.4 illustrates the contribution of various participants during all the phases of the construction project life cycle for the design/bid/build type of contracting system.

4.3.3 Data Collection

After signing of the contract and understanding the project objectives, the designer has to collect data. The purpose of data collection is to gather all the relevant information on existing conditions, both on the project site

TABLE 4.4

Contribution of Various Participants (Design/Bid/Build Type of Contracts)

Phase	Example of Contribution		
	Owner	**Designer**	**Contractor**
Conceptual Design	Identification of need Selection of alternative Selection of team members Approval of time schedule Approval of budget TOR	Feasibility Development of alternatives Cost estimates Schedule Development of concept design	
Preliminary Design	Approval of preliminary (schematic) design	Develop general layout/scope of facility/project Regulatory approval Budget Schedule Contract terms and conditions Value engineering	
Detail Design	Approval of budget Approval of time schedule Approval of design Contract negotiation Signing of contract	Development of detail design Authorities, approval Detail plan Budget Schedule BOQ Tender documents Evaluation of bids	Collection of tender documents Preparation of proposal Submission of bid
Construction	Approve subcontractors Approve contractor's core staff Legal/regulatory clearance SWI V.O. Payments	Supervision Approve plan Monitor work progress Approve shop drawings Approve material Recommend payment	Execution of work Contract management Selection of subcontractors Planning Resources Procurement Quality Safety
Testing Commissioning and Handover	Training Acceptance of project Substantial completion certificate Payments	Witness tests Check closeout requirements Recommend takeover Recommend issuance of substantial completion certificate	Testing Commissioning Authorities' approval Documents Training Handover

Source: Abdul Razzak Rumane (2010), *Quality Management in Construction Projects*. CRC Press, Boca Raton, FL. Reprinted with permission from Taylor & Francis Group.

and surrounding areas that will impact the planning and design of the project. The data related to the following major elements is required to be collected by the designer:

1. Certificate of title
   a. Site legalization
   b. Historical records
2. Topographical survey
   a. Location plan
   b. Site visits
   c. Site coordinates
   d. Photographs
3. Geotechnical investigations
4. Field and laboratory test of soil and soil profile
5. Existing structures in/under the project site
6. Existing utilities/services passing through the project site
7. Existing roads, structure surrounding the project site
8. Shoring and underpinning requirements with respect to adjacent area/structure
9. Requirements to protect neighboring area/facility
10. Environmental studies
11. Daylighting requirements
12. Wind load, seismic load, dead load, and live load
13. Site access/traffic studies
14. Applicable codes, standards, and regulatory requirements
15. Usage and space program
16. Design protocol
17. Scope of work/client requirements

4.3.4 Owner/Client Requirements

Based on the scope of work/requirement mentioned in the terms of reference (TOR), a detailed list of requirements is prepared by the designer (consultant). The project design is developed based on these requirements.

Figures 4.3 to 4.7 are checklists for the owner's preferred requirements related to architectural works; structural works; mechanical works; heating, ventilation, and cooling (HVAC) works; and electrical works, respectively.

Owner's Preferred Requirements (Architectural)				
Sr. No.	**Description**	**Yes**	**No**	**Notes**
1	Building Type			
	a. Commercial			
	b. Residential			
	c. Public			
	d. Other type (specify)			
2	Multistoried			
3	Basement			
	Floor number:			
	a. Mezzanine			
	b. Ground			
	c. Typical number (...)			
4	Type of Façade System			
	a. Curtain wall/glass			
	b. Fair face concrete			
	c. Painting plaster			
	d. Stone			
	e. Ceramic			
	f. Aluminum composite panels			
	g. Precast panels			
	h. Brick			
	i. GRG			
	j. Louvers			
	k. Façade cleaning system			
5	Type of roof			
	a. Space frame			
	b. Brick			
	c. Skylight			
	d. Louvers			
	e. Canopies			
6	Type of Partition			
	a. Block work			
	b. Concrete			
	c. Gypsum board			
	d. Demountable partition			
	e. Operable partition			
	f. Glass partition			
	g. Wood partition			
	h. Sandwich panels			

FIGURE 4.3
Checklist for owner requirements (architectural).

7	Type of Doors a. Wooden b. Steel c. Aluminum d. Glazing e. Fire rated f. Rolling shutter g. Sliding			
8	Type of Windows a. Wooden b. Aluminum c. Metal			
9	Wall Finishes a. Wood cladding b. Painting plaster c. Stone d. Ceramic e. Curtain glass f. Mirror g. Wallpaper h. Polished plaster			
10	Type of Ceiling a. Acoustic b. Wooden c. Metallic d. Plain Gypsum panels e. Curve light f. Metal tiles g. Gypsum tiles h. Perforated tiles i. GRG tiles j. Spider glass			
11	Floor Finishes a. Stone b. Carpet c. Wood parquet d. Ceramic e. Epoxy f. Terrazzo g. Rubber h. Raised floor			

FIGURE 4.3 (*Continued*)

12	Toilet			
	Wash basin			
	Solid surface			
	Mirror			
	WC			
	Accessories			
13	Bathroom			
	Wash basin			
	Mirror			
	Solid surface			
	WC			
	Shower			
	Accessories			
14	Kitchen			
	Rough carpentry			
	Solid surface			
	Sink type			
	Accessories			
15	Staircase			
16	Elevator			
17	Entrance			
	a. Front			
	b. Rear			
	c. Service			
18	Terraces			
	a. Indoor			
	b. Outdoor			
19	Landscape garden			
20	Parking			
	a. Car shade			
	b. Open area			
21	Swimming pool			
	a. Indoor			
	b. Outdoor			
22	Other services			

FIGURE 4.3 (*Continued*)

Owner's Preferred Requirements (Structural)				
Sr. No.	Description	Yes	No	Notes
1	Building Type			
	a. Commercial			
	b. Residential			
	c. Other type (specify)			
2	Multistoried			
3	Basement			
4	Type of Construction			
	a. Reinforced cast in situ			
	b. Reinforced precast			
	c. Steel structure			
5	Type of roof			
	a. Concrete			
	b. Steel			
	c. Other type (specify)			
6	Type of basement			
	a. Heated and cooled			
	b. Unconditioned			
7	Type of partition walls			
	a. Block			
	b. Concrete			
	c. Other type			

FIGURE 4.4
Checklist for owner's preferred requirements (structural).

4.3.5 Development of Alternatives

Once the owner/client defines the project objectives and priorities through the TOR or client brief, the designer starts development of alternatives. The TOR serves as a guide for the development of alternatives. The designer develops several alternative schemes and solutions. Each alternative is based on the predetermined set of performance measures to meet the owner's requirements. In case of construction projects, it is mainly the extensive review of development options that are discussed between the owner and the designer/consultant. The consultant engineer provides engineering advice to the owner to enable him/her to assess its feasibility and the relative merits of various alternative schemes to meet their requirements. The social, economical, and environmental impact; functional capability; and safety and reliability should be considered when evaluating alternatives. Each alternative is compared by considering the advantages and disadvantages of each systematically to meet the predetermined set of performance measures and the owner's requirements. The designer makes a brief presentation to

Owner's Preferred Requirements (Mechanical)				
Sr. No.	**Description**	**Yes**	**No**	**Notes**
1	Equipment Type: Pumps			
	a. Normal			
	b. Standby unit			
	c. Preferential country of origin			
2	Water storage tank			
3	Water system type			
	a. Cold water only			
	b. Cold/hot water			
4	Type of Pipe Water Supply			
	a. Metallic			
	b. Nonmetallic			
5	Type of Pipe Drainage			
	a. Metallic			
	b. Nonmetallic			
6	Boiler System			
	a. Central			
	b. Individual unit			
	c. Preferential type of boiler			
7	Preferential fittings and accessories			
8	Water pumps on emergency			
9	Fuel storage type			
10	Interface with BMS			

FIGURE 4.5
Checklist for owner's preferred requirements (mechanical).

the owner, and the project is selected based on the preferred conceptual alternatives.

4.3.5.1 Functional/Technical

While developing project alternatives, the designer has to consider alternate materials, systems, and equipment by comparing the following elements and developing a conceptual alternative:

1. Economy
2. Cost efficiency
3. Sustainable
4. Suitability to the purpose and objectives
5. Performance parameters

Owner's Preferred Requirements (HVAC)				
Sr. No.	**Description**	**Yes**	**No**	**Notes**
1	Equipment Type			
	a. Normal			
	b. Special			
	c. Preferential country of origin			
2	Standby unit			
3	System Type			
	a. Cooling only			
	b. Cooling/heating			
4	BMS System			
	a. Manual operation			
	b. Automatic			
	c. Addressable			
	d. Web based			
5	Power for Equipment			
	a. Normal			
	b. Emergency			
6	Thermostat Type			
	a. Analog			
	b. Digital			

FIGURE 4.6
Checklist for owner's preferred requirements (HVAC).

6. Environmental impact
7. Environment-friendly material and products
8. Physical properties, thermal comfort, insulation, fire resistance
9. Utilization of space
10. Accessibility
11. Ventilation
12. Indoor air quality
13. Water efficiency
14. Power consumption (energy-saving measures, renewable energy)
15. Daylighting
16. High-performance lighting
17. Green building concept
18. Aesthetics
19. Safety and security
20. Any other critical issues

Owner's Preferred Requirements (Electrical)				
Sr. No.	**Description**	**Yes**	**No**	**Notes**
1	Lighting Type			
	a. Normal			
	b. LED			
	c. Fiber optic			
2	Lighting Switching			
	a. Normal switching			
	b. Central switching			
	c. Smart building			
	d. Dimming			
3	Power			
	a. Normal			
	b. Emergency			
	c. Through UPS			
4	Diesel Generator Set			
5	UPS			
6	Fire Alarm System			
	a. Analog			
	b. Addressable			
7	Communication System			
	a. Analog			
	b. Digital			
	c. IP telephony			
8	IT System			
	a. Passive network			
	b. Active components			
9	Security System			
	a. Analog cameras			
	b. Digital cameras			
	c. IP cameras			
	d. Analog system			
	e. Digital system			
	f. IP integration			
	g. Guard tour system			
	h. X-ray machine			
	i. Metal detectors			
	j. Other requirements			

FIGURE 4.7
Checklist for owner's preferred requirements (electrical).

10	Access Control System			
11	Parking Control System			
12	Road Blocker			
13	Public Address System a. Analog b. Digital c. IP			
14	Audiovisual System			
15	Conference System			
16	Satellite Antenna System a. Digital b. IP			
17	Central Clock System			
18	System Integration			
19	Any Other System			

FIGURE 4.7 ***(Continued)***

4.3.5.2 Time Schedule

The duration of a construction project is finite and has a definite beginning and a definite end. Therefore, the expected time schedule for the completion of the project/facility is worked out during the conceptual phase. The expected time schedule is important from both financial and acquisition of the facility by the owner/end user. It is the owner's goal and objective that the facility be completed in time for occupancy.

Figure 4.8 illustrates a typical time schedule for a construction project.

4.3.5.3 Financial Implication/Resources

The next step is to refine cost estimates for the conceptual alternatives, as the owner must know the capital cost of construction. Only then can the owner arrange the financial resources. It is the owner's responsibility to provide an approved financial upper limit to complete the facility. It is required that the owner formulate his thoughts on project financing, as the financial conditions will affect the possible options from the beginning. Normally, the following three points should be considered:

1. What are sources of funding?
2. What criteria or rules apply?
3. How could the project best respond to those rules?

In case any funding agency is involved in financing the project, it may impose certain conditions that affect the project feasibility and implementation.

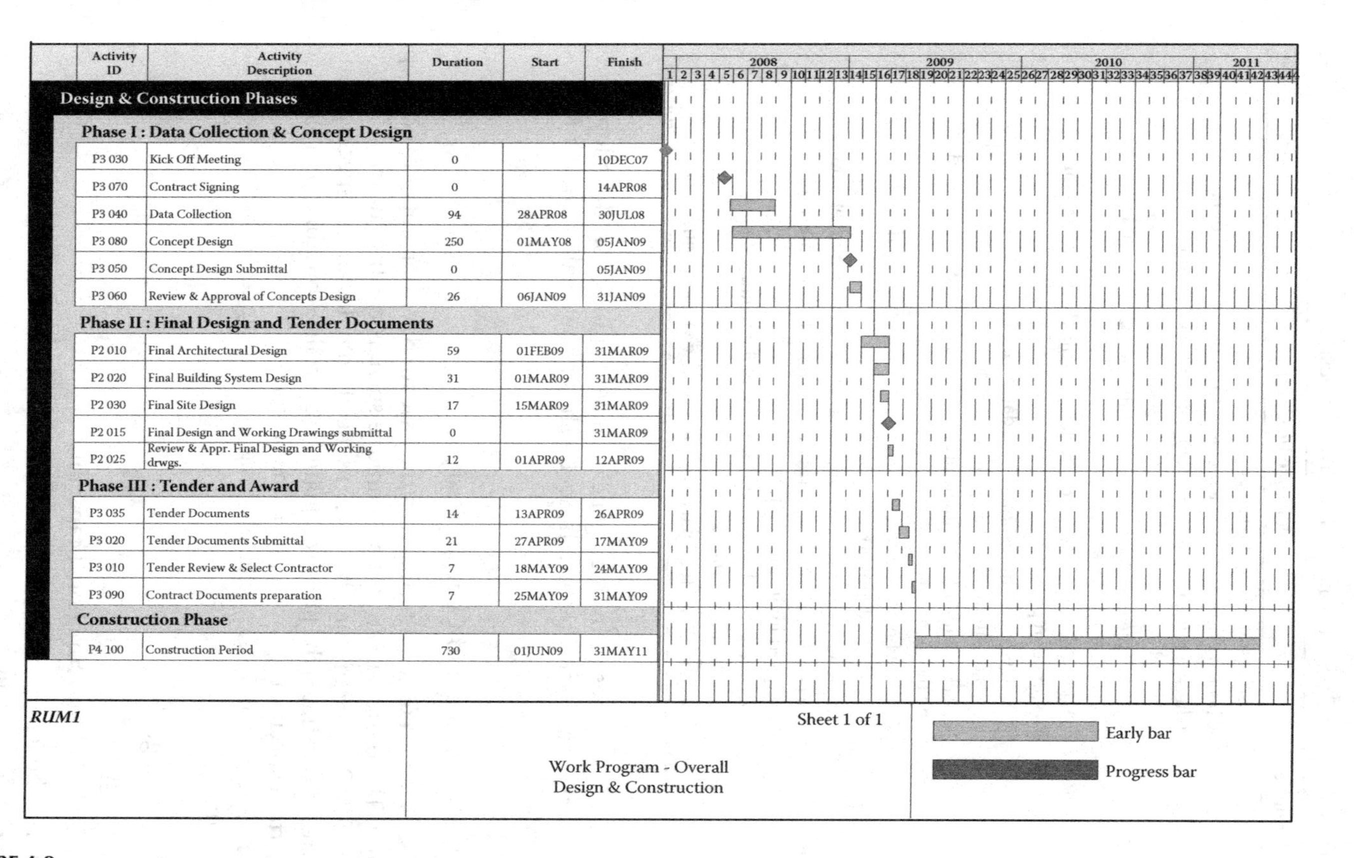

FIGURE 4.8
Typical time schedule for construction project.

TABLE 4.5

Quality Check for Cost Estimate during Concept Design

Serial Number	Points to be Checked	Yes/No
1	Check for use of historical data	
2	Check if estimate factors were used to adjust historical data	
3	Check if the estimate is updated with revision or update of concept	
4	Check if the updates/revisions are chronologically listed and cost estimate updated	
5	Check whether the scope of work is described well enough for estimation purposes	
6	Check whether the estimate is updated taking into consideration feedback from each trade	
7	Check whether the estimate is based on area schedule provided by the architect	
8	Check whether the cost estimate clearly identifies the quantities and associated work	
9	Check whether all the assumptions are per current market data	
10	Check whether cost estimates from all the relevant trades are included in the final sum	
11	Check whether the estimate includes all the requirements of specialist consultants	
12	Check whether the total estimate is reviewed and verified	

It is likely that these funding agencies may also insist on the adoption of a particular contract strategy. Table 4.5 shows the quality checks for cost estimation.

4.3.6 Development of Concept Design

The selected preferred alternative is the base for development of the concept design. The designer can use a technique such as quality function deployment (QFD) to translate the owner's need into technical specifications. Figure 4.9 illustrates the house of quality for a college building project based on certain specific requirements by the customer.

While developing the concept design, the designer must consider the following:

1. Project goals
2. Usage
3. Technical and functional capability
4. Aesthetics
5. Constructability
6. Sustainability (environmental, social, and economical)

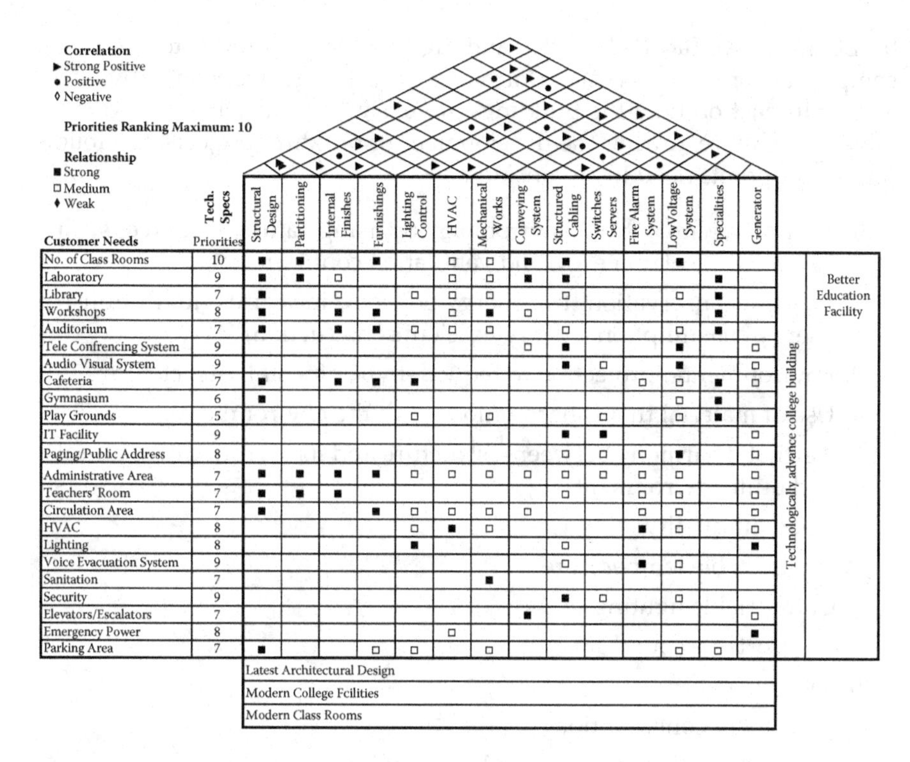

FIGURE 4.9
Quality function deployment.

7. Health and safety
8. Reliability
9. Environmental compatibility
10. Fire protection measures
11. Supportability during maintenance/maintainability
12. Cost-effectiveness over the entire life cycle (economy)

It is the designer's responsibility to pay greater attention to improving environment and achieving sustainable development. Numerous UN meetings (such as the first United Nations conference on Human Development held in Stockholm in 1972; the 1992 Earth Summit in Rio de Janeiro; the 2002 Earth Summit in Johannesburg; the 2005 World Summit), and the Brundtland Commission on Environment and Development in 1987 emphasizes "sustainability," whether it is sustainable environment, sustainable economic development, sustainable economical development, sustainable agricultural and rural development, and so on. Accordingly the designer has to address environmental and social issues and comply with local environmental protection codes. A number of tools and rating systems have been created by LEED (USA), BREEAM (UK), and HQE (France) in order to assess and compare the environmental performance of buildings. These initiatives have a great impact on how the buildings are designed, constructed, and maintained. Therefore, during the development of building projects, the following twelve points need to be considered:

1. Accretion with natural environment by using natural resources such as sunlight, solar energy, and ventilation configuration
2. Energy conservation by energy-efficient measures to diminish energy consumption (energy-efficient construction)
3. Environmental protection to reduce environmental impact
4. Use of material that is harmonious with the environment
5. Aesthetic harmony between a structure and its surrounding natural and built environment
6. Good air quality
7. Comfortable temperature
8. Comfortable lighting
9. Comfortable sound
10. Clean water
11. Less water consumption
12. Integration with social and cultural environment

During the design stage, the designer must work jointly with the owner to develop details on the owner's need and objectives and give due consideration to each part of the requirements. The owner on his part should ensure that the project objectives are

- Specific
- Measurable
- Agreed upon by all the team members
- Realistic
- Possible to complete with in defined time limits
- Within the budget

4.3.6.1 Concept Design Deliverables

The following are the concept design deliverables:

1. Concept design report (narrative/descriptive report)
   a. Space program
   b. Building exterior

   c. Building interior
   d. Structural system
   e. MEP System
   f. Conveyance system
2. Drawings
   a. Overall site plan
   b. Floor plan
   c. Elevations
   d. Sketches
   e. Sections (indicative to illustrate overall concept)
3. Concept schedule of material and finishes
4. Data collection, studies, reports
5. Lighting/daylight studies
6. Cost estimate
7. Project schedule estimate
8. Models

4.3.7 Cost of Quality during Design Phase

Construction projects are unique and nonrepetitive in nature, and so need specific attention to maintain quality. To a great extent, each project has to be designed and built to serve a specific need. It is the designer's (consultant) responsibility to develop project documents to ensure

1. Conformance to the owner's requirements
2. Conformance to requirements listed under the TOR
3. Compliance with applicable standards, codes, regulations and practices
4. Compliance with regulatory requirements
5. Development of design drawings and specifications adhere to the economic objectives
6. Great accuracy to avoid any disruption/stoppage/delay of work during construction
7. Completion within the stipulated time to avoid delay in starting of construction
8. Minimal design errors
9. Minimal omissions
10. Reduction of risk and liabilities

In order to achieve a zero defect policy during the construction phase and reduce rework during the design phase, the designer (consultant) has to take necessary steps to reduce the cost of quality. Categories of quality costs related to the design phase of construction projects can be summarized as follows:

Internal Failure Cost:

- Redesign/redraw to meet requirements of the owner and other trades
- Redesign/redraw to meet a fully coordinated design
- Rewrite specifications/documents to meet the requirements of all other trades

External Failure Cost:

- Incorporate design review comments by client/project manager
- Incorporate specifications/documents review comments by client/project manager
- Incorporate review comments from regulatory authority(ies)
- Resolve RFI (Request for Information) during construction

Appraisal Cost:

- Review of design drawings
- Review of specifications
- Review of contract documents to ensure meeting owner's needs, quality standards, constructability, and functionality
- Review for regulatory requirements, codes

Prevention Cost:

- Conduct technical meetings for proper coordination
- Follow the quality system
- Meet submission schedule
- Train project team members
- Update of software used for design

In order to improve the design and reduce hidden costs, a PDCA cycle model can be developed as a process improvement tool.

PDCA activities during the design phase are listed as follows:

- PLAN (Establish Scope)
- Establish owner's requirements

- Determine/define scope of work
- Establish standards, codes
- Establish project schedule
- Establish project budget
- DO (develop design)
- Develop design drawings
- Develop specifications
- Coordinate with other disciplines/trades
- Prepare contract documents
- CHECK (review)
- Conformance to client needs/requirements
- Check for project schedule
- Check for project budget
- Check for conflict with other disciplines
- Conformance to regulatory requirements, standards, codes
- Check for constructability
- Check for environmental compatibility
- ACT (implement comments)
- Implement review comments (if any)
- Take corrective action (if required)

  or
- Release documents for construction bid

Figure 4.10 illustrates the PDCA cycle for the design phase.

4.4 Schematic Design

Schematic design is mainly a refinement of the elements in the conceptual design phase. Schematic design is also known as *preliminary design*. It refers to design intent documents that quantify functional performance expectations and parameters for each system to be commissioned. It is traditionally labeled as 30% design. Schematic design adequately describes information about all proposed project elements in sufficient detail for obtaining regulatory approvals, necessary permits, and authorization. In this phase, the project is planned to the level where sufficient details are available for the initial cost and schedule. This phase also include the initial preparation of all documents necessary to build the facility/construction project. The central

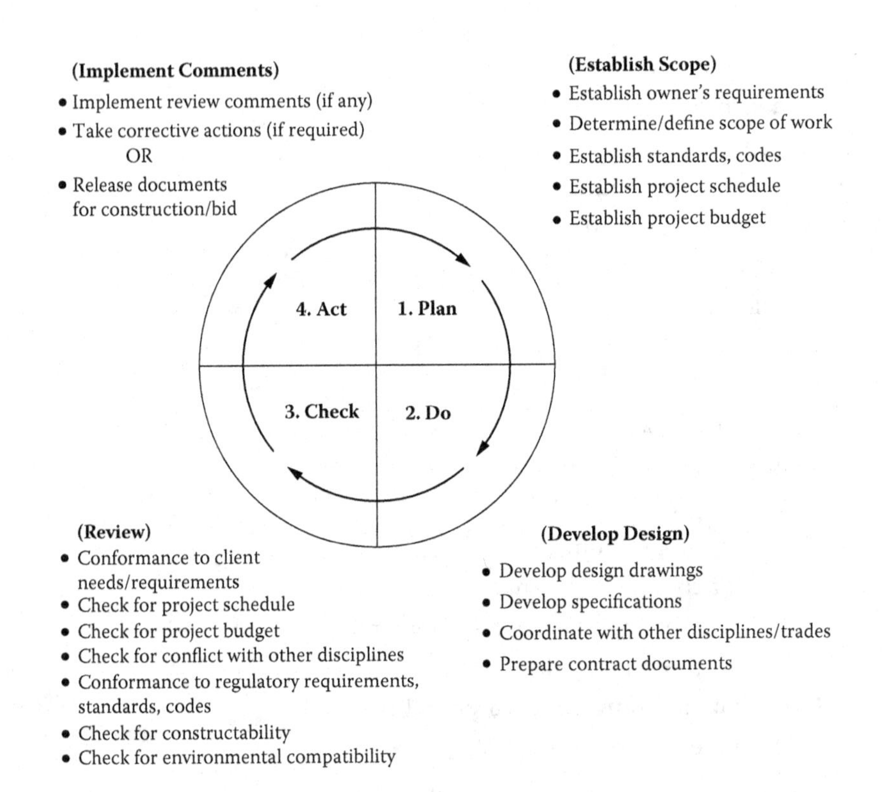

FIGURE 4.10
PDCA cycle for design phase.

activity of preliminary design is the architect's design concept of the owner's objective that can help draw up the detailed engineering and design for the required facility. Preliminary design is the subjective process of transforming ideas and information into plans, drawings, and specifications of the facility to be built. Component/equipment configurations, material specifications, and functional performance are decided during this stage. Design is a complex process. Before design is started, the scope must adequately define deliverables, that is, what will be provided. These deliverables are design drawings, contract specifications, type of contracts, construction inspection record drawings, and reimbursable expenses.

For the development of the preliminary design phase, the designer has to

- Investigate site conditions
- Collect and analyze the required data
- Analyze building code requirements
- Analyze energy conservation requirements
- Study fire and other regulatory codes and requirements

Based on the evaluation and analysis of the required data, the designer can proceed with development of preliminary design by considering the following 15 major points:

1. Concept design deliverables
2. Calculations to support the design
3. System schematics for electromechanical system
4. Coordination with other members of the project team
5. Authorities' requirements
6. Availability of resources
7. Constructability
8. Health and safety
9. Reliability
10. Energy conservation issues
11. Environmental issues
12. Selection of systems and products that support functional goals of the entire facility
13. Sustainability
14. Requirements of all stakeholders
15. Optimized life-cycle cost (value engineering)

4.4.1 Schematic Design Deliverables

Preliminary design is the basic responsibility of the architect (designer/consultant—A/E). In the case of a building construction project, the preliminary design determines the

1. General layout of the facility/building/project (preliminary design drawings)
2. Required number of buildings, number of floors in each building, area of each floor
3. Different types of functional facilities required, such as offices, stores, workshops, recreation, training centers, parking, etc.
4. Type of construction, such as reinforced concrete or steel structure, precast or cast in situ
5. Type of electromechanical services required
6. Type of infrastructure facilities inside the facilities area
7. Type of landscape
8. Preliminary specifications

Table 4.6 lists the deliverables of the schematic design phase for building projects.

TABLE 4.6

Schematic Design Deliverables

Serial Number	Deliverables
1	General 1.1 Preliminary/outline specification 1.2 Zoning 1.3 Permits and regulatory approvals 1.4 Energy code requirements Construction methodology narration Descriptive report of environmental, health, and safety requirements Estimated cost Estimate construction period Value engineering suggestions and resolutions Life safety requirements Sketches/perspective a. Interior b. Exterior Graphic presentation
2	Preliminary Design Drawings
2.1	Architectural
2.1.1	Overall site plans a. Location of building, roads, parking, access, and landscape b. Project boundary limits c. Site utilities d. Water supply, drainage, storm water lines e. Zoning f. Reference grids and axis
2.1.2	Floor plans a. Floor plans of all floors b. Structural grids c. Vertical circulation elements d. Vertical shafts e. Partitions f. Doors g. Windows h. Floor elevations i. Designation of rooms j. Preliminary finish schedule k. Services closets l. Raised floor, if required
2.1.3	Roof plans a. Roof layout b. Roof material c. Roof drains and slopes
2.2	Structural 2.2.1 Building structure 2.2.2 Floor grade and system 2.2.3 Foundation system

TABLE 4.6 (*Continued*)

Schematic Design Deliverables

Serial Number	Deliverables
	2.2.4 Tentative size of columns, beams 2.2.5 Stairs 2.2.6 Roof
2.3	Elevator 1. Traffic studies 2. Elevator/escalator location 3. Equipment room
2.4	HVAC 1. Ducting layout plan 2. Piping layout plan 3. Development of preliminary system schematics 4. Calculations to allow preliminary plant selections 5. Establishment of primary building services distribution routes 6. Establishment of preliminary plant location and space requirements 7. Determine heating and cooling requirements based on heat dissipation of equipment, lighting loads, type of wall, roof, glass, etc. 8. Estimation of HVAC electrical load 9. Development of BMS schematics showing interface 10. Location of plant room, chillers, cooling towers
2.5	Plumbing and fire suppression 1. Sprinkler layout plan 2. Piping layout plan 3. Water system layout plan 4. Water storage tank location 5. Development of preliminary system schematics 6. Location of mechanical room
2.6	Electrical 1. Lighting layout plan 2. Power layout plan 3. System design schematic without any sizing of cables and breakers 4. Substation layout and location 5. Total connected load 6. Location of electrical rooms and closets 7. Location of MLTPs, MSBs, SMBs, EMSB, SEMBs, DBs, EDBs, etc. 8. Location of starter panels, MCC panels, etc. 9. Location of generator, UPS 10. Raceway routes 11. Riser requirements 12. Information and communication technology (ICT) a. Information technology (computer network) b. IP telephone system (telephone network) c. Smart building system 13. Loss prevention systems a. Fire alarm system b. Access control security system layout c. Intrusion system

(*Continued*)

TABLE 4.6 (*Continued*)

Schematic Design Deliverables

Serial Number	Deliverables
	14. Public address, audiovisual system layout
	15. Schematics for F.A and other loss prevention systems
	16. Schematics for ICT system, and other low-voltage systems
	17. Location of LV equipment
2.7	Landscape
	1. Green area layout
	2. Selection of plants
	3. Irrigation system
2.8	External
	1. Street/road layout
	2. Street lighting
	3. Bridges (if any)
	4. Security system
	5. Location of electrical panels (feeder pillars)
	6. Pedestrian walkways

During this phase, the owner gets an opportunity to alter the scope and consider the alternatives. The owner seeks to optimize certain facility features within the constraints of other factors such as cost, schedule, and vendor capabilities.

4.4.2 General Scope of Work/Basic Design

The purpose of development of general scope of work documents during this phase is to provide sufficient information to identify the work to be performed and to allow the detailed design to proceed without significant changes that may adversely affect the project budget and schedule. At the preliminary design stage, the scope must define deliverables, that is, what will be furnished. It should include a schedule of dates for delivering drawings, specifications, calculations and other information, forecasts, estimates, contracts, materials, and construction. The designer develops a design concept with a plan, elevation, and other related information that meets the owner's requirements. The designer also develops the concept of how various systems, such as the heating and cooling system, communication system, etc., will fit into the system.

4.4.3 Regulatory Approval

Once the preliminary design is approved, it should be submitted to regulatory bodies for their review and approval for compliance with the regulations, codes, and licensing procedure.

4.4.4 Schedule

After the preliminary scope of work, the preliminary design, and budget for the facility/project are finalized, the logic of the construction program is set. On the basis of this logic, the Critical Path Method (CPM) schedule/(bar chart) is prepared to determine the critical path and set the contract milestones.

4.4.5 Budget

Based on the preliminary design, the budget is prepared by estimating the cost of activities and resources. The preparation of the budget is an important activity that results in a timed phased plan summarizing the expected expenses toward the contract and also the income or the generation of funds necessary to achieve the milestone. The budget for the construction project is the maximum amount the owner is willing to spend for design and construction of a facility that meets the owner's need. The budget is determined by estimating the cost of activities and resources and is related to the schedule of the project. If the cash flow or resulting budget is not acceptable, the project schedule should be modified. It is required that while preparing the budget, a risk assessment of the project is also performed.

4.4.6 Contract Terms and Conditions

Normally it is the consultant/design team that is responsible for developing a set of contract documents that meet the owner's needs, required level of quality, budget, and schedule. At this stage, the contract exists between the consultant and the client for the development of the project, and any good management test will demand that the contract be clearly understood by all parties associated with it. There are numerous combinations of contract arrangements for handling construction projects; however, the design/bid/build contract is predominantly used for most construction projects. This delivery system has been chosen by owners for many centuries and is called the traditional contracting system. In this system, the detailed design for the project is completed before tenders for construction are invited. The detailed engineering is carried out by the consultant/design professional to make the project qualitative and economical.

Based on the type of contracting arrangements the owner prefers, necessary documents are prepared by establishing a framework for execution of the project. Generally, FIDIC (Federation International des Ingénieurs–Counseils) model conditions for international civil engineering contracts are used as a guide to prepare these contract documents. Preliminary specifications and documents are prepared in line with the model contract documents.

4.4.7 Value Engineering Study

Value engineering (VE) studies can be conducted at various phases of a construction project; however, the studies conducted in the early stage of a project tend to provide the greatest benefit. In most projects, VE studies are performed during the schematic phase of the project. At this stage, the design professionals have considerable flexibility to implement the recommendations made by the VE team without significant impact to the project schedule or design budget. In certain countries, for a project over $5 million USD, a VE study must be conducted as part of the schematic design process. The team members perform a VE study depending on the client's/owner's requirements. It is advisable that a SAVE International–registered Certified Value Specialist be assigned to lead this study. Figure 4.11 illustrates VE process activities.

4.5 Design Development

Design development is the third phase of the construction project life cycle. It follows the preliminary design phase and takes into consideration the configuration and the allocated baseline derived during the preliminary phase. The design development phase is also known as *detail design/detailed engineering*. The client-approved preliminary design is the base for preparation of the detail design. All the comments and suggestions on the preliminary design from the client and regulatory bodies are reviewed and resolved to ensure that changes will not detract from meeting the project design goals/objectives. Detailed design involves the process of successively breaking down, analyzing, and designing the structure and its components so that it complies with the recognized codes and standards of safety and performance while rendering the design in the form of drawings and specifications that will tell the contractors exactly how to build the facility to meet the owner's need. During this phase, the following activities are performed:

1. Review of comments on the preliminary design by the client (project manager)
2. Review of comments on the preliminary design by the regulatory authorities
3. Preparation of detail design for all the works
4. Interdisciplinary coordination to resolve the conflict
5. Obtaining regulatory approval
6. Preparing the project schedule
7. Preparing the project budget

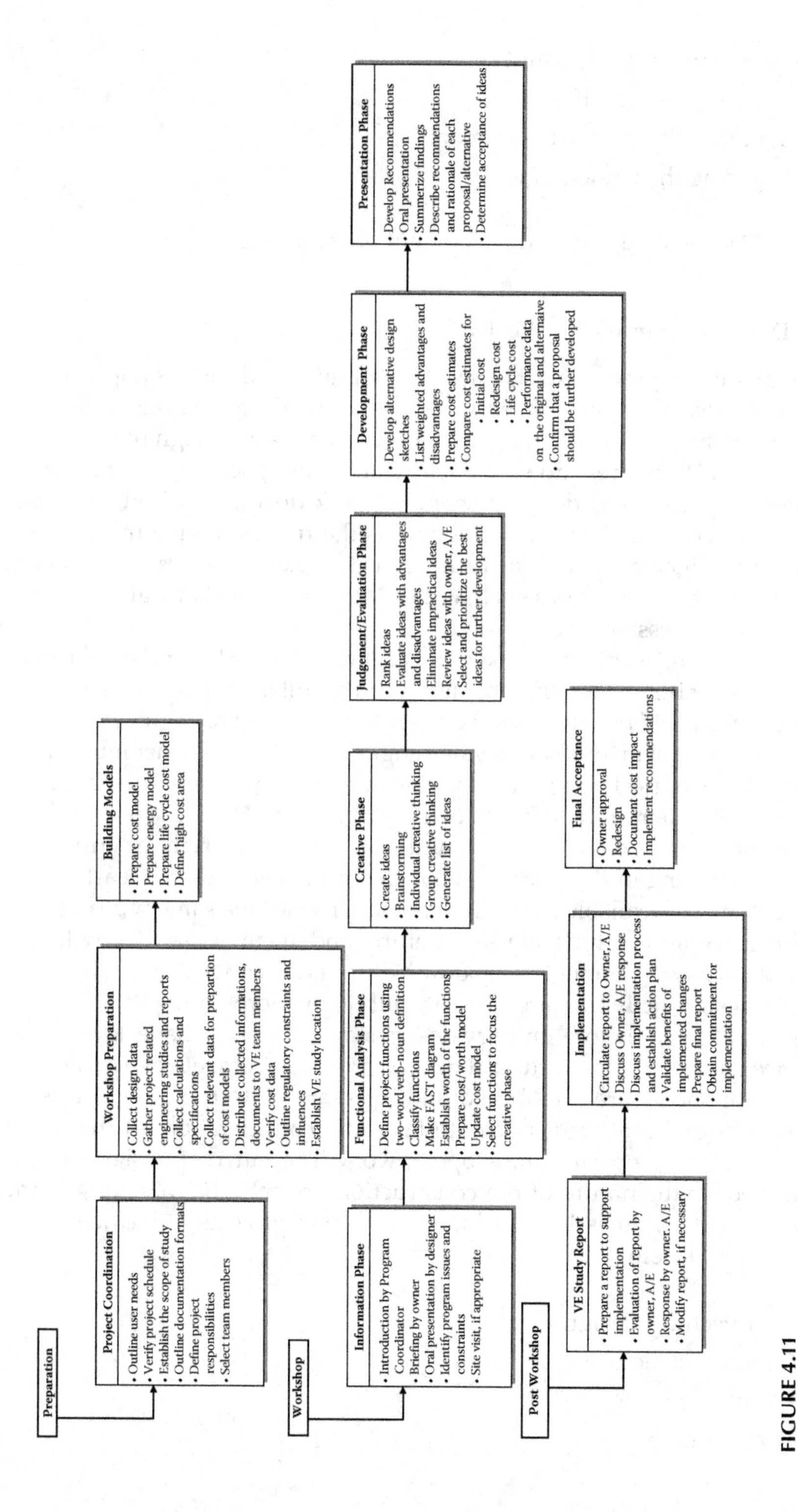

FIGURE 4.11
Value engineering process.

8. Prepare the Bill of Quantity.
9. Preparing the specifications.
10. Preparing the contract documents.
11. Preparing the tender/bidding documents.

Figure 4.12 shows the design development phase stages.

4.5.1 Detail Design of the Works

The detail design process starts once the preliminary design is approved by the owner. Detail design is an enhancement of work carried out during the preliminary stage. During this phase, a comprehensive design of the works with detailed Work Breakdown Structure and work packages are prepared. In general, specific and detailed scope of work documents lead to better-quality projects. The detail design phase is the traditional realm of design professionals, including architects, interior designers, landscape architects, and several other disciplines such as civil, electrical, mechanical, and other engineering professionals as needed.

Accuracy in project design is a key consideration for the life cycle of the project; therefore, it is required that the designer/consultant not only be an expert in the technical field but also have a broad understanding of engineering principles, construction methods, and value engineering. The designer must know the availability of the latest product in the market and use proven technology, methods, and materials to meet the owner's objectives. He or she must refrain from using a monopolistic product, unless its use is important or critical for proper functioning of the system. He or she must ensure that at least two or three sources are available in the market that produce the same type of product which complies with all required features and intent of use. This will help the owner in getting competitive bidding during the tender stage.

Detail design activities are similar, although more in-depth, than the design activities in the preliminary design stage. The size, shape, levels, performance characteristics, technical details, and requirements of all the individual components are established and integrated into the design. Design engineers of different trades have to take into consideration all these at a minimum while preparing the scope of work. The range of design work is determined by the nature of the construction project. The following is the list of major disciplines in building construction projects for which detail drawings are developed:

1. Architectural design
2. Concrete structure
3. Elevator
4. HVAC works

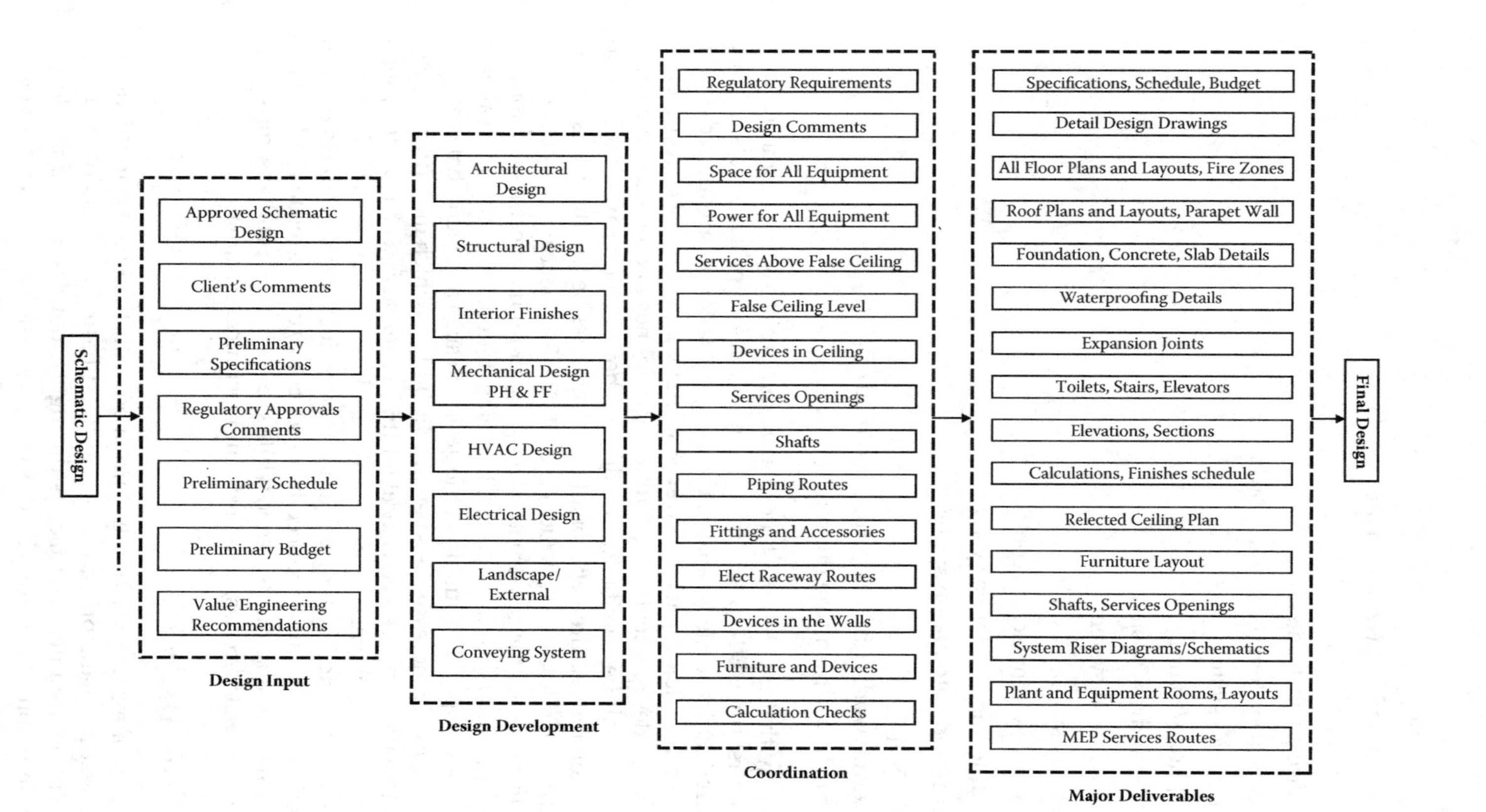

FIGURE 4.12
Design development stages.

5. Mechanical
   a. Cold and hot water systems
   b. Drainage system
   c. Fire protection system
6. Electrical system (light and power)
7. Fire alarm system
8. Telephone/communication system
9. Public address system
10. Audio visual system
11. Security system/CCTV
12. Security system/access control
13. Satellite/main antenna system
14. Integrated automation system
15. Landscape
16. External works (infrastructure and road)
17. Furnishings/furniture (loose)

It is unlikely that the design of a construction project will be right in every detail the first time. Effective management and design professionals who are experienced and knowledgeable in the assigned task will greatly reduce the chances of error and oversight. However, so many aspects must be considered, especially for designs involving multiple disciplines and enfaces to ensure, that changes will be inevitable. The design should be reviewed taking into consideration the requirements of all the disciplines before the release of the design drawings for the construction contract. Engineering design has significant importance for construction projects and must meet the customer's requirement from the start of project implementation. Engineering weakness can adversely impact the quality of the design to such an extent that marginal changes can easily increase costs beyond the budget, which may affect project schedule. Some areas deemed critical for the proper design of a product, such as explicit design, material specification, and grades of the material specified in documentation, have great importance. Most of the products used in construction projects are produced by other construction-related industries/manufacturers, and so the designer, while specifying the products, must give details of the related codes, standards, and technical compliance.

4.5.1.1 Design Quality Check

The success of a project is highly correlated with the quality and depth of the engineering plans prepared during this phase. Coordination and conflict resolution are important factors during the development of the design to avoid omissions and errors. Figure 4.13 shows the design review steps for

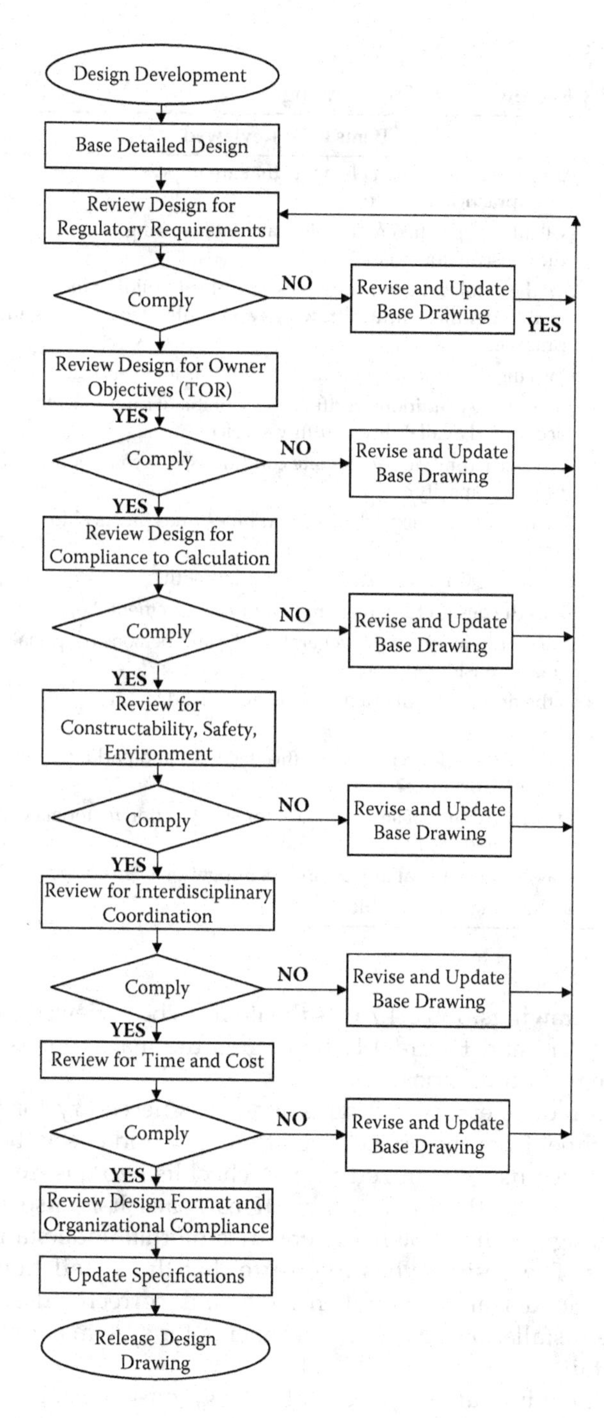

FIGURE 4.13
Design review steps for detail design.

TABLE 4.7

Constructability Review for Design Drawings

Serial Number	Items to be Reviewed	Yes/No
1	Are there construction elements that are impossible or impractical to build?	
2	Will all the specified material be available during the construction phase?	
3	Are the available labor resources capable of building the facility per the contract drawings and contracted methods and practices?	
4	Does the design follow industry standards and practices?	
5	Are the site conditions verified and suitable with respect to access and availability of utility services?	
6	Is the structural design per site conditions, soil conditions, and bearing capacity?	
7	Are the construction schedule and milestones practical to achieve?	
8	Can QA/QC requirements be complied with?	
9	Does the design fully meet regulatory requirements?	
10	Are requirements of the general public and persons of special needs considered?	
11	Is the design coordinated with adjacent land and its accessibility?	
12	Is there space for temporary office facilities and parking space for the workforce?	
13	Has availability of storage space for construction material been considered?	
14	Has environmental impact and its mitigation been considered?	
15	Is the design sustainable?	

detail design drawings, Table 4.7 lists the items to be reviewed for constructability of design, and Figure 4.14 highlights the major points facilitating interdisciplinary coordination.

In order to reduce errors and omissions, it is necessary for the quality control personnel from the project team to review and check the design for quality assurance using itemized review checklists to ensure that design drawings fully meet the owner's objectives/goals. It is also required to review the design with the owner to ensure a mutual understanding of the build process. The designer has to ensure that the installation/execution specification details are comprehensively and correctly described, and also that the installation quality requirements for systems are specified in sufficient detail.

Before the drawings are released for bidding/construction, it is necessary to check them for formatting, annotation, and interpretation. Table 4.8 lists the items to be checked for correctness of design drawings.

Serial Number	Discipline	Discipline					
		Architectural	Structural	Mechanical	HVAC	Electrical	External Works
1	Architectural		1. Structural framing plans 2. Axis, grids, levels 3. Location of columns, beams 4. Modules to match with structural plan 5. Location of stairs, fire exits 6. Expansion joints 7. Building dimensions	1. Pump room location and size of room 2. Void above false ceiling for piping 3. Sprinkler in false ceiling 4. Location of sanitary fixtures and accessories 5. Location of fire hydrant, cabinet, and landing valves 6. Location of water tank 7. Shaft for water supply, sanitary and drainage pipes 8. Location of fuel filling point for fuel carrying tanker 9. Location of manholes	1. Plant room location and size 2. Void above false ceiling for HVAC equipment, duct, and piping 3. Access for maintenance of equipment 4. K-value of thermal insulation, type of external glazing and U-value 5. Location of louvers, grills, diffusers 6. Location of thermostat and other devices 7. Staircase pressurization system with respect to HVAC 8. Location of HVAC equipment on roof 9. HVAC shaft requirement	1. Location and size of substation and door sizing 2. Trenches for cables in substation, electrical room, and generator room 3. Location and size of electrical room and closets 4. Location of electrical devices 5. Location of light fittings in the false ceiling 6. Void above false ceiling for cable tray and trunking 7. Cable tray, cable trunking route 8. Location and size of low voltage rooms 9. Location and size of generator room	1. Property limits 2. Location of outdoor equipment 3. Location of plants 4. Location of seating/ recreation area 5. Location of maintenance room/area 6. Location of manholes 7. Location of generator exhaust pipe

FIGURE 4.14
Interdisciplinary coordination.

2	Structural	1. Structural framing plans		1. Opening for pipe crossing in the walls and slab	1. Shaft for pipings and duct	1. Base for transformers	1. Manholes
		2. Axis, grids, levels		2. Shaft for pipe risers (water supply, sanitary, drainage	2. Openings/sleeves for duct and pipings	2. Base for generator	2. Foundation for light poles
		3. Location of columns, beams		3. Opening for roof drain	3. Operating weight of all HVAC equipment	3. Trenches for electrical cables	3. Manhole/ foundation for electrical panels
		4. Modules to match with structural plan		4. Openings/ sleeves for piping	4. Floor height to accommodate equipment	4. Openings/sleeves for cable tray, electrical bus duct	4. Manhole/ foundation for feeder pillars
		5. Location of stairs, fire exits		5. Opening for main circulation drain	5. Expansion joints requirements	5. Shaft for cable trays	5. Underground services tunnel
		6. Expansion joints		6. Water tank inlet location	6. Pump room equipment loads with HVAC equipment	6. Foundation for light poles	
		7. Building dimensions		7. Sanitary manholes		7. Manhole/ foundation for electrical panels	

FIGURE 4.14 ***(Continued)***

3	Mechanical	1. Pump room location and size of room 2. Void above false ceiling for piping 3. Sprinkler in false ceiling 4. Location of sanitary fixtures and accessories 5. Location of fire hydrant, cabinet, and landing valves 6. Location of water tank 7. Shaft for water supply, sanitary and drainage pipes 8. Location of fuel filling point for fuel-carrying tanker 9. Location of manholes	1. Opening for pipe crossing in the walls and slab 2. Shaft for pipe risers (water supply, sanitary, drainage) 3. Opening for roof drain 4. Openings/sleeves for piping 5. Opening for main circulation drain 6. Water tank inlet location 7. Sanitary manholes		1. Make up water requirements for HVAC 2. Connection of chilled water for plumbing works 3. Interface with building management system 4. HVAC/AHU drain with drainage system	1. Power supply for pumps and other equipment 2. Location of isolators for power supply 3. Interface with fire alarm system	1. Irrigation system with external works 2. Area drain, road gully with external/asphalt work 3. Storm water manholes with external works 4. External service to be hooked up with municipality route

FIGURE 4.14 ***(Continued)***

4	HVAC	1. Plant room location and size 2. Void above false ceiling for HVAC equipment duct and 3. Access for maintenance of equipment 4. K-value of thermal insulation, type of external glazing, and U-value 5. Location of louvers, grills, diffusers 6. Location of thermostat and other devices 7. Staircase pressurization system 8. Location of HVAC equipment on roof 9. HVAC shaft requirement	1. Shaft for ducts and piping 2. Opening for ducts and pipings in the wall and roof	1. Make up water requirements for HVAC 2. Connection of chilled water for plumbing works 3. Interface with building management system 4. HVAC/AHU drain with drainage system		1. Power supply for chillers, pumps, AHUs, and other equipment 2. Location of isolators for power supply 3. Heat dissipation from lighting and other electrical panels 4. 3-phase/single-phase power requirements 5. Power supply load during summer/winter 6. Electrical power supply for equipment connected to generator 7. Interface with fire alarm system 8. Interface with building management system	1. Access for underground services 2. Location of exhaust for underground ventilation system

FIGURE 4.14 *(Continued)*

5	Electrical	1. Location and size of substation and door sizing 2. Trenches for cables in substation, electrical room, and generator room 3. Location and size of electrical room and closets 4. Location of electrical devices 5. Location of light fittings and other devices in the false ceiling 6. Void above false ceiling for cable tray and trunking 7. Cable tray, cable trunking route 8. Location and size of low-voltage rooms 9. Location and size of generator room 10. Ventilation of substation and generator room	1. Base for transformers 2. Base for generator 3. Trenches for electrical cables 4. Openings/sleeves for cable tray, electrical 5. Shaft for cable trays 6. Foundation for light poles 7. Manhole/foundation for electrical	1. Power supply for pumps and other equipment 2. Location of isolators for power supply 3. Interface with fire alarm system	1. Power supply for chillers, pumps, AHUS, and other equipment 2. Location of isolators for power supply 3. Heat dissipation from lighting and other electrical panels 4. 3-phase/single-phase power 5. Power supply load during summer/winter 6. Electrical power supply for equipment connected to generator 7. Interface with fire alarm system 8. Interface with building management system		1. Location of lighting poles 2. Location of earth pits 3. Location of electrical manholes, handholes 4. Underground cable routes 5. Location of bollards 6. Location pf electrical panels, feeder pillars

FIGURE 4.14 (*Continued*)

6	Landscape/ External	1. Property limits 2. Location of outdoor equipment 3. Location of plants 4. Location of seating/relax area 5. Location of maintenance room/area 6. Location of manholes 7. Location of generator exhaust pipe	1. Manholes 2. Foundation for light poles 3. Manhole/ foundation for electrical panels 4. Manhole/ foundation for feeder pillars 5. Underground services tunnel	1. Irrigation system with external works 2. Area drain, road gully with external/ asphalt work 3. Storm water manholes with external 4. External services to be hooked up with	1. Access for underground 2. Location of exhaust for underground ventilation system	1. Location of lighting poles 2. Location of earth pits 3. Location of electrical manholes, handholes 4. Underground cable routes 5. Location of bollards 6. Location of electrical panels, feeder pillars 7. Location of generator exhaust pipe	

FIGURE 4.14 (*Continued*)

TABLE 4.8
Quality Check for Design Drawings

Serial Number	Points to Be Checked	Yes/No
1	Check for use of approved version of AutoCAD	
2	Check drawing for Title frame Attribute North orientation Key plan Issues and Revision Number	
3	Client name and logo	
4	Designer (consultant name)	
5	Drawing title	
6	Drawing number	
7	Contract reference number	
8	Date of drawing	
9	Drawing scale	
10	Annotation: Text size Dimension style Fonts Section and elevation marks	
11	Layer standards including line weights	
12	Line weights, line type	
13	Drawing continuation reference and match line	
14	Plot styles (CTB-color dependent plot style tables)	
15	Electronic CAD file name and project location	
16	XREF (X Reference) attachments (if any)	
17	Image reference (if any)	
18	Section references	
19	Symbols	
20	Legends	
21	Abbreviations	
22	General notes	
23	Drawing size per contract requirements	
24	List of drawings	

4.5.2 Regulatory/Authorities' Approval

Government agency regulatory requirements have considerable impact on pre-contract planning. Some agencies require that the design drawings be submitted for their preliminary review and approval to ensure that the designs are compatible with local codes and regulations. These include submission of drawings to electrical authorities showing the anticipated electrical load required for the facility, approval of fire alarm and firefighting system drawings, and approval of drawings for water supply and drainage system. The technical details of the conveying system are also required to be submitted for approval by the concerned authorities.

4.5.3 Contract Documents and Specifications

Preparation of detailed documents and specifications per the master format is one of the activities performed during this phase of the construction project. The contract documents must specify the scope of work, location, quality, and duration for completion of the facility. As regards the technical specifications of the construction project, master format specifications are included in the contract documents. Normally, construction documents are prepared per master format contract documents produced jointly by the Construction Specification Institute (CSI) and Construction Specifications Canada (CSC), which are widely accepted as standard practice for preparation of contract documents.

The master format is a master list of section titles and numbers for organizing information about construction requirements, products, and activities into a standard sequence. It is a uniform system for organizing information in project manuals, organizing cost data, filling product information and other technical data, identifying drawing objects, and presenting construction market data. *MasterFormat*, 2004 edition, consists of 48 divisions (division 49 is reserved).

4.5.3.1 Particular Specifications

Specifications of work quality are an important feature of construction project design. Specifications of required quality and components represent part of the contract documents and are detailed under various sections of particular specifications. Generally, the contract documents include all the details as well as references to generally accepted quality standards published by international standards organizations. Proper specifications and contract documentation are extremely important, as these are used by the contractor as a measure of quality compliance during the construction process.

Particular specifications consist of many sections related to a specific topic. Detailed requirements are written in these sections to enable the contractor to understand the product or system to be installed in the construction project. The designer has to interact with project team members and the owner while preparing the contract documents.

The generalized writing of these sections is as follows:

Section No.

Title

Part 1 General

1.01 General Reference/Related Sections

1.02 Description of Work

1.03 Related Work Specified Elsewhere in Other Sections

1.04 Submittals
1.05 Delivery, Handling, and Storage
1.06 Spare Parts
1.07 Warranties

In addition to the above, a reference is made for items such as preparation of mockup, quality control plan, and any other specific requirement related to the product or system specified herein.

Part 2 Product

2.01 Materials
2.02 Manufacturer's Qualification/List of Recommended Manufacturers

Part 3 Execution

3.01 Installation
3.02 Site Quality Control

4.5.4 Detail Plan

The project plan is a formal, approved document used to manage project execution. It is an evaluation of the time and effort needed to complete the project. Based on the detailed engineering, design drawings, and contract documents, the design team (consultant) prepares a detail plan for construction. The plan is based on following:

- Assessment of the owner's capabilities and final estimated cost (budget)
- Scheduling information
- Resource management, which includes availability of financial resources, expected cash flow statement, supplies, and human resources

A typical preliminary work program prepared based on the contracted construction documents is illustrated in Figure 4.15.

4.5.5 Budget

The budget for a project is the maximum amount of money the owner is willing to spend for the design and construction of the project. The preparation of a budget is an important activity that results in a time-phased plan summarizing expected expenditure, income, and milestones. Normally, project

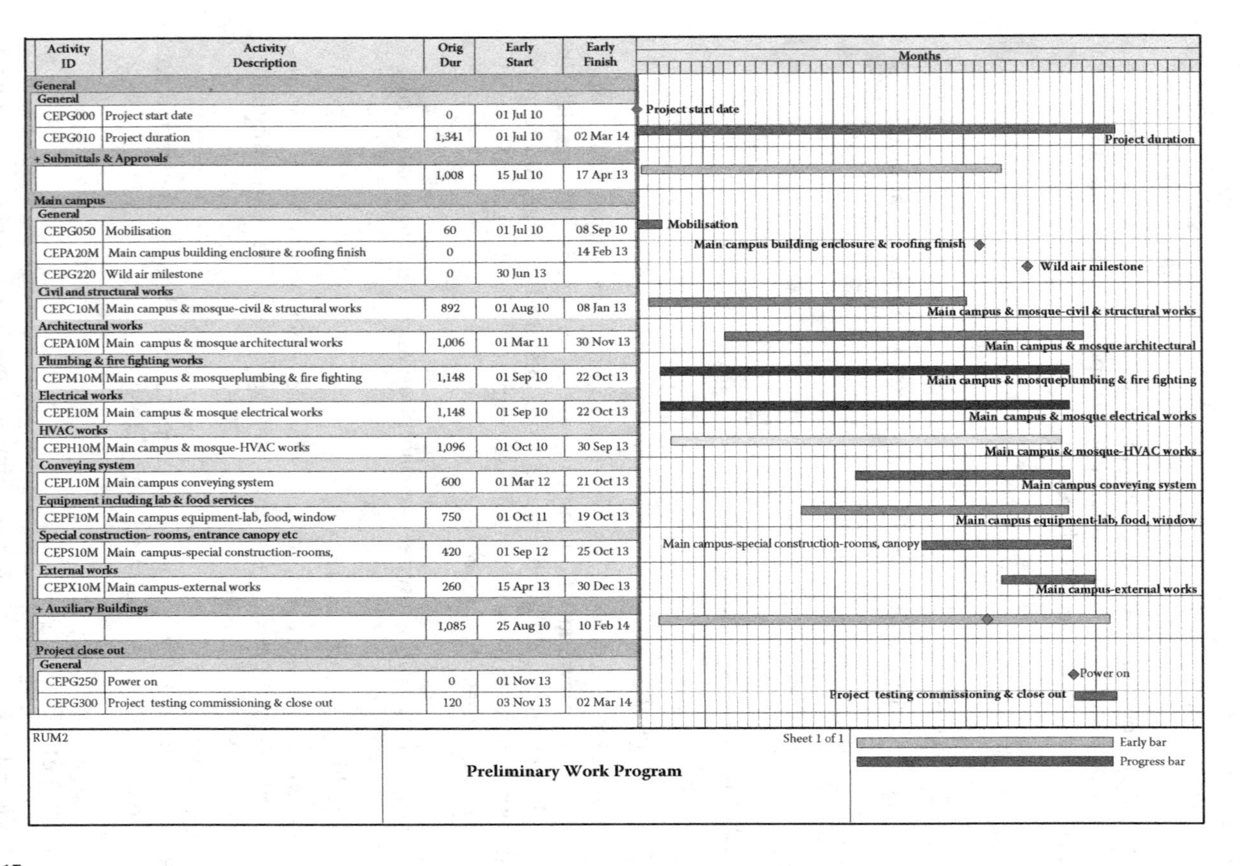

Activity ID	Activity Description	Orig Dur	Early Start	Early Finish
General				
General				
CEPG000	Project start date	0	01 Jul 10	
CEPG010	Project duration	1,341	01 Jul 10	02 Mar 14
+ Submittals & Approvals				
		1,008	15 Jul 10	17 Apr 13
Main campus				
General				
CEPG050	Mobilisation	60	01 Jul 10	08 Sep 10
CEPA20M	Main campus building enclosure & roofing finish	0		14 Feb 13
CEPG220	Wild air milestone	0	30 Jun 13	
Civil and structural works				
CEPC10M	Main campus & mosque-civil & structural works	892	01 Aug 10	08 Jan 13
Architectural works				
CEPA10M	Main campus & mosque architectural works	1,006	01 Mar 11	30 Nov 13
Plumbing & fire fighting works				
CEPM10M	Main campus & mosqueplumbing & fire fighting	1,148	01 Sep 10	22 Oct 13
Electrical works				
CEPE10M	Main campus & mosque electrical works	1,148	01 Sep 10	22 Oct 13
HVAC works				
CEPH10M	Main campus & mosque-HVAC works	1,096	01 Oct 10	30 Sep 13
Conveying system				
CEPL10M	Main campus conveying system	600	01 Mar 12	21 Oct 13
Equipment including lab & food services				
CEPF10M	Main campus equipment-lab, food, window	750	01 Oct 11	19 Oct 13
Special construction- rooms, entrance canopy etc				
CEPS10M	Main campus-special construction-rooms,	420	01 Sep 12	25 Oct 13
External works				
CEPX10M	Main campus-external works	260	15 Apr 13	30 Dec 13
+ Auxiliary Buildings				
		1,085	25 Aug 10	10 Feb 14
Project close out				
General				
CEPG250	Power on	0	01 Nov 13	
CEPG300	Project testing commissioning & close out	120	03 Nov 13	02 Mar 14

FIGURE 4.15
Preliminary work program.

budgeting starts with the identification of need; however, a detailed cost estimate is done during the engineering phase. On the basis of the work packages, the consultant/designer starts computing the project budget. The bill of material or bill of quantities is prepared based on the approved design drawings. The BOQ is considered as a base for computing the budget. If the budget exceeds the owner's capability for financing the project, then the designs are reviewed to ensure that they meets the owner's estimated cost to build the facility.

Figure 4.16 illustrates the Planned Project S-Curve for a building construction project.

4.5.6 Cash Flow

The estimate of cash flow requirements for the project is prepared from the preliminary estimate and preliminary work program. An accurate cash flow projection helps the owner to plan the payments on time according to the project schedule. A simple cash flow projection based on prior planning helps the owner to make available all the resources required from his or her side. The cash flow is used as part of the control package during the construction.

4.5.7 Final Design

The design and documents prepared during this phase are submitted to the client/owner. Upon approval from the client, the designer (consultant) prepares the final drawings and specifications. Any feedback received from the client/owner is incorporated in the final design.

4.5.8 Six Sigma DMADV Tool for Design Development

Six Sigma is, basically, a process quality goal. It is a process quality technique that focuses on reducing variation in processes and preventing deficiencies in product. It is primarily used in manufacturing, process, and services having a large population where it is possible to collect a large amount of data for rigorous analysis to minimize variation and prevent deficiencies in product. Quality for construction is different from quality in manufacturing or the services industries as the product (project) is nonrepetitive and is a unique piece of work with specific requirements. Construction projects do not pass through a series of processes where the output can be repeatedly monitored by inspection and testing at various stages of production and analyzed as is the case in manufacturing. Six Sigma methodology involves utilizing rigorous data analysis to minimize variation in the process. It focuses on reducing variation in a process and preventing defects in the product.

Appendix A discusses the systematic approach of the Six Sigma methodology concept of teamwork, team roles, and the analytical tool DMADV (Define, Measure, Analyze, Design, and Verify) to develop the designs of construction projects.

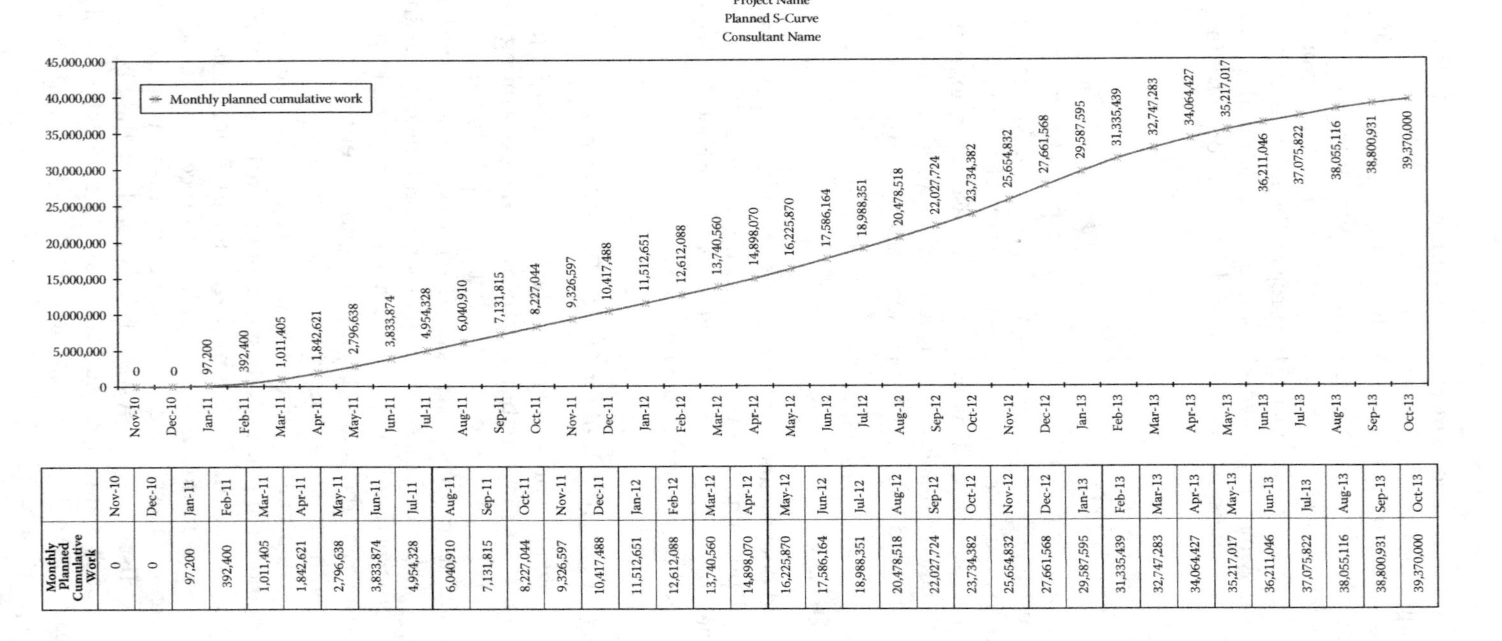

Month	Monthly Planned Cumulative Work
Nov-10	0
Dec-10	0
Jan-11	97,200
Feb-11	392,400
Mar-11	1,011,405
Apr-11	1,842,621
May-11	2,796,638
Jun-11	3,833,874
Jul-11	4,954,328
Aug-11	6,040,910
Sep-11	7,131,815
Oct-11	8,227,044
Nov-11	9,326,597
Dec-11	10,417,488
Jan-12	11,512,651
Feb-12	12,612,088
Mar-12	13,740,560
Apr-12	14,898,070
May-12	16,225,870
Jun-12	17,586,164
Jul-12	18,988,351
Aug-12	20,478,518
Sep-12	22,027,724
Oct-12	23,734,382
Nov-12	25,654,832
Dec-12	27,661,568
Jan-13	29,587,595
Feb-13	31,335,439
Mar-13	32,747,283
Apr-13	34,064,427
May-13	35,217,017
Jun-13	36,211,046
Jul-13	37,075,822
Aug-13	38,055,116
Sep-13	38,800,931
Oct-13	39,370,000

FIGURE 4.16
S-curve.

4.6 Construction Documents

The designer (consultant) proceeds with preparation of construction documents upon approval of final design and specifications by the owner. A contract for construction commits the contractor to construct the works and the owner to pay. Once the contract is signed, it commits all the parties to obligations and liabilities and is enforceable in law. A break of contract by either party may make that party liable for payment of damages to the other.

There are a standard set of conditions of contracts published by engineering institutes/societies and other bodies. Depending on the need of construction projects and the type of contract arrangements, an appropriate set of contracts is selected. The contract document must include the health and safety programs to be followed by the contractor during the construction process.

The construction documents consist of the following four major elements:

1. Design drawings in detail and fully coordinated
2. Particular specifications for all the trades
3. Bill of Quantity (BOQ) document
4. Schedule

Based on the above, the consultant prepares the tender documents. Table 4.9 lists the contents of the contract documents followed in Kuwait for tendering purposes.

4.7 Bidding/Tendering

Construction documents prepared by the consultant are sent to the client for tendering purposes.

In many countries, it is a legal requirement for government-funded projects to employ the competitive bidding method. This requirement gives an opportunity to all the qualified contractors to participate in the tender, and normally the contract is awarded to the lowest bidder. Privately funded projects have more flexibility in evaluating the tender proposal. Private owners may adopt the competitive bidding system, or the owner may select a specific contractor and negotiate the contract terms. The negotiated contract systems have flexibility of pricing arrangement as well as the selection of contractor based on his expertise or the owner's past experience with the contractor

TABLE 4.9

Contents of Contract Documents

Serial Number	Document	
Document-I	Tendering Procedure	
	I.1	Tendering Invitation
	I.2	Instruction for Tenderers
	I.3	Form of Tender and Appendix
	I.4	Initial Bond (Form of Bank Guarantee)
	I.5	Performance Bond (Form of Bank Guarantee)
	I.6	Form of Agreement
	I.7	List of Tender Documents
	I.8	Declaration No. (1)
Document-II	General Conditions of Contract	
	II.1	Legal Clauses and Conditions
	II.2	Particular Conditions of Contract
	II.3	Special Conditions for Public Services, Safety Regulations for Individuals, Properties, and Public Utilities
Document-III	Technical Conditions	
	III-1	General Specifications and Rules and Regulations Issued by Various Ministries and Regulatory Authorities
	III-2	Particular Specifications
	III-3	Drawings
	III-4	Bill of Quantities (BOQ)
	III-5	Price Analysis Schedule
	III-6	Addendum (if any)
	III-7	Technical Requirements (if any), and Any Other Instructions Issued by the Employer

successfully completing one of his or her projects. Figure 4.17 illustrates the tendering process.

For most construction projects, selection of the tenderer is based on the lowest tender price. Tenders received are opened and evaluated by the owner/owner's representative. Normally, tender results are declared in the official gazette or by some sort of notification. The successful tenderer is informed of the acceptance of the proposal and is invited to sign the contract. The tenderer has to submit the performance bond before a formal contract agreement is signed. If a successful tenderer fails to submit the performance bond within the specified period or withdraws his tender, then the contractor loses the initial bond and may be subjected to other applicable regulatory conditions.

The signing of the contract agreement between the owner/owner's representative and the contractor bind both parties to fulfilling their contractual obligations.

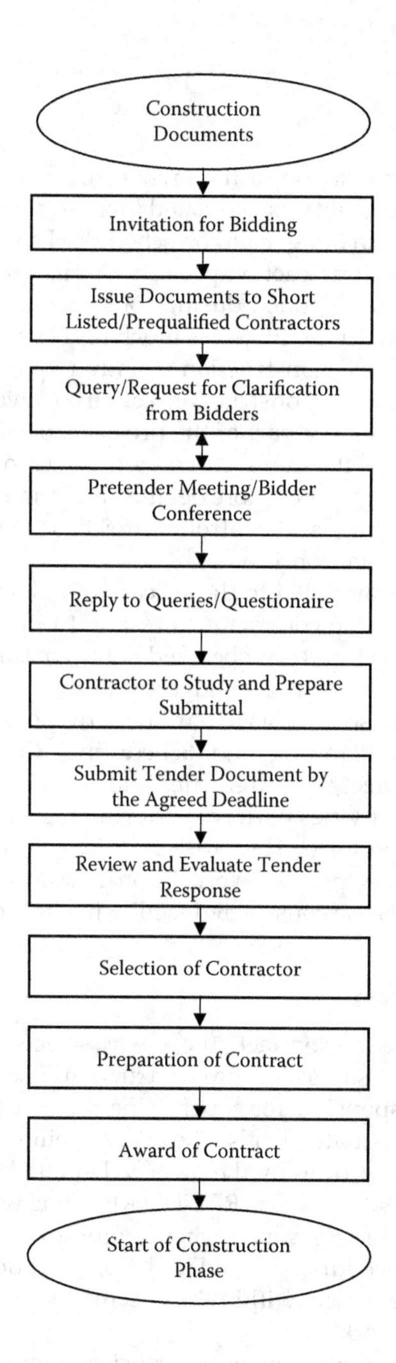

FIGURE 4.17
Tendering process.

4.8 Construction

Construction is the translation of the owner's goals and objectives, by the contractor, to build the facility as stipulated in the contract documents, plans, and specifications within budget and on schedule. The construction phase is an important phase in construction projects. A majority of the total project budget and schedule is expended during construction. Similar to costs, the time required to construct the project is much higher than the time required for the preceding phases. Construction usually requires a large workforce and a variety of activities. Construction activities involve erection, installation, or construction of any part of the project. Construction activities are actually carried out by the contractor's own forces or by subcontractors. Construction therefore requires more detailed attention regarding planning, organizations, monitoring, and control of project schedule, budget, quality, safety, and environmental concerns.

Once the contract is awarded to the successful bidder (contractor), then it is the responsibility of the contractor to respond to the needs of the client (owner) by building the facility as specified in the contract documents, drawings, and specifications within the budget and time.

The construction phase consists of various activities such as Mobilization, Execution of Works, Planning and Scheduling, Control & Monitoring, Management of Resources/Procurement, Quality, and Inspection. Figure 4.18 illustrates the major activities performed during the construction phase.

These activities are performed by various parties having contractual responsibilities to complete the specified work. Coordination among these parties is essential to ensure that the constructed facility meets the owner's objectives.

4.8.1 Supervision Team

In a traditional type of contract, the client selects the same firm that designed the project to supervise the construction. The firm is known as the consultant, and is responsible for supervising the construction process and achieving the project's quality goals. The firm appoints a representative, who is acceptable to and approved by the owner/client, to be on site, and who is often called the resident engineer (RE). The RE, along with supervision team members, is responsible for supervising, monitoring and controlling, and implementing the procedure specified in the contract documents and ensuring completion of the project within the specified time and budget, and per the defined scope of work.

In order to ensure smooth flow of supervision activities, the RE has to follow the organization's supervision manual and contractual requirements. Table 4.10 illustrates an example checklist listing the items to be verified by the RE to ensure availability of all the necessary documents and information to facilitate smooth flow of supervision work.

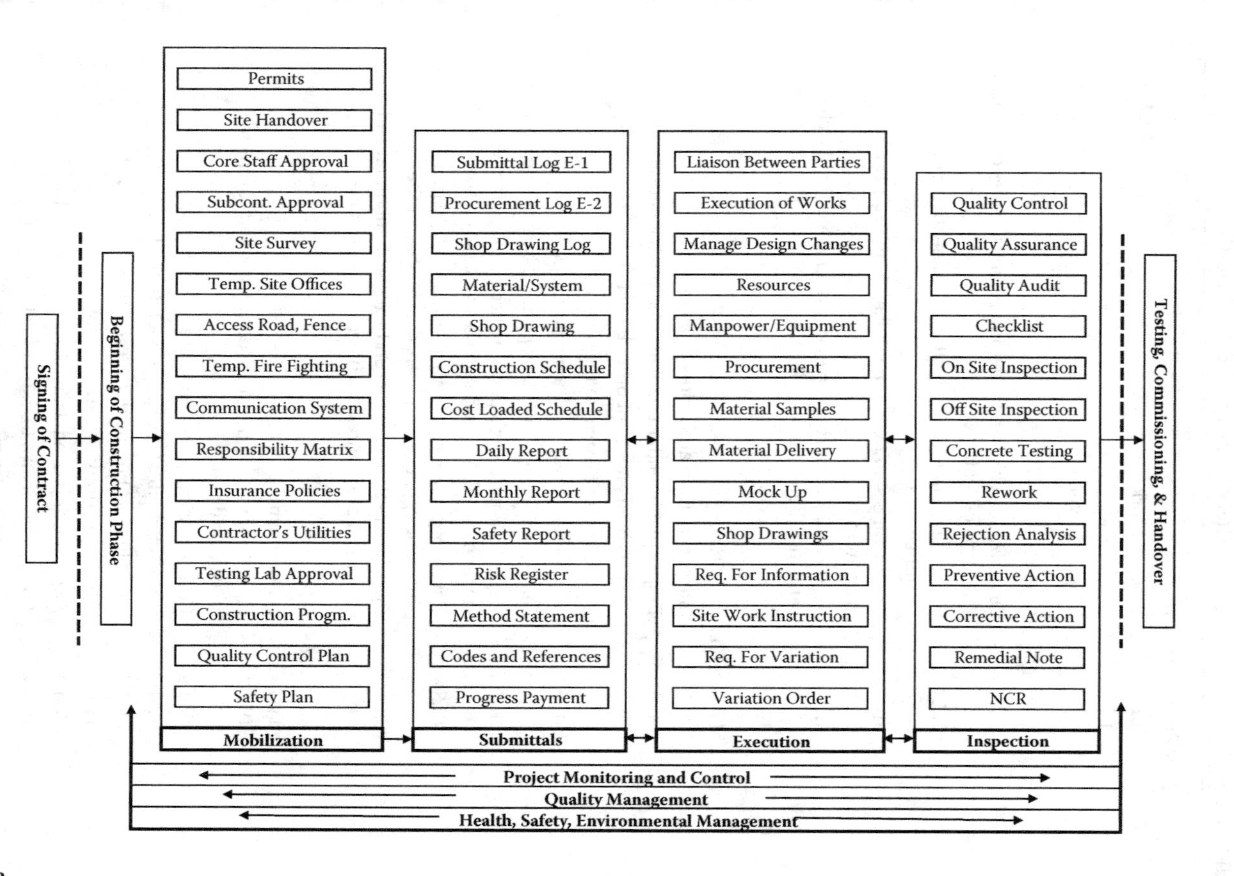

FIGURE 4.18
Major activities during construction phase.

TABLE 4.10

Consultant's Checklist for Smooth Functioning of Project

Serial Number	Items to be Checked/Verified	
I	*Project Details*	
	I.1	Scope of work
	I.2	Project objectives
	I.3	Project deliverables
II	*Project Organization*	
	II.1	Organization chart and roles and responsibilities of defined supervision staff
	II.2	Supervision staff deployment matching with project requirements
	II.3	Contractor's staff deployment plan approved per contract requirements
	II.4	Responsibility matrix prepared and approved by the client and distributed among all project parties
	II-5	Project directory
III	*Mobilization*	
	III.1	Site permit from authorities available
	III.2	Project plot boundaries are marked per the permit
	III.3	Project commencement order issued
	III.4	Copy of permit issued to the contractor
	III.5	Temporary site offices drawings approved
	III.6	Temporary firefighting plan approved by respective authority
	III.7	Copies of contractor's performance bond, guarantees, insurance policies, and licenses available on site
	III.8	Copies of consultant's performance bond, guarantees, insurance policies and licenses available on site
	III.9	Pre-construction meeting conducted and submittal and approval procedures discussed and agreed upon
IV	*Project Administration*	
	IV-1 Contract Documents	
	IV-1.1	Signed copy of contract between owner and contractor available on site
	IV-1.2	Copies of contract documents available on site
	IV-1.3	Contracted bill of quantity (BOQ) is available
	IV-1.4	All volumes of particular specifications are available
	IV-1.5	Contracted drawings are available
	IV-1.6	Authority approved drawings, duly stamped, are available
	IV-1.7	Addendum, if any, to the contract is available
	IV-1.8	Replies to tender queries are available
	IV-1.9	Copy of signed contract documents and drawings handed over to the contractor and have been acknowledged
	IV-1.10	Log for Codes and Standards available

TABLE 4.10 (*Continued*)

Consultant's Checklist for Smooth Functioning of Project

Serial Number		Items to be Checked/Verified
	IV-2 Document Management	
	IV-2.1	Document control system is in place
	IV-2.2	Filing index is available
	IV-2.3	Material submittal log is available
	IV-2.4	Shop drawing submittal log is available
	IV-2.5	Logs for correspondence between various parties available
	IV-2.6	Log for checklist (Request for Inspection) available
	IV-2.7	Log for JSI (Job Site Instruction) available
	IV-2.8	Log for SWI (Site Work Instruction) available
	IV-2.9	Log for RFI (Request for Information) available
	IV-2.10	Log for VO (Variation Order) available
	IV-2.11	Log for NCR (Nonconformance Report) available
	IV-2.12	Material sample log and place identified
	IV-2.13	Log for equipment test certificate available
	IV-2.14	Log for visitors at site
	IV-2.15	Contractor's staff approval log in place
	IV-2.16	Subcontractor's approval log in place
	IV-2.17	Consultants staff approval in place
	IV-2.18	Overtime request log available
V	*Communication*	
	V-1	Communication matrix established and agreed by all the parties
	V-2	Distribution system for transmittals/submittals agreed
VI	*Project Monitoring and Control*	
	VI-1	Daily Report log in place
	VI-2	Weekly Report log in place
	VI-3	Monthly Report log in place
	VI-4	Progress Meetings log in place
	VI-5	Minutes of Meetings log in place
	VI-6	Progress Payment log in place
	VI-7	Construction Schedule log in place
VII	*Construction*	
	VII-1	Quality Control Plan log in place
	VII-2	Safety Management Plan log in place
	VII-3	Risk Management Plan log in place
	VII-4	Method Statement Submittal log in place
	VII-5	Accident and fire report
	VII-6	Off-site inspection visits
	VII-7	Location of gathering point established

(*Continued*)

TABLE 4.10 (*Continued*)

Consultant's Checklist for Smooth Functioning of Project

Serial Number		Items to be Checked/Verified
VIII	*General*	
	VIII-1	Correspondence between site and head office
	VIII-2	Staff-related matters
	VIII-3	Copy of Supervision Manual available
	VIII-4	Emergency contact telephones and contact details displayed on site

4.8.2 Mobilization

The contractor is given a few weeks to start the construction works after the signing of the contract. The activities to be performed during the mobilization period are defined in the contract documents. During this period, the contractor is required to perform many of the activities before the beginning of actual construction work at the site. Necessary permits are obtained from the relevant authorities to start the construction work. Upon receipt of the construction site from the owner, the contractor starts mobilization works, which consist of preparation of site offices/field offices for the owner, supervision team (consultant), and the contractor himself. This includes all the necessary on-site facilities and services necessary to carry out specific works and tasks. Mobilization activities usually occur at the beginning of a project but can occur anytime during a project when specific on-site facilities are required.

Examples of mobilization activities include the following:

1. Selection of core staff as mentioned in the contract documents
2. Selection of subcontractor (this may be an ongoing activity per the approved schedule)
3. Setting up of temporary facilities consisting of
    a. Site offices
    b. Storage facilities
    c. Parking
    d. Fences and gates
    e. Project identification sign boards
4. Construction of temporary access roads
5. Installation of the necessary utilities for construction
6. Testing laboratory
7. Temporary firefighting system
8. Performing site survey and testing

9. Insurance policies covering the following areas:
   a. Contractor's all risks and third party insurance policy
   b. Contractor's plant and equipment insurance policy
   c. Workmen's compensation insurance policy
   d. Site storage insurance policy
10. Advance payment bank guarantee
11. Performance bond guarantee
12. Preliminary construction program including resource schedule
13. Contractor's quality control plan
14. Safety management plan (health, safety, and environment)
15. Site communication plan
16. Responsibility matrix
17. Designating authority-approved dumping area for waste material

In anticipation of the award of contract, the contractor begins the following two activities much in advance, but these are part of the contract documents and the contractor's action is required immediately after the signing of the contract in order to start construction:

1. Mobilization of construction equipment and tools
2. Manpower to execute the project

4.8.2.1 Communication System

For smooth implementation of the project, a proper communication system is established clearly identifying the submission process for correspondence and transmittals. Correspondence between the consultant and contractor is normally done through job site instructions; whereas correspondence between the owner, consultant, and contractor is normally done through letters or via email. Figure 4.19 shows a sample job site instruction form used by the consultant to communicate with the contractor.

In construction projects, there are different types of forms used to assist project monitoring and control, quality management, and to track the progress of work. These forms can mainly be divided into the following four categories:

1. PCS (Project Control and Scheduling) Forms
2. Engineering Submittal Forms
3. Field and Quality Control Forms
4. Handover and Closeout Forms

Project Name	
Consultant Name	
JOB SITE INSTRUCTION (JSI)	
CONTRACTOR:	JSI No.:
CONTRACT No.:	DATE:

The work shall be carried out in accordance with the Contract Documents without change in Contract Sum or Contract Time. Proceeding with the work in accordance with these instructions indicates your acknowledgement that there will be no change in the Contract Sum or Contract Time.	
Subject: SAMPLE FORM ATTACHMENTS: (List attached documents that support description.)	
Signed: Resident Engineer	Received by Contractor: Date:

Distribution: ☐ Owner ☐ A/E ☐ Contractor

FIGURE 4.19
Job site instruction form.

Table 4.11 lists typical forms used in construction projects, and Figure 4.20 is a sample transmittal form.

4.8.2.2 Responsibility Matrix

For smooth flow of construction process activities, proper communication and submittal procedures need to be established between all the concerned parties at the beginning of the construction activities. Table 4.12 shows an example matrix for site administration of a building construction project.

4.8.2.3 Selection of Subcontractor

In most construction projects, the contractor engages special subcontractors to execute certain portions of the contracted project works. Areas of sub-contracting are generally listed in the Particular Conditions section of the

TABLE 4.11
List of Transmittal Forms

Serial Number		Description of Form
I	*General Forms*	
	I-1	Job Site Instruction
	I-2	Request for Staff Approval
	I-3	Request for Subcontractor Approval
	I-4	Request for Overtime
	I-5	Site Entry Permit
	I-6	Visitor Entry Permit
	I-7	Material Entry Permit
	I-8	Material Removal Permit
	I-9	Vehicular Entry Permit
	I-10	Accident Report
	I-11	Theft & Damage Report
	I-12	Site Works Instruction
	I-13	Attachment to Site Works Instruction
	I-14	Variation Order
	I-15	Attachment to Variation Order
	I-16	Minutes of Meeting
	I-17	Transmittal Form
II	*PCS Reporting Forms*	
	II-1	Contractor's Submittal Status Log E-1
	II-2	Contractor's Procurement Log E-2
	II-3	Contractor's Shop Drawing Status Log
	II-4	Daily Progress Report
	II-5	Daily Progress Report
	II-6	Daily Checklist Status
	II-7	Progress Payment Request
	II-8	Look Ahead Schedule
	II-9	Schedule Update Report
III	*Engineering Submittal Forms*	
	III-1	Site Transmittal for Material Approval
	III-2	Specification Comparison Statement
	III-3	Site Transmittal for Workshop Drawings
	III-4	Request for Information
	III-5	Request for Modification
	III-6	Variation Order (Proposal)
	III-7	Request for Alternative or Substitution
	III-8	Sample Tag

(*Continued*)

TABLE 4.11 (*Continued*)

List of Transmittal Forms

Serial Number	Description of Form	
IV	*Quality Control Forms*	
	IV-1	Checklist (Request for Inspection)
	IV-2	Checklist for Form Work
	IV-3	Notice for Daily Concrete Casting
	IV-4	Checklist for Concrete Casting
	IV-5	Quality Control of Concreting
	IV-6	Report on Concrete Casting
	IV-7	Notice for Testing at Lab
	IV-8	Concrete Quality Control Form
	IV-9	Remedial Note
	IV-10	Non-Conformance Report
	IV-11	Material Inspection Report
	IV-12	Safety Violation Notice
	IV-13	Notice of Commencement of New Activity
	IV-14	Removal of Rejected Material
V	*Closeout Forms*	
	V-1	Handing Over Certificate
	V-2	Handing Over of Spare Parts
	V-3	Defect Liability Certificate

contract document. Generally, the contractor has to submit subcontractors/specialist contractors to execute the following types of work:

1. Precast concrete works
2. Metal works
3. Space frame, roofing works
4. Woodwork
5. Aluminum works
6. Internal finishes such as painting, false ceiling, tiling, cladding
7. Furnishings
8. Waterproofing and insulation works
9. Mechanical works
10. HVAC works
11. Electrical works
12. Low-voltage systems/smart building system
13. Landscape
14. External works
15. Any other specialized works

Project Name			
Consultant Name			
TRANSMITTAL FORM			
Contractor Name:			
Contract No.:			
To.	Resident Engineer		
Transmittal No.:		Date :	

Type:		Action Requested:	
DG	Drawings	1	For Incorporation Within the Design
SP	Specifications	2	For Information
SK	Sketches	3	For Review and Comment
S	Samples	4	For Costing
MM	Minutes of Meeting	5	For Approval
MD	Manufacturer's Data	6	For Tender
R	Reports	7	For Construction
L	Logs		
O	Others (please specify)		

We are sending herewith the following:					
ENCLOSURES					
Item	**Qty**	**Ref. No.**	**Description**	**Type**	**Action**

Comments:

Issued by:
Signature:
Date:

Received by:
Signature:
Date:

FIGURE 4.20
Transmittal form.

The contractor has to submit the names of the subcontractors for approval to the owner prior to their engagement to perform any work at the site. The subcontractors for various subprojects have to be submitted for approval in a timely and orderly manner, as planned in the approved work program, so that work progresses in a smooth and efficient manner. This log helps the project manager to remind the main contractor to submit the subcontractors' names on time.

TABLE 4.12

Matrix for Site Administration and Communication

Sr. No.	Description of Activities	Contractor	Consultant	Owner
1	General			
	1.1 Notice to Proceed	—	—	P
	1.2 Bonds and Guarantees	P	R	A
	1.3 Consultant Staff Approval	—	P	A
	1.4 Contractor's Staff Approval	P	R/B	A
	1.5 Payment Guarantee	P	R	A
	1.6 Master Schedule	P	R	A
	1.7 Stoppage of Work	—	P	A
	1.8 Extension of Time	—	P	A
	1.9 Deviation from Contract Documents	P	R	A
	a. Material			
	b. Cost			
	c. Time			
2	Communication			
	2.1 General Correspondence	P	P	P
	2.2 Job Site Instruction	D	P	C
	2.3 Site Works Instruction	D	P/B	A
	2.4 Request for Information	P	A	C
	2.5 Request for Modification	P	B	A
3	Submittals			
	3.1 Subcontractor	P	B/R	A
	3.2 Materials	P	A	C
	3.3 Shop Drawings	P	A	C
	3.4 Staff Approval	P	B	A
	3.5 Pre-Meeting Submittals	P	D	C
4	Plans and Programs			
	4.1 Construction Schedule	P	R	C
	4.2 Submittal Logs	P	R	C
	4.3 Procurement Logs	P	R	C
	4.4 Schedule Update	P	R	C
5	Monitor and Control			
	5.1 Progress	D	P	C
	5.2 Time	D	P	C
	5.3 Payments	P	R/B	A
	5.4 Variations	P	R/B	A
	5.5 Claims	P	R/B	A
6	Quality			
	6.1 Quality Control Plan	P	R	C
	6.2 Checklists	P	D	C
	6.3 Method Statements	P	A	C
	6.4 Mock Up	P	A	B
	6.5 Samples	P	A	B
	6.6 Remedial Notes	D	P	C
	6.7 Nonconformance Report	D	P	C
	6.8 Inspections	P	D	C
	6.9 Testing	P	A	B

TABLE 4.12 (*Continued*)

Matrix for Site Administration and Communication

Sr. No.	Description of Activities	Contractor	Consultant	Owner
7	Site Safety			
	7.1 Safety Program	P	A	C
	7.2 Accident Report	P	R	C
8	Meetings			
	8.1 Progress	E	P	E
	8.2 Coordination	E	P	C
	8.3 Technical	E	P	C
	8.4 Quality	P	C	C
	8.5 Safety	P	C	C
	8.6 Closeout	—	P	
9	Reports			
	9.1 Daily Report	P	R	C
	9.2 Monthly Report	P	R	C
	9.3 Progress Report	—	P	A
	9.4 Progress Photographs	—	P	A
10	Closeout			
	10.1 Snag List	P	P	C
	10.2 Authorities Approvals	P	C	C
	10.3 As-Built Drawings	P	D/A	C
	10.4 Spare Parts	P	A	C
	10.5 Manuals and Documents	P	R/B	A
	10.6 Warranties	P	R/B	A
	10.7 Training	P	C	A
	10.8 Handover	P	B	A
	10.9 Substantial Completion Certificate	P	B/P	A

P Prepare/Initiate
B Advise/Assist
R Review/Comment
A Approve
D Action
E Attend
C Information

Source: Abdul Razzak Rumane (2010), *Quality Management in Construction Projects*. CRC Press, Boca Raton, FL. Reprinted with permission from Taylor & Francis Group.

The request includes all the related information to prove the subcontractor's ability to provide the services to meet the project quality, sufficient available resources to meet the specified schedule, past performance, and whether any quality system is implemented.

Sometimes the owner/consultant nominates subcontractors to execute a portion of a contract; such a subcontractor is known as a nominated subcontractor. Table 4.13 shows an example subcontractor selection questionnaire.

TABLE 4.13

Subcontractor Prequalification Questionnaire

Instructions

Please type or write all your replies legibly. Attach additional sheets, if required.

PART I

I.1 Company Information

I.1.1 Name of Organization:

I.1.2 Commercial Registration No.:

I.1.3 Year of Establishment:

I.1.4 Type of Company:

I.1.5 Company Address:

I.1.6 Affiliate Company Name(s) and Address:

I.2 Subcontract Works (Please check all interested)

Sr. No.	Work Description	Detail Design	Preparation of Shop Drawing	Construction	Inspection/ Auditing
1	Architectural Work				
2	Structural Work				
3	Precast Work				
4	Internal Finishes				
5	External Finishes				
6	Plumbing				
7	Drainage				
8	Firefighting				
9	HVAC				
10	Elevator				
11	Escalator				
12	Electrical Work (Power)				

13	Electrical Work (Low Voltage): Specify
14	Instrumentation
15	Building Management System
16	Irrigation
17	Landscape Work
18	Pavements
19	Streets/Roads
20	Waterproofing

PART II

II.1 Financial Information

II.1.1 Provide copy of Audited Balance Sheet:

II.1.2 Provide Bonding Capacity:

II.1.3 Provide Insurance Capacity:

II.1.4 Provide Bank Reference:

PART III

III.1 Organization Details

III.1.1 Core Business Area:

III.1.2 Organization Chart:

III.1.3 ISO Certification:

III.1.4 Years of Experience:

III.2 Project Details

III.2.1 Project History for the last 10 years

Sr. No.	Name of Project	Type of Work	Value	Peak Workforce	Start Date	Finish Date
1						
2						

(*Continued*)

TABLE 4.13 (*Continued*)

Subcontractor Prequalification Questionnaire

3						
4						

III.2.2 Current Projects

Sr. No.	Name of Project	Type of Work	Value	Peak Workforce	Start Date	Expected Finish Date
1						
2						
3						

PART IV

IV.1 Management Staff

IV.1.1 Provide list of Project Managers, Project Engineers, Engineers

IV.2 Workforce

Sr. No.	Work Description	Technicians	Foreman	Skilled	Unskilled
1	Architectural Work				
2	Structural Work				
3	Precast Work				
4	Internal Finishes				
5	External Finishes				
6	Plumbing				
7	Drainage				
8	Firefighting				
9	HVAC				
10	Elevator				
11	Escalator				
12	Electrical Work (Power)				

13	Electrical Work (Low Voltage): Specify
14	Instrumentation
15	Building Management System
16	Irrigation
17	Landscape Work
18	Pavements
19	Streets/Roads
20	Waterproofing

PART V

V.1 Quality Management System

V.1.1 Provide copy of ISO Certificate:

V.I.2 Person in Charge of QA/QC Activities:

V.I.3 Number of Quality Auditors:

PART VI

VI.1 HSE System

VI.1 Does the company have an HSE Policy?

VI.1.2 Provide site accident records for the last 2 years.

Declaration

We hereby declare that the information provided herein is true to our knowledge.

Note:

All relevant documents attached.

Signature of Authorized Person

4.8.3 Execution of Works

Construction activities mainly consist of the following:

- Site work such as cleaning and excavation of project site
- Construction of foundations including footings and grade beams
- Construction of columns and beams
- Forming, reinforcing, and placing the floor slab
- Laying up masonry walls and partitions
- Installation of roofing system
- Finishes
- Furnishings
- Conveying system
- Installation of firefighting system
- Installation of water supply, plumbing, and public health system
- Installation of heating, ventilating, and air conditioning system
- Integrated automation system
- Installation of electrical lighting and power system
- Emergency power supply system
- Fire alarm system
- Communication system
- Electronic security and access control system
- Landscape works
- External works

Figure 4.21 illustrates the sequence of construction activities.

4.8.3.1 Materials and Shop Drawing Submittals

Prior to the start of execution/installation of work, the contractor has to submit a specified material/product/system to the consultant for approval. The consultant reviews the submittal and returns the transmittal to the contractor, mentioning one of the following actions on the transmittal:

- **A** Approved
- **B** Approved As Noted
- **C** Revise and Resubmit OR Not Approved
- **D** For Information OR More Information Required

In case of deviation from specified items, the contractor has to submit a schedule of such deviations, listing all the points that do not conform to the specifications.

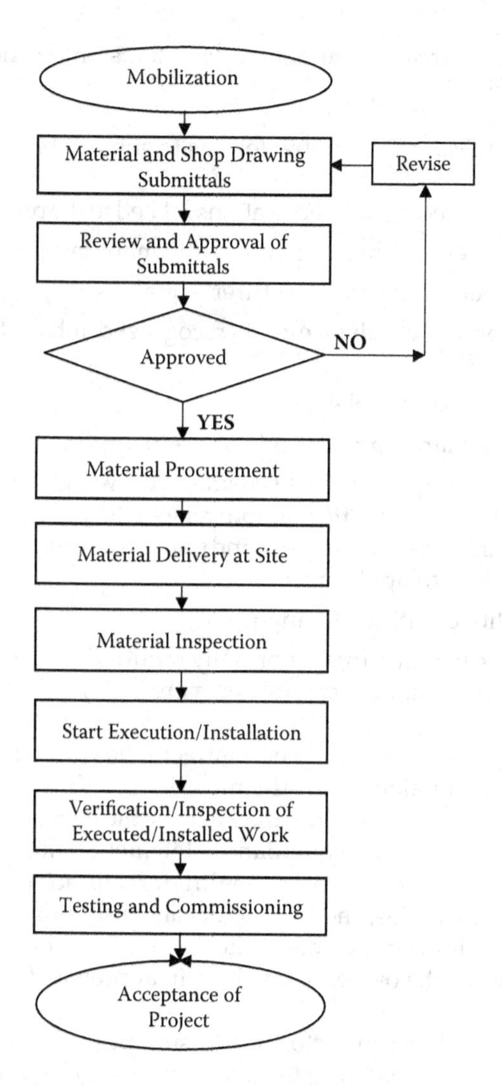

FIGURE 4.21
Sequence of construction activities.

The detailed procedure for submitting materials/products/systems, samples, and shop drawings is specified under the section "Submittal" of the contract specifications. The contractor has to submit the same to the owner/consultant for review and approval. The following subsections provide the details of preparation of the shop drawings and materials.

4.8.3.1.1 Materials

The contractor has to submit the following, at a minimum, to the owner/consultant to get their review and approval of materials, products, equipment, and systems. The contractor cannot use these items unless they are approved for use in the project.

4.8.3.1.1.1 Product Data The contractor has to submit the following details:

- Manufacturer's technical specifications related to the proposed product
- Installation methods recommend by the manufacturer
- Relevant sheets of the manufacturer's catalogs
- Confirmation of compliance with recognized international quality standards
- Mill reports (if applicable)
- Performance characteristics and curves (if applicable)
- The manufacturer's standard schematic drawings and diagrams to supplement standard information related to project requirements and configuration of the same to indicate product application for the specified works (if applicable)
- Compatibility certificate (if applicable)
- Single source liability (this is normally required for systems approval where different manufacturers' items are used)

4.8.3.1.1.2 Compliance Statement The contractor has to submit a specification comparison statement along with the material transmittal. The compliance statement form is normally included as part of the contract documents. The information provided in the compliance statement helps the consultant to review and verify product compliance with the contracted specifications.

In case of any deviations, the contractor has to submit a schedule of such deviations listing all the points not conforming to the specification.

In certain projects, the owner is involved in approval of materials.

4.8.3.1.1.3 Samples The contractor has to submit (if required) the samples from the approved material to be used for the work. The samples are mainly required to

- Verify color, texture, and pattern
- Verify that the product is physically identical to the proposed and approved material
- Be used for comparison with products and materials used in the works

At times it may be required to install the samples in such a manner as to facilitate a review of qualities indicated in the specifications.

Figure 4.22 shows the procedure for selection of materials/products, whereas Figure 4.23 shows the material approval procedure.

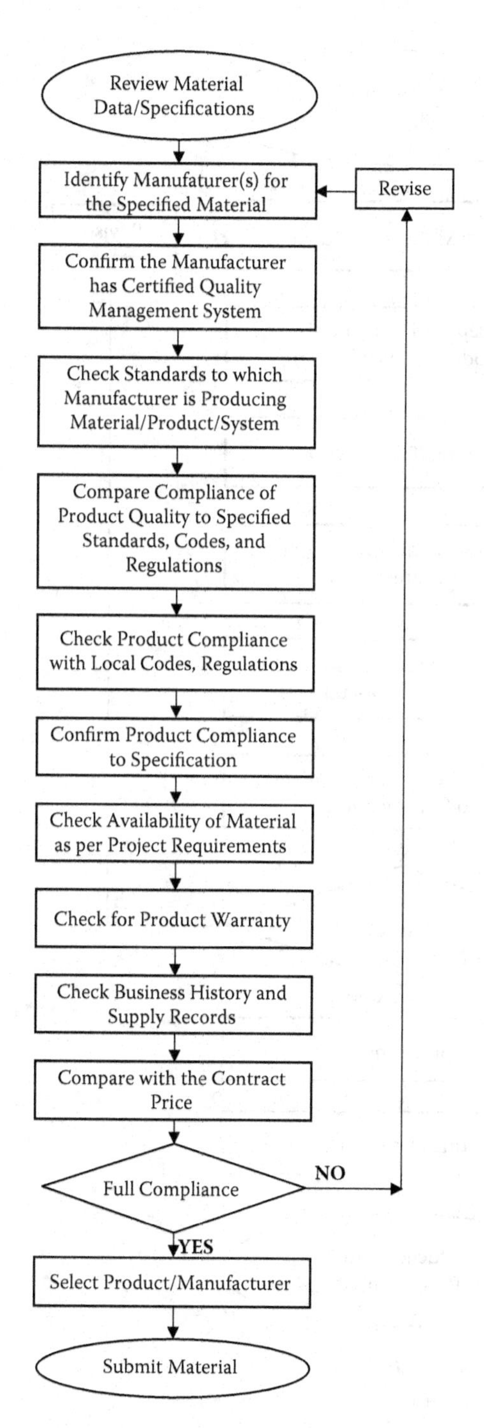

FIGURE 4.22
Material/manufacturer selection procedure.

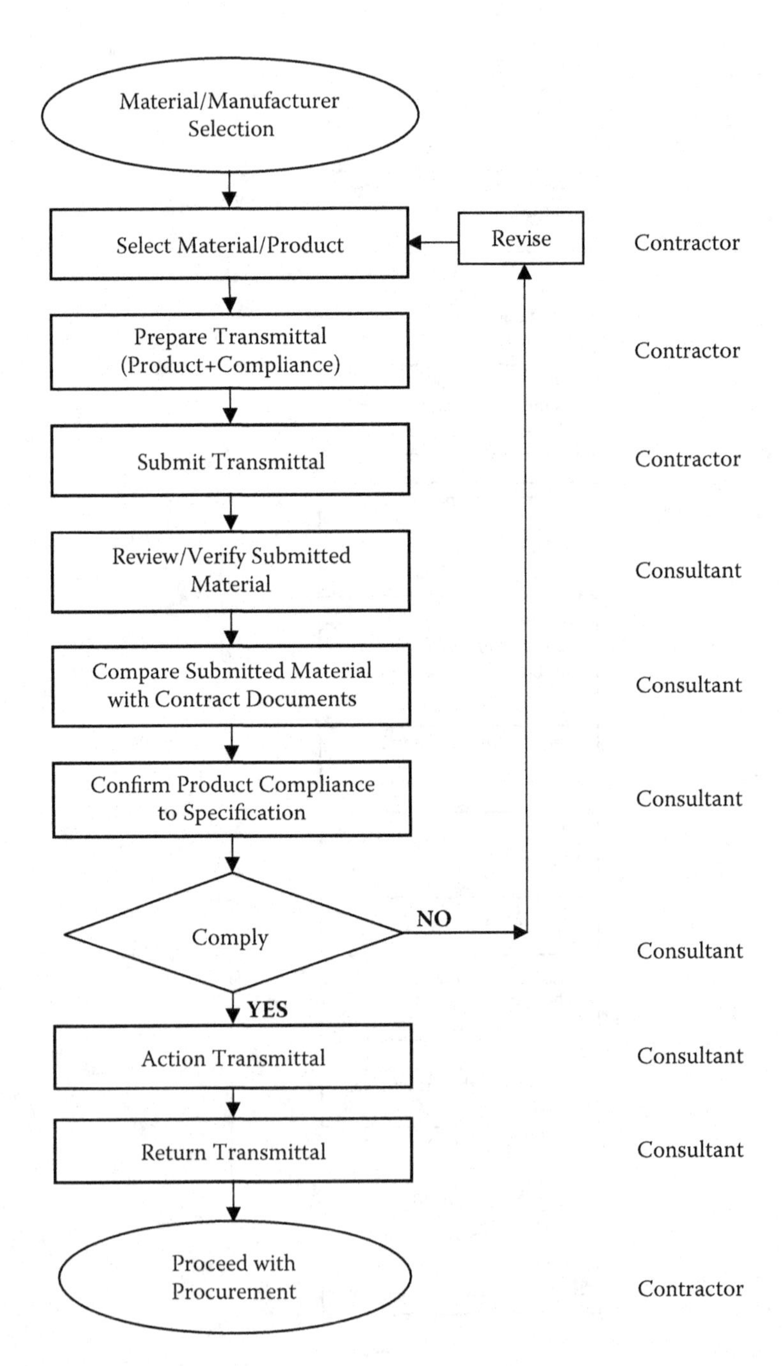

FIGURE 4.23
Material approval procedure.

4.8.3.1.2 Shop Drawings

The contractor is required to prepare shop drawings taking into consideration following as a minimum but not limited to

1. Reference to contract drawings. This help the A/E (consultant) to compare and review the shop drawing with the contract drawing.
2. Detailed plans and information based on the contract drawings.
3. Notes of changes or alterations from the contract documents.
4. Detailed information about fabrication or installation of works.
5. All dimensions need to be verified at the job site.
6. Identification of product.
7. Installation information about the materials to be used.
8. Type of finishes, color, and textures.
9. Installation details relating to the axis or grid of the project.
10. Roughing in and setting diagram.
11. Coordination certification from all other related trades (subcontractors).

The shop drawings are to be drawn accurately to scale and should have project-specific information in it. The shop drawings should not be reproductions of contract drawings.

Immediately after approval of individual trade shop drawings, the contractor has to submit the builder's workshop drawings and composite/coordinated shop drawings taking into consideration the following at a minimum. Figure 4.24 shows the shop drawing preparation and approval procedure.

4.8.3.1.2.1 Builder's Workshop Drawings Builders workshop drawings indicate the openings required in the civil or architectural work for services and other trades. These drawings indicate the size of openings, sleeves, and level references with the help of detailed elevation and plans. Figure 4.25 shows the builder's workshop drawing preparation and approval procedure.

4.8.3.1.2.2 Composite/Coordination Shop Drawings The composite drawings specify the interrelationships between the components shown on the related shop drawings and indicate the required installation sequence. Composite drawings shall show the interrelationship of all services with each other and with the surrounding civil and architectural work. Composite drawings should also show the detailed coordinated cross sections, elevations, reflected plans, etc., resolving all conflicts in levels, alignment, access, space, etc. These drawings should be prepared taking into consideration the actual physical dimensions required for installation within the available space. Figure 4.26 illustrates the composite drawing preparation and approval procedure.

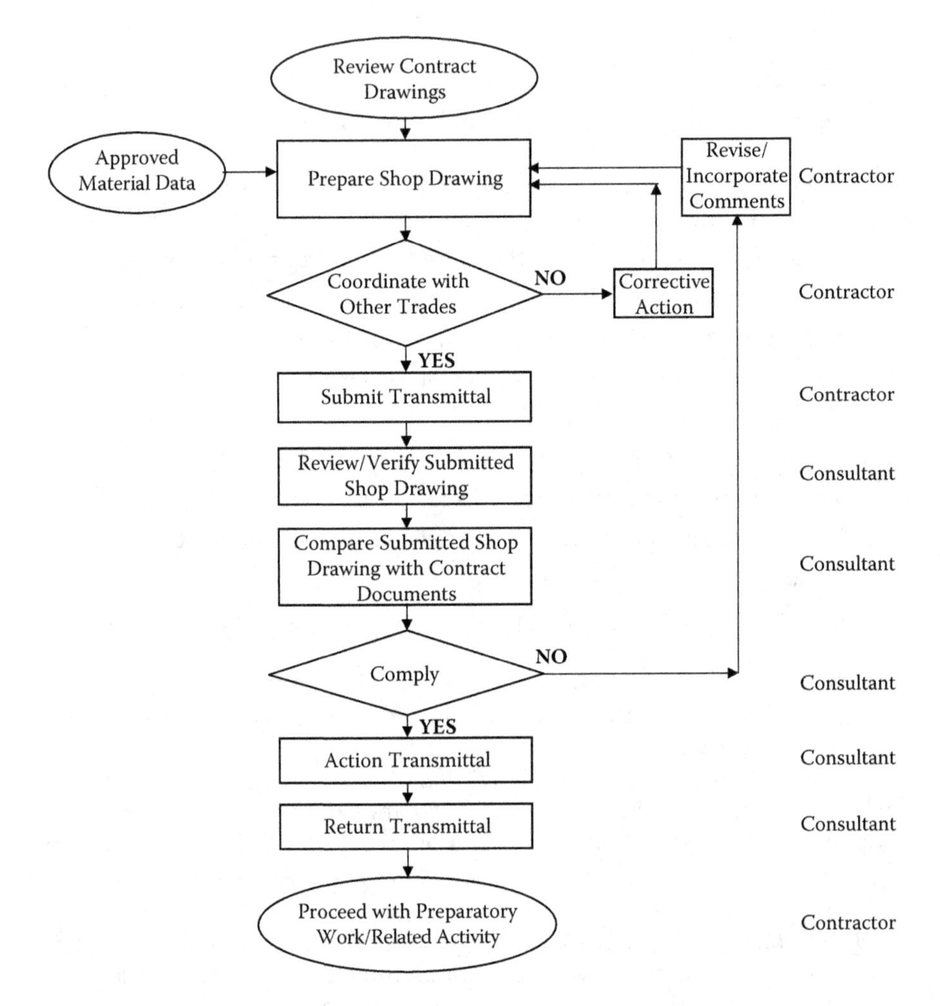

FIGURE 4.24
Shop drawing preparation and approval procedure.

4.8.3.2 Contractor's Quality Control Plan (CQCP)

The contract documents specify the details of the Quality Control Plan (QCP) to be prepared by the contractor for the construction project and submitted to the consultant for approval. The following is the outline for preparation of such QCPs:

1. Purpose of the quality control plan
2. Project description
3. Site staff organization chart for quality control
4. Quality control staff and their responsibilities

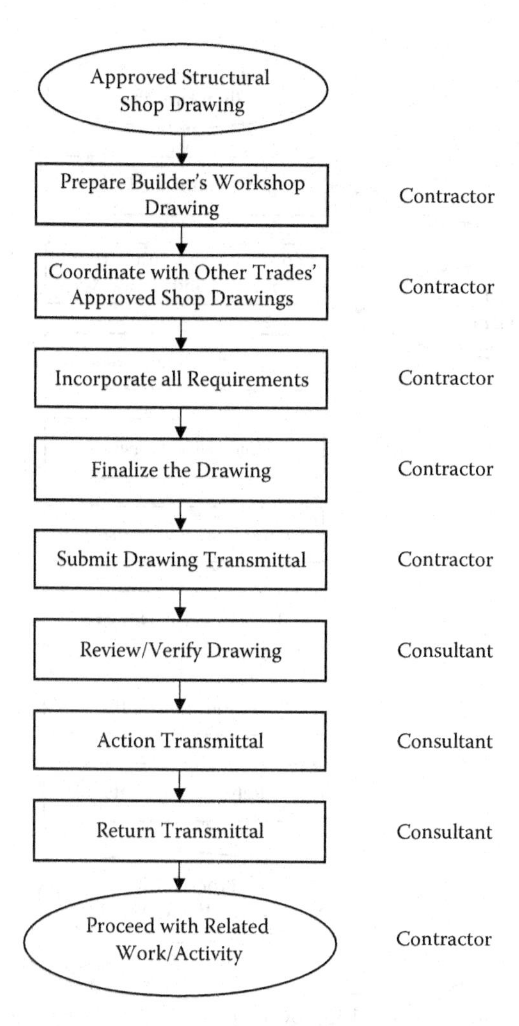

FIGURE 4.25
Builder's workshop drawing preparation and approval procedure.

5. Construction program and subprograms
6. Schedule for submission of subcontractors, manufacturer of materials, and shop drawings
7. QC procedure for all the main activities such as
    - Procurement (direct bought-out items)
    - Off-site manufacturing, inspection, and testing
    - Inspection of site activities (checklists)
    - Inspection and testing procedure for systems
    - Procedure for laboratory testing of material

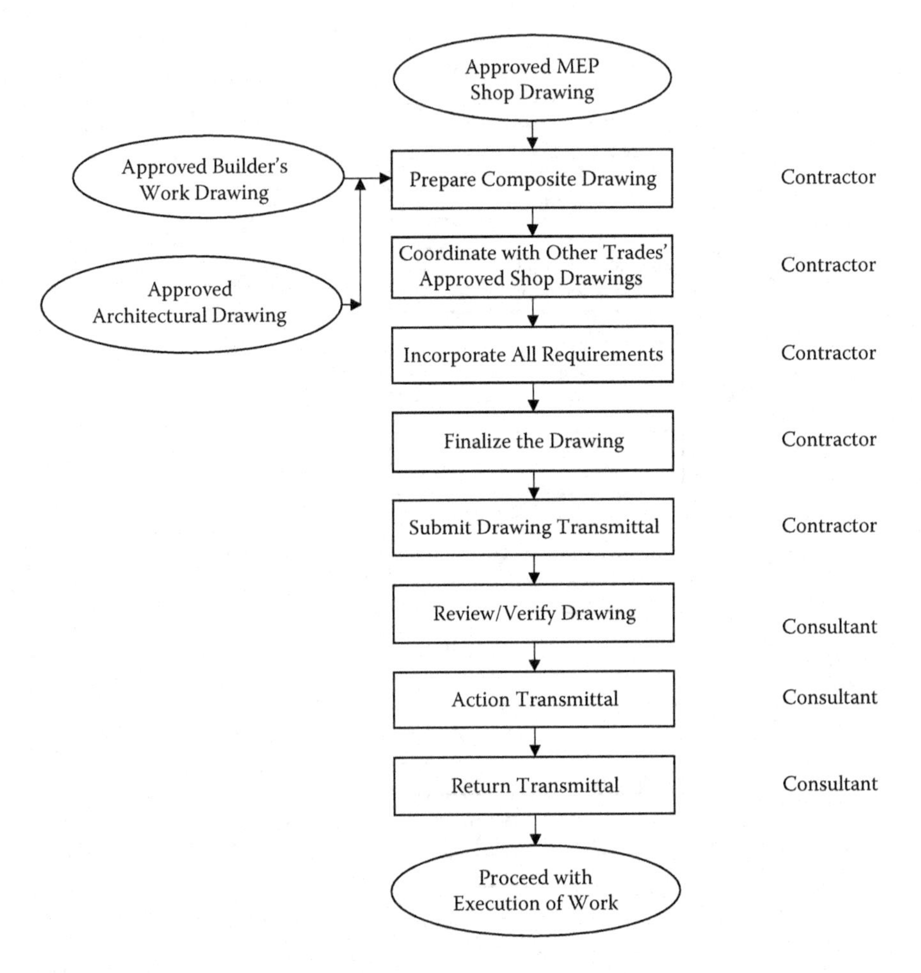

FIGURE 4.26
Composite drawing preparation and approval procedure.

- Inspection of material received at site
- Protection of works

8. Method statement for various installation activities
9. Project-specific procedures for site work instructions and remedial notes
10. Quality control records
11. List of quality procedures applicable to project in reference to company's Quality Manual and Procedure
12. Periodic testing procedure for construction equipment and tools
13. Quality updating program

14. Quality auditing program
15. Testing
16. Commissioning
17. Handover
18. Health, Safety, and Environment

4.8.3.3 Method Statement

The contractor has to submit a method statement to the consultant for approval per the contract documents. The method statement describes the steps involved in the execution/installation of work by ensuring safety at each stage. It should have the following information:

1. Scope of Work: Brief description of work/activity
2. Documentation: Relevant technical documents to undertake this work/activity
3. Personnel Involved
4. Safety Arrangement
5. Equipment and Plant Required
6. Personal Protective Equipment
7. Permits/Authorities' Approval to Work
8. Possible Hazards
9. Description of the Work/Activity: Detailed method and sequence of each operation/key steps to complete the work/activity

Figures 4.27 to 4.34 are illustrative examples of the work sequences of major activities, for different trades, performed during the construction process.

4.8.4 Planning and Scheduling

Project planning is a logical process to determine what work must be done to achieve project objectives and ensure that the work of the project is carried out

- In an organized and structured manner
- Reducing uncertainties to a minimum
- Reducing risk to a minimum
- Establishing quality standards
- Achieving results within the budget and scheduled time

Prior to the execution of a project or immediately after the actual project starts, the contractor prepares the project construction plans based on

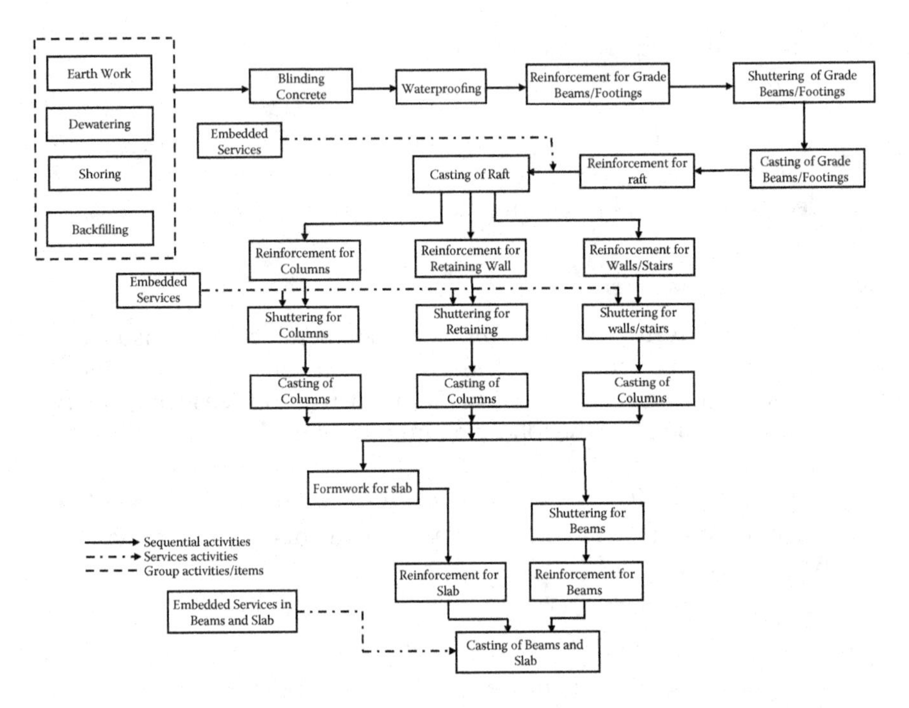

FIGURE 4.27
Work sequence for concrete work.

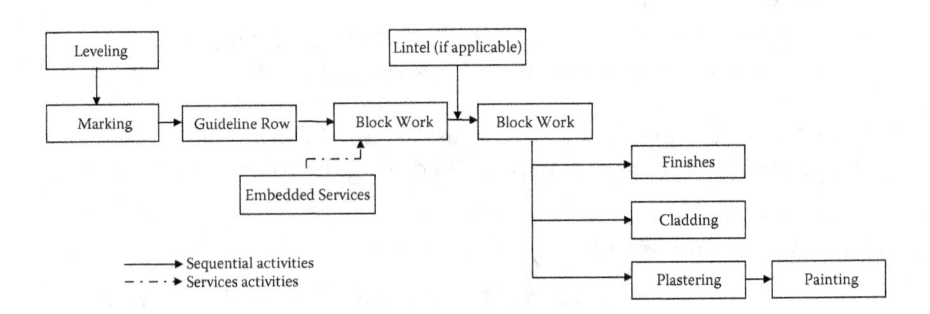

FIGURE 4.28
Work sequence for block work.

the contracted time schedule of the project. Detailed planning is needed at the start of construction in order to decide how to use the resources such as labor, plant, materials, finance, and subcontractors economically and safely to achieve the specified objectives. The plan shows the periods for all sections of the works and activities, indicating that everything can be completed by the date specified in the contract and the facility will be ready for use or for installation of equipment by other contractors.

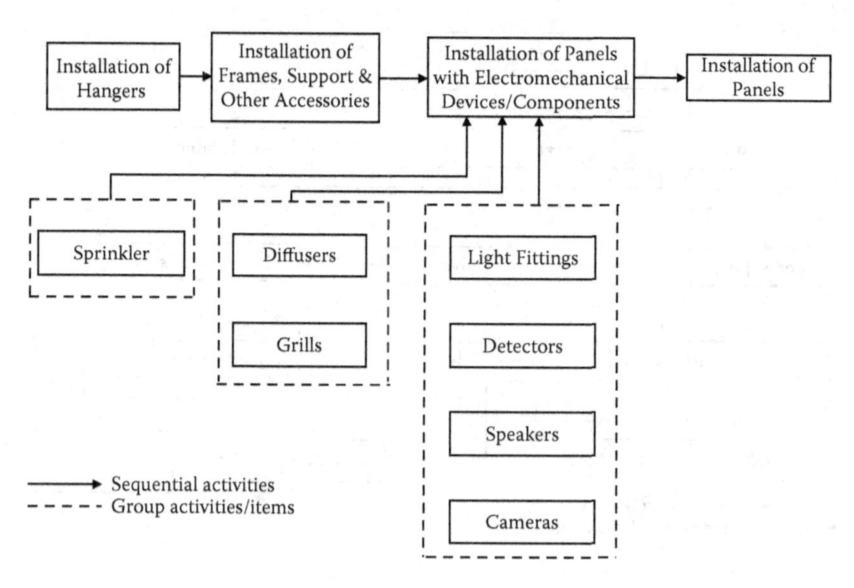

FIGURE 4.29
Work sequence for false ceiling work.

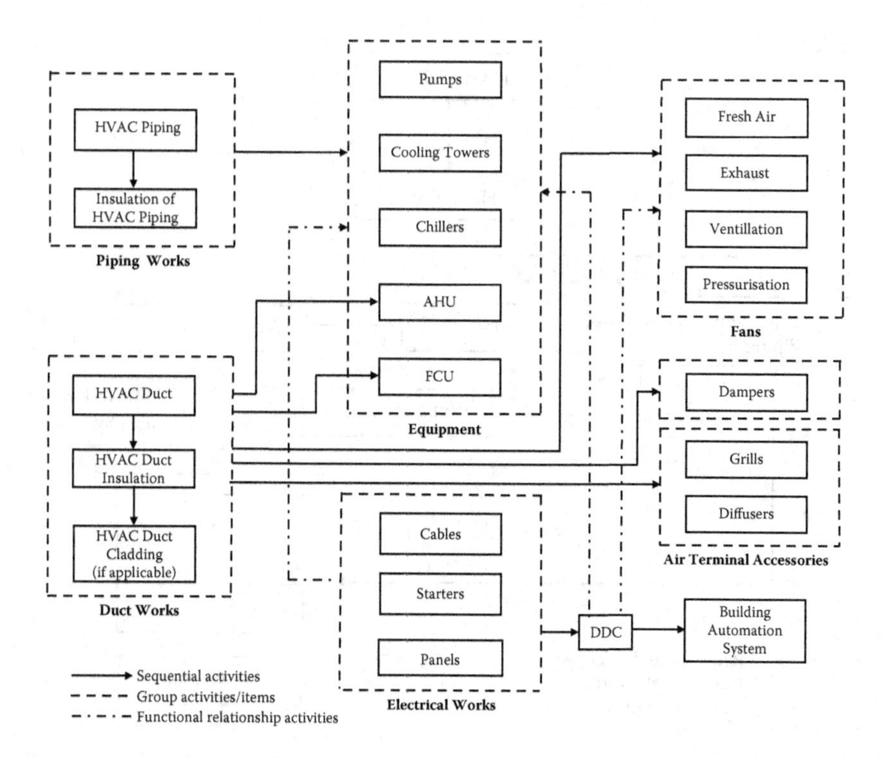

FIGURE 4.30
Work sequence for HVAC work.

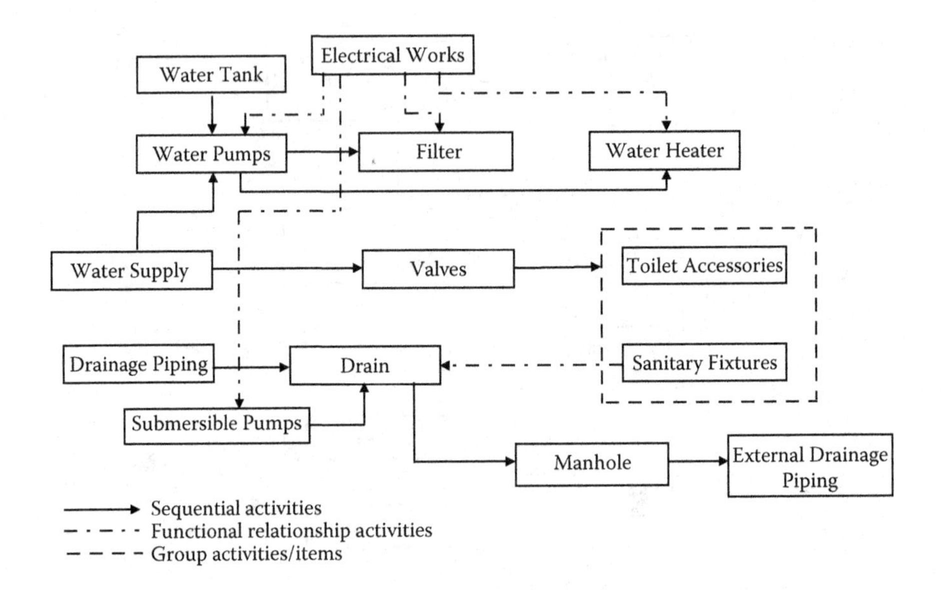

FIGURE 4.31
Work sequence for plumbing work.

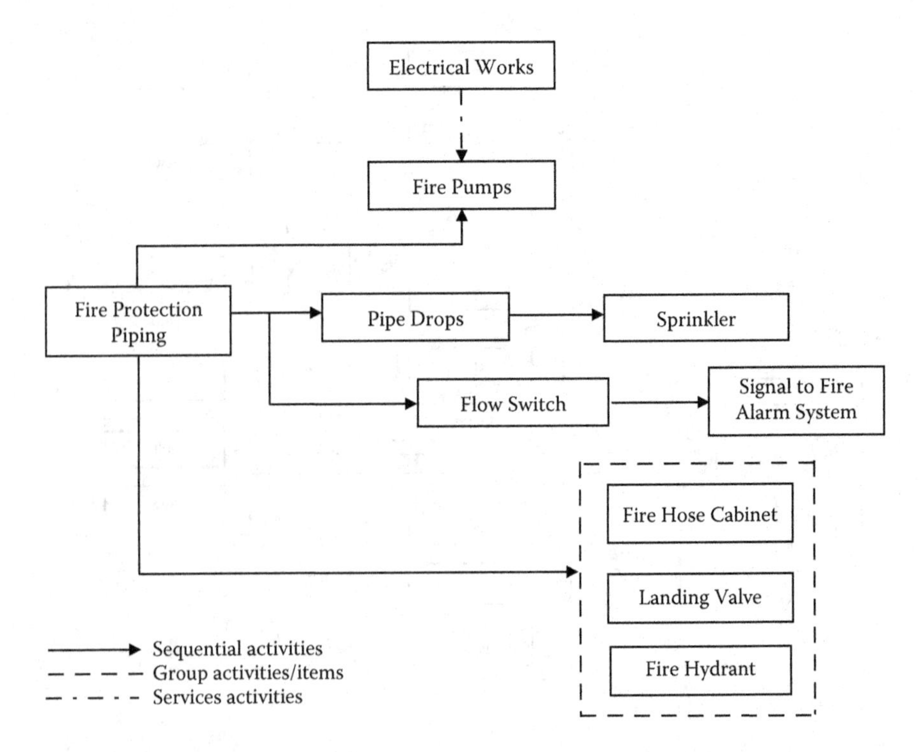

FIGURE 4.32
Work sequence for firefighting work.

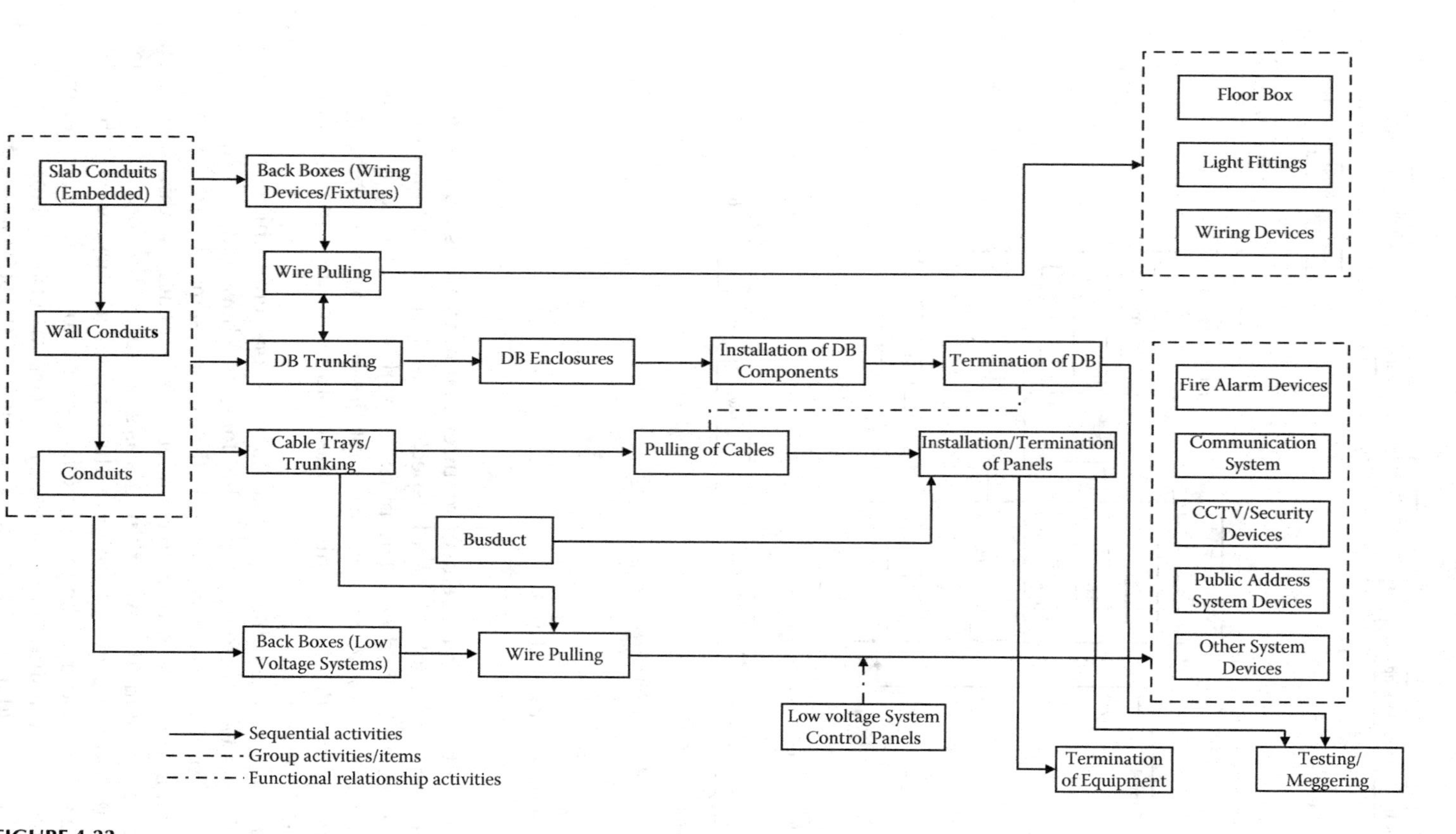

FIGURE 4.33
Work sequence for electrical work.

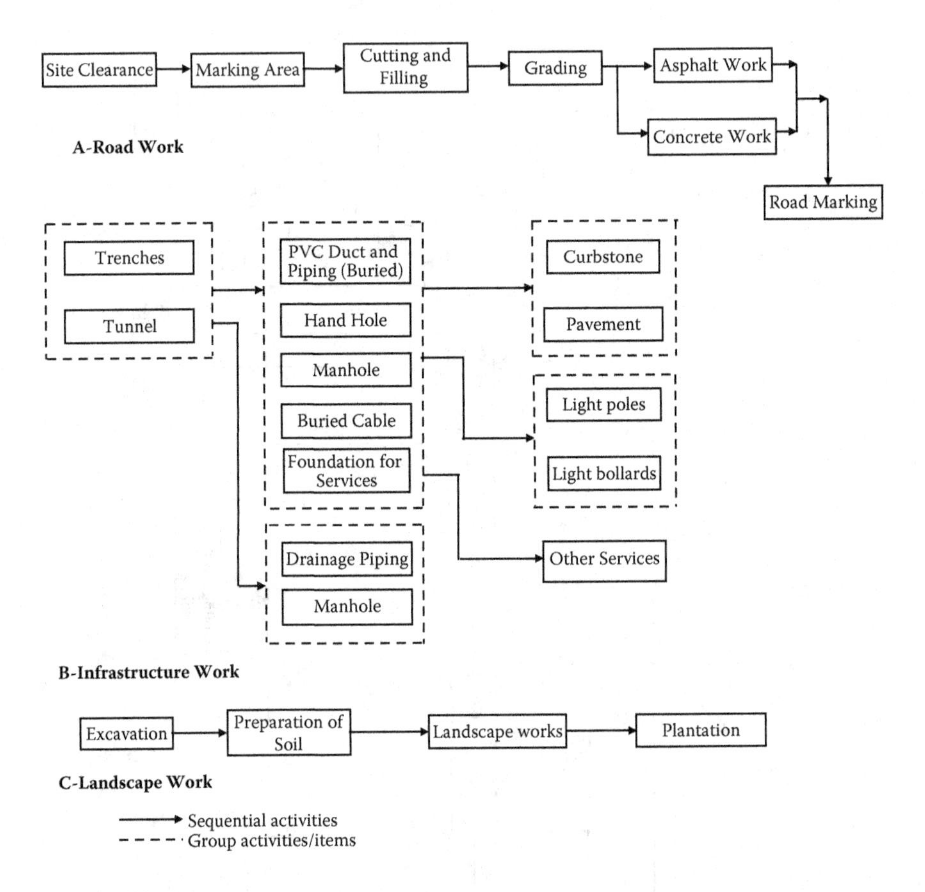

FIGURE 4.34
Work sequence for external work.

Effective project management requires planning, measuring, evaluating, forecasting, and controlling all aspects of project quality and quality of work, cost, and schedules. The purpose of the project plan is to successfully control the project to ensure completion within the budget and schedule constraints. Project planning is the evolution of the time and efforts to complete the project.

Planning is a mechanism that conveys or communicates to project participants what activity should be done, and how and in what order to meet the project objectives by scheduling the same. Project planning is required to bring the project to completion on schedule, within budget, and in accordance with the owner's needs as specified in the contract. The planning process considers all the individual tasks, activities, or jobs that make up the project and must be performed. It takes into account all the resources available, such as human resources, finances, materials, plant, equipment, etc. It also considers the works to be executed by the subcontractors.

Planning and scheduling are often used synonymously for preparing a construction program because both are performed interactively. Planning is the process of identifying the activities necessary to complete the project, while scheduling is the process of determining the sequential order of the planned activities and the time required to carry out and complete the activities. Scheduling is the mechanical process of formalizing the planned functions, assigning the starting and completion dates to each part or activity of the work in such a manner that the whole work proceeds in a logical sequence and in an orderly and systematic manner. Figure 4.35 illustrates the project management planning process.

Construction projects are unique and nonrepetitive in nature. Construction projects consist of many activities directed toward the accomplishment of a desired objective. Activities are those operations of the plan that take time to carry out and on which resources are expended. In order to achieve the quality objectives of the project, each activity has to be completed within the specified limit, using the specified product and approved method of installation. Construction projects consist of a number of related activities that are dependent on other activities and cannot be started until others are completed, and some that can run in parallel. The most important point while starting the planning is to establish all the activities that constitute the project.

Planning involves defining the objectives of the project; listing the tasks or jobs that must be performed; determining gross requirements for material, equipment, and manpower; and preparing costs and durations for the various jobs or activities to bring about the satisfactory completion of the project. The techniques for planning vary depending on the projects' size, complexity, duration, personnel, and owner's requirements. The techniques used during the construction phase of the project should make possible the evaluation of the project's progress against the plan. There are many different analytical and graphical techniques that are commonly used for planning of the project. Some of these are:

- Bar Charts
- Critical Path Method (CPM)
- Progress Curves
- Matrix Schedules

The most widely used techniques are bar charts and network diagrams. The bar chart is the oldest planning method used in project management. It is a graphical representation of the estimated duration of each project activity, the planned sequence of activities, and the planned sequence of activity performance. The horizontal axis represents the time schedule, whereas the project activities are shown along the vertical axis.

Network diagrams such as the Program Evaluation and Review Technique (PERT) and Critical Path Method (CPM) are used for scheduling

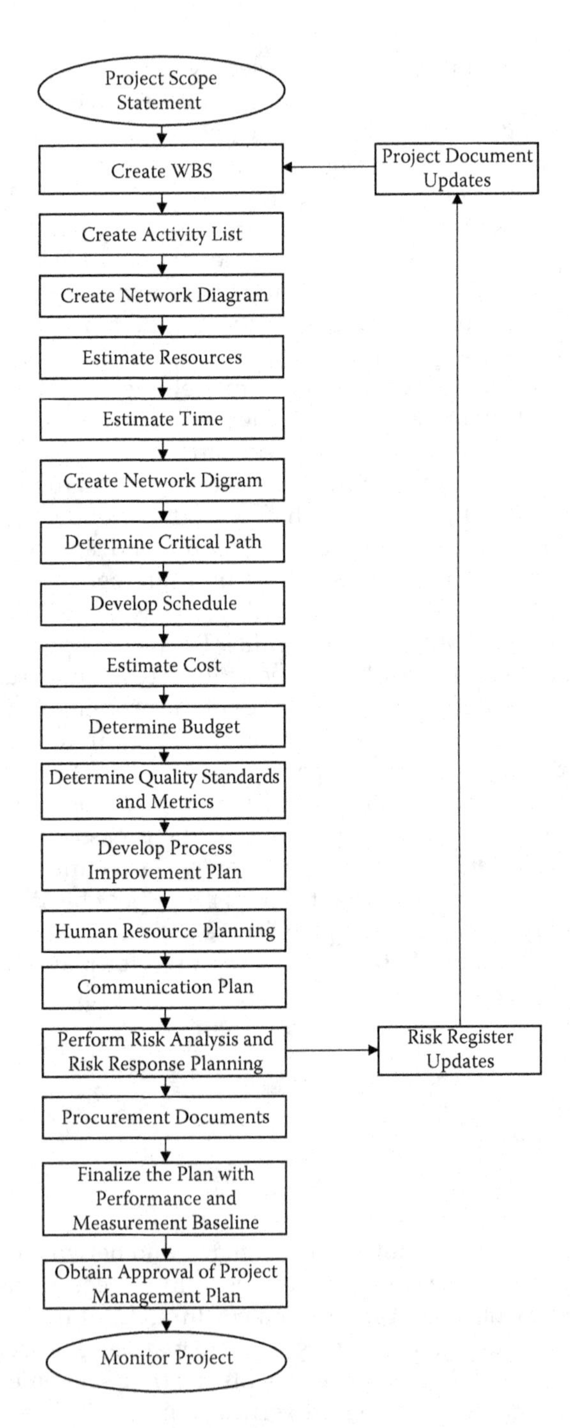

FIGURE 4.35
Project management planning process.

complex projects. PERT/CPM diagrams consist of nodes and links and represent the entire project as a network of arrows (activities) and nodes (events). In order to draw network diagrams, work activities have to be identified, relationships among the activities specified, and precedence relationships between the activities in a particular sequence established.

The most widely used scheduling technique is CPM. The CPM analysis represents the set of sequence of predecessor/successor activities that will take the longest time to complete. The duration of the critical path is the sum of all the activities' duration along the path. Thus, the critical path is the longest possible path of the project activities network. The duration of the critical path represents the minimum time required to complete the project.

There are many computer-based programs available for preparing the network and critical path of activities for construction projects. These programs can be used to analyze the use of resources, review project progress, and forecast the effects of changes in the schedule of works or other resources. Most computer programs automate preparation and presentation of various planning tools such as bar charts, PERT, and CPM analysis. The programs are capable of storing huge data and help process and update the program quickly. They manipulate data for multiple usages from the planning and scheduling perspectives.

In order to manage and control the project at different levels in the most effective manner, the project is broken down into a group of smaller subprojects/subsystems and then to small well-defined activities. This breakdown is necessary because of the size and complexity of a construction project and is referred to as Work Breakdown Structure (WBS). To begin the preparation of a detailed construction program, the contractor prepares a WBS. Its purpose is to define the various activities that must be executed to complete the project. WBS help the construction project planner to

- Plan and schedule the work
- Estimate costs and budget
- Control schedule, cost, and quality

Depending on the size of the project, the project is divided into multiple zones, and relevant activities are considered for each zone to prepare the construction program. While preparing the program, the relationships between project activities and their dependency and precedence are considered by the planner. These activities are connected to their predecessor and successor activity based on the way the task is planned to be executed. There are four possible relationships that exist between various activities: finish–to-start relationship, start-to-start relationship, finish-to-finish relationship, and start–to–finish relationship.

Once all the activities are established by the planner and the estimated duration of each activity has been assigned, the planner prepares the detailed program fully coordinating all the construction activities.

The Critical Path Method calculates the minimum completion time for a project along with the possible start and finish times for the project activities. The critical path is the longest in the network, whereas the other paths may be equal to or shorter than that path. Therefore, there is a possibility that some of the events and activities can be completed before they are actually needed, and accordingly it is possible to develop a number of activity schedules from the Critical Path Method analysis to delay the start of each activity as long as possible but still finish the project with the minimum possible time without extending the completion date of the project. To develop such a schedule, it is required to find out when each activity needs to start and when it needs to be finished. There may be some activities in the project with some leeway in which some activity can start and finish. This is called slack time or float in an activity. For each activity in a project, there are four points in time; early start, early finish, late start, and late finish. The early start and early finish are the earliest times an activity can start and be finished. Similarly, the late start and late finish are the latest times the activities can start and finish. The difference between the late start time and early start time is the slack time or float.

With the advent of powerful computer-based programs like Primavera and Microsoft Project, it is possible for the details of the work breakdown to be fed to these software programs. The software is capable of producing network diagrams, schedules, and countless different reports that also help in the efficient monitoring of the project schedule by comparing actual with planned progress. The software can be used to analyze the project for use of resources, forecasting the effects of changes in the schedule, and cost control.

The first step in preparation of a construction program is to establish the activities, and the next step is to establish the estimated time duration of each activity. The deadline for each activity is fixed, but it is often possible to reschedule by changing the sequence in which the tasks are performed, while retaining the original estimated time.

The activities to be performed during execution of a project are grouped in a number of categories. Each of these categories has a number of activities. The following are the major categories of construction project schedules:

A. General Activities
   1. Mobilization
B. Engineering
   1. Subcontractor Submittal and Approval
   2. Materials Submittal and Approval
   3. Shop Drawing Submittal and Approval
   4. Procurement
C. Site Activities
   a. Site Earth Works
   b. Dewatering and Shoring

c. Excavation and Backfilling
d. Raft Works
e. Retaining Wall Works
f. Concrete Foundation and Grade Beams
g. Waterproofing
h. Concrete Columns and Beams
i. Casting of Slabs
j. Wall Partitioning
k. Interior Finishes
l. Furnishings
m. External Finishes
n. Equipment
o. Conveying Systems Works
p. Plumbing and Public Health Works
q. Firefighting Works
r. HVAC Works
s. Electrical Works
t. Fire Alarm System Works
u. Communication System Works
v. Low Voltage Systems Works
w. Landscape Works
x. External Works

D. Closeout
1. Testing and Commissioning
2. Completion and Handover

The contractor also submits the following along with the construction schedule:

- Resources (equipment and manpower) schedule
- Cost loading (schedule of item pricing based on bill of quantities)

Figure 4.36 shows the logic flow diagram for development of the construction schedule.

4.8.4.1 Six Sigma for Development of Construction Schedule

The contractor's construction schedule (CCS) is an important document used during the construction phase. It is used to plan, monitor, and control project activities and resources. The document is voluminous

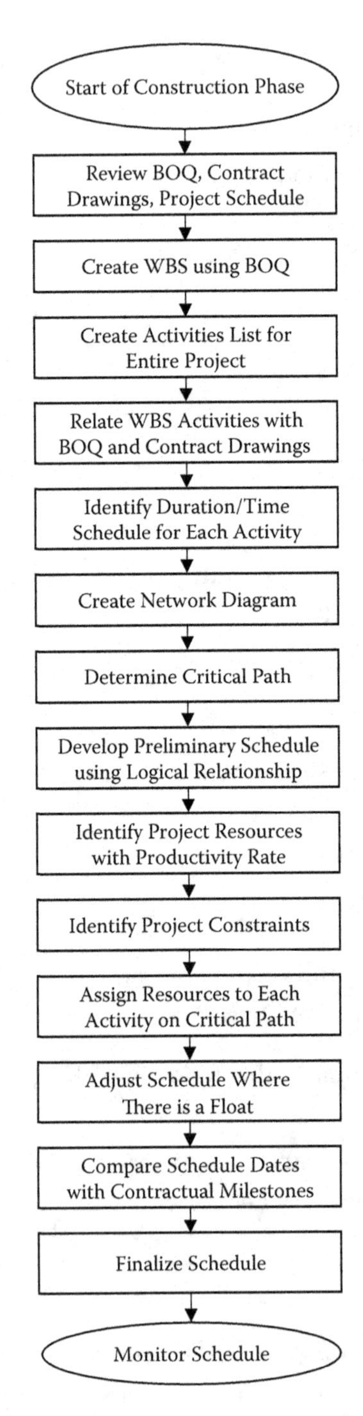

FIGURE 4.36
Logic flow diagram for development of construction schedule.

and important. It has to be prepared accurately in order to follow up work progress without deviation from the milestones set up in the contract documents. In most cases, contractors experience problems with the approval of the CCS, at the very first submission, from the construction manager/project manager/consultant. The CCS is rejected if it does not meet the specification requirements. The contractor is not paid unless the CCS is approved.

Appendix B is an example procedure to develop a contractor's CCS using the systematic approach of the Six Sigma methodology concept of the DMADV analytic tool set.

4.8.5 Management Resources/Procurement

In most construction projects, the contractor is responsible for engaging subcontractors, specialist installers, and suppliers and arranging for materials, equipment, construction tools, and all types of human resources to complete the project per contract documents and to the satisfaction of the owner's appointed supervision team. Workmanship is one of the most important factors in achieving quality in construction, and therefore the construction workforce should be fully trained and be knowledgeable about all the related activities to be performed during the construction process.

Once the contract is awarded, the contractor prepares a detailed plan for all the resources he needs to complete the project. The contractor also prepares a procurement log based on the project completion schedule.

4.8.5.1 Project Core Team Members

Contract documents normally specify a list of the minimum number of core staff to be available on site during the construction period. The absence of any of these staff may result in penalties on the contractor by the owner.

The following is a typical list of contractor's minimum core staff needed during the construction period for execution of work of a major building construction project:

1. Project Manager
2. Site Senior Engineer for Civil Works
3. Site Senior Engineer for Architectural Works
4. Site Senior Engineer for Electrical Works
5. Site Senior Engineer for Mechanical Works
6. Site Senior Engineer for HVAC Works
7. Site Senior Engineer for Infrastructure Works
8. Planning Engineer

9. Senior Quantity Surveyor/Contract Administrator
10. Civil Works Foreman
11. Architectural Works Foreman
12. Electrical Works Foreman
13. Mechanical Works Foreman
14. HVAC Works Foreman
15. Laboratory Technician
16. Quality Control Engineer
17. Safety Officer

The contractor has to submit the names of the staff for these positions for approval from the owner/consultant to work on the project. The staff approval request form is used by the contractor to propose the names of the staff to work for these positions and is submitted along with the qualification and experience certificates of the proposed staff. Those who get approval are allowed to work on the project under the core staff requirements. Their absence from the project site makes the contractor liable for penalties per the contract documents.

The contractor's human resources can be divided into two categories:

1. The contractor's own staff and workers
2. Subcontractors' staff and workers

The main contractor has to manage all these personnel by

- Assigning the daily activities
- Observing their performance and work output
- Ensuring daily attendance
- Ensuring safety during the construction process

Figure 4.37 shows a sample contractor's manpower chart of a construction project.

4.8.5.2 Equipment Schedule

The contract documents specify that a minimum set of equipment is to be available on site during the construction process to ensure smooth operation of all the construction activities. These are normally

- Tower crane
- Mobile crane
- Normal mixture
- Concrete mixing plant
- Dump trucks

Manpower Histogram

Month	Jan-12	Feb-12	Mar-12	Apr-12	May-12
Planned Manpower	813	968	1129	1117	1188
Actual Manpower "Campus A"	295	300	325	383	450
Actual Manpower "Campus B"	305	60	100	214	423
Actual Manpower "Total"	600	360	425	597	873

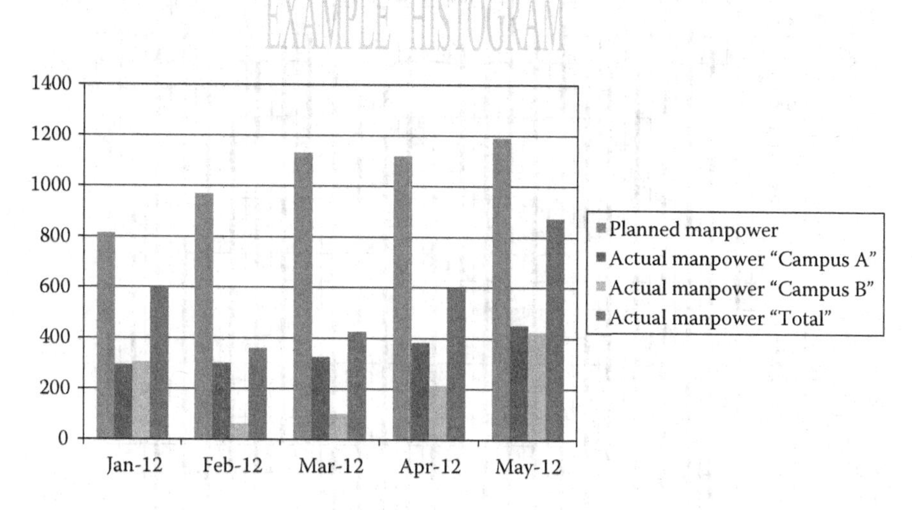

FIGURE 4.37
Contractor's manpower chart.

- Compressor
- Vibrators
- Water pumps
- Compactors
- Concrete pumps
- Trucks
- Concrete trucks
- Diesel generator sets

Figure 4.38 shows a sample equipment schedule listing equipment the contractor has to make available at a major building construction project.

4.8.5.3 Procurement

In most construction projects, the contractor is responsible for procurement of material, equipment, and systems to be installed on the project. The contractors have their own procurement strategies. While submitting the bid,

Owner Name
Project Name
Equipment Schedule

Contractor Name

Equipment Description	Feb-09	Mar-09	Apr-09	May-09	Jun-09	Jul-09	Aug-09	Sep-09	Oct-09	Nov-09	Dec-09	Jan-10	Feb-10	Mar-10	Apr-10	May-10	Jun-10	Jul-10	Aug-10	Sep-10	Oct-10	Nov-10	Dec-10	Jan-11	Feb-11	Mar-11	Apr-11	May-11	Jun-11
	1	2	3	4	5	6	7	8	9	10	11	12	13	14	15	16	17	18	19	20	21	22	23	24	25	26	27	28	29
Batching Plant																													
Air Compressor																													
Bar Bending Machine																													
Bar Cutting Machine																													
Bulldozer																													
Buses																													
Fuel Tanker																													
Generator																													
Jeep																													
Half Loory																													
Mobile Crane																													
Pick Up																													
Flat Compactor																													
Roller Compactor																													
Bob Cat																													
Tipper Trailer																													
Tower Light																													
Asphalt Finisher																													
Bitumen Sprayer																													
Wheel Loader																													
Welding Machine																													
Water Truck																													
Silo Cement																													
Transit Mixer (Concrete)																													
Foam Pump																													
Marble Cutter																													

FIGURE 4.38
Equipment schedule.

the contractor obtains quotations from various suppliers/subcontractors. The contractor has to consider the following at a minimum while finalizing the procurement:

- Contractual commitment
- Specification compliance
- Statutory obligations
- Time
- Cost
- Performance

Figure 4.39 illustrates the material procurement procedure.

4.8.6 Monitoring and Control

Once planning and scheduling are complete and the project is under way, progress on the project must be monitored on an ongoing basis to ensure that the goals and objectives on the project are being met. The monitoring and control of a construction project occurs during the execution of the project, and its aim is to recognize any obstacles encountered during execution and to apply measures to mitigate these difficulties.

Monitoring is collecting, recording, and reporting information concerning any and all aspects of project performance that the project manager or others in the organization need to know about. Monitoring of the construction project is normally done by collecting and recording the status of various activities and compiling them in the form of progress reports. These are prepared by the consultant and contractor, and distributed to the concerned members of the project team.

Monitoring involves not only tracking time but also resources and budget. Monitoring in construction projects is normally done by compiling the status of various activities in the form of progress reports. These are prepared by the contractor, supervision team (consultant), and construction/project management team. The objectives of project monitoring and control are

- To report the necessary information in detail and in an appropriate form that can be interpreted by management and other concerned personnel to assess how resources are being used to achieve project objectives
- To provide an organized and efficient means of measuring, collecting, verifying, and quantifying data reflecting the progress and status of the execution of project activities with respect to schedule, cost, resources, procurement, and quality

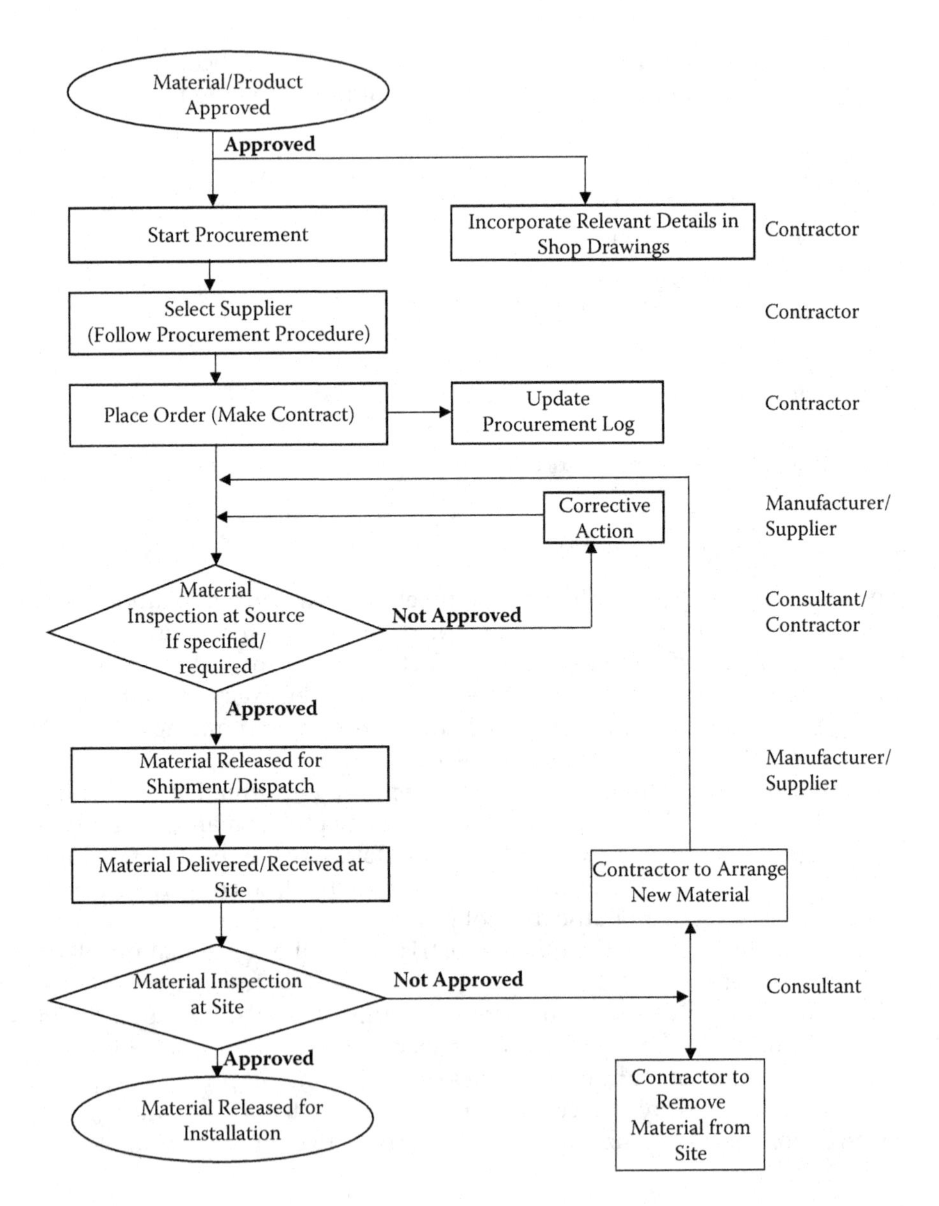

FIGURE 4.39
Material procurement procedure.

- To provide an organized, efficient, and accurate means of converting the data from the execution process into information
- To identify and isolate the most important and critical information about the project activities to enable decision-making personnel to take corrective action for the benefit of the project

- To forecast and predict about future progress of activities to be performed

Figure 4.40 shows the logic flow diagram for the monitoring and control process, and Figure 4.41 shows monitoring and control references for construction projects.

Construction project control is exercised through knowing where to focus the main efforts at a given time and maintaining good communication. There are three main areas where project control is required: (1) Budget, (2) Schedule, and (3) Quality.

All of these areas have to be balanced to achieve the project objectives. In order to accomplish the project objectives in construction projects, monitoring and control is done through various tools and methods.

4.8.6.1 Cost Control

Monitoring and control of project payments is essential to keep within the budgeted amount. This is done through monitoring the cash flow with the help of S-curves and progress curves that give the exact status of payments and also identify when the budget is being exceeded. Uninterrupted cash flow is one of the most important elements in the overall success of the project. Figure 4.42 shows a planned S-curve.

4.8.6.1.1 Project Payment/Progress Curve (S-Curve)

Cash flow is a simple comparison of when revenue will be received and when the financial obligations must be paid. It is also an indication of the progress of the work to be completed in a project. This is obtained by loading each activity in the approved schedule with the budgeted cost in the bill of quantities (BOQ). The process of inputting the schedule of values is known as cost loading. The graphical representation of the above is obtained as a curve and is known as the S-curve. This also represents the planned progress of a project.

Figure 4.43 shows a progress S-curve.

4.8.6.1.2 Earned Value Management

Earned Value Management (EVM) is a methodology used to measure and evaluate project performance against cost, schedule, and scope baseline. It compares the amount of planned work with what is actually accomplished to determine whether the project is progressing as planned. EVM is used to measure the progress of the project budget, schedule, and scope to

- Know the percentage of budget spent
- Know the percentage of time elapsed
- Know the percentage of work done

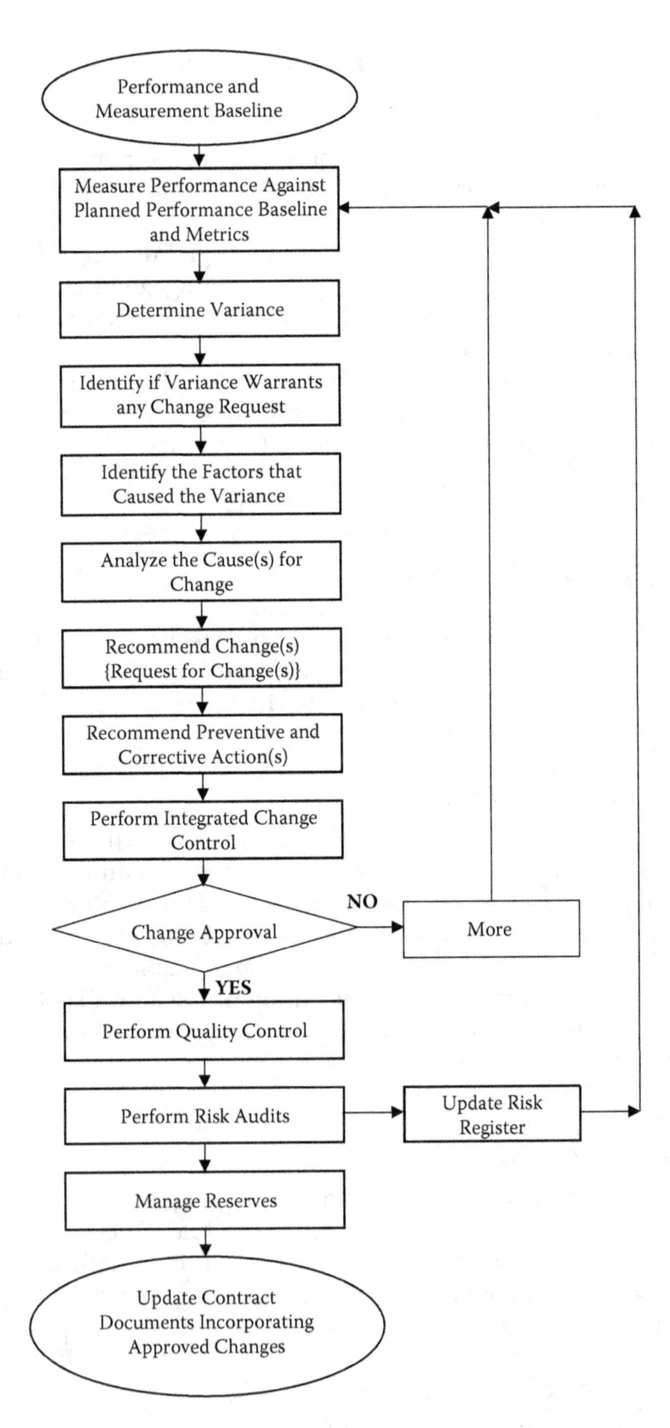

FIGURE 4.40
Logic flow diagram for monitoring and control process.

Serial Number	Elements	Contract Reference	Contractor Reference
1	Performance Baseline	Schedule	1. Contractor's Construction Schedule
		Specifications, Drawings	1. Approved Materials 2. Approved Shop Drawings 3. Approved Composite Drawings
		Cost	1. Approved S-Curve
2	Data Collection Methods	Reports	1. Daily Report 2. Weekly Report 3. Monthly Report 4. Safety Report 5. Risk Report 6. Accident Report 7. Checklist 8. Risk Report
3	Frequency of Data Collection	1. Daily 2. Weekly 3. Monthly	1. Daily Report 2. Weekly Report 3. Monthly Report
4	Status Information Collection	1. Logs 2. Reports 3. Meetings 4. Checklists	1. Logs 2. Reports 3. Meetings 4. Checklists
5	Comparison Between Planned and Actual (Variance)	1. S-Curves 2. Milestones	1. Progress Reports 2. Progress Payment 3. Milestones
6	Analysis	1. Price Analysis	1. Construction Schedule Attachment 2. Progress Payment
7	Corrective Action	1. Comments by Consultant	1. Incorporate Comments
8	Change Order	1. Variation Order	1. Request for Information 2. Request for Modification 3. Request for Variation
9	Document Updates	1. On Regular Basis	1. Reports

FIGURE 4.41
Monitoring and control references for construction projects.

- Forecast its completion date and cost
- Provide budget and schedule variances

The following are the basic terminologies used in EVM:

- BCWS → Budgeted Cost of Work Scheduled or Planned Value (PV)
  This is the planned cost of the total amount of work scheduled to be performed by the milestone date.

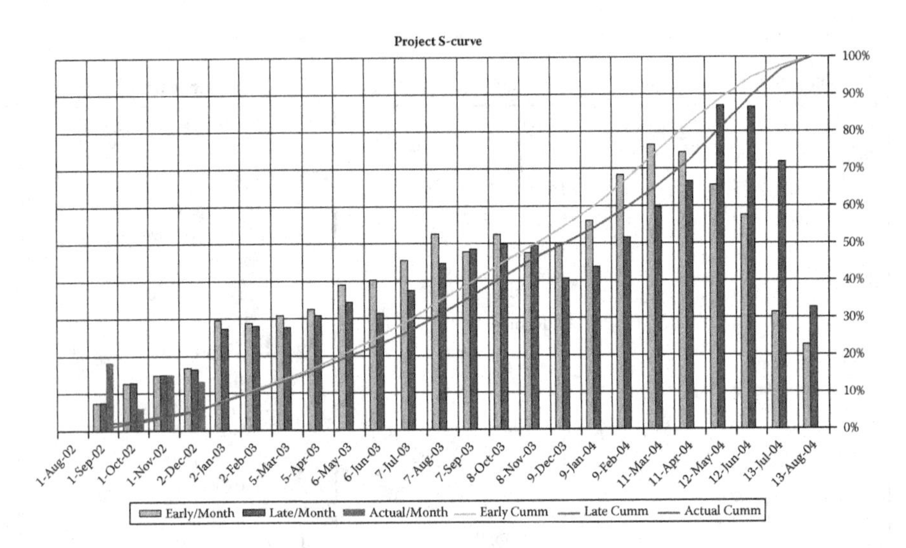

FIGURE 4.42
Planned S-curve.

- BCWP → Budgeted Cost of Work Performed or Earned Value (EV)

  This is the actual cost incurred to accomplish the work that has been done to date.
- ACWP → Actual Cost of Work Performed or Actual Cost (AC)

  This is the planned cost to complete the work that has been done.

Figure 4.44 shows the diagram for the earned value method.

These three key values are used in various combinations to determine cost and schedule performance and provide an estimated cost of the project at its completion.

Table 4.14 illustrates the terms used in EVM and their interpretation, whereas Table 4.15 illustrates the formulas used in EVM and their interpretation.

4.8.6.1.3 Variation Orders

It is common that during the construction process there will be some changes to the original contract. Even under most ideal circumstances, contract documents cannot provide complete information about every possible condition or circumstance that the construction team may encounter. Construction projects involve mainly the owner, consultant, and contractor. Any of these parties can originate the changes. Table 4.16 lists the causes of changes in construction projects.

These changes help build the facility to achieve the project objective. These changes are identified as the construction proceeds. Prompt identification of such requirements helps both the owner and contractor to avoid unnecessary

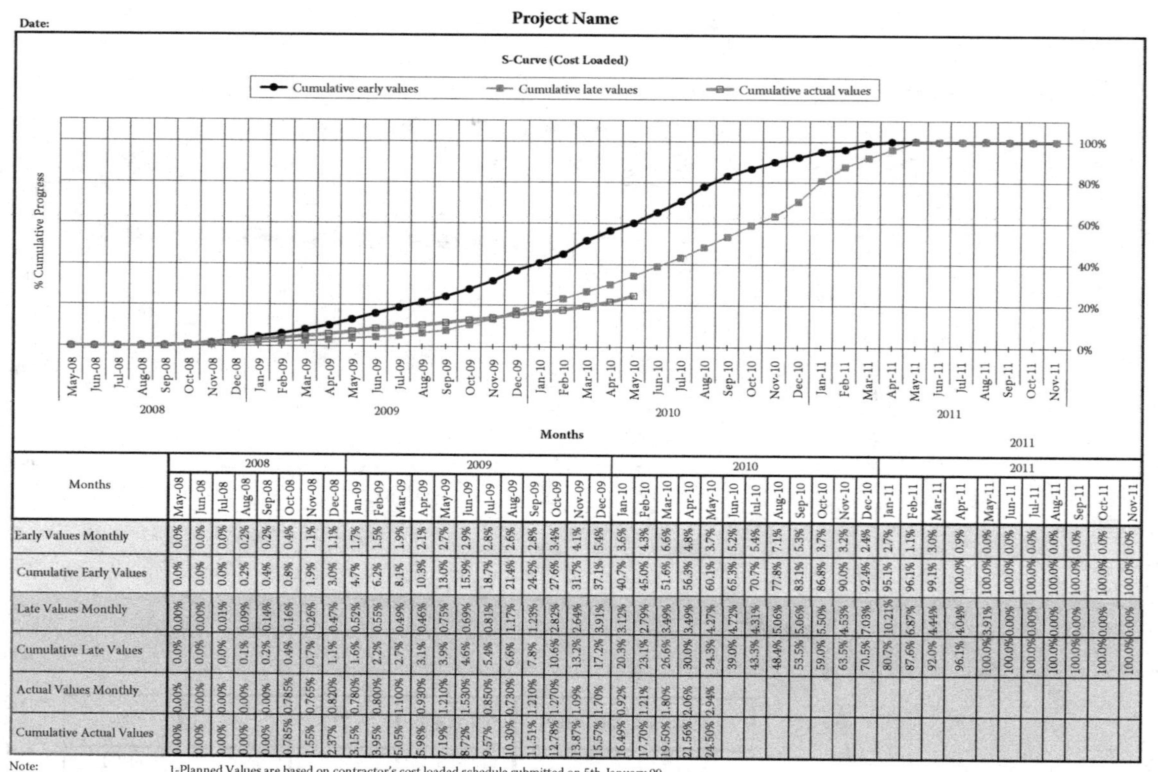

Months	2008							
	May-08	Jun-08	Jul-08	Aug-08	Sep-08	Oct-08	Nov-08	Dec-08
Early Values Monthly	0.0%	0.0%	0.0%	0.2%	0.2%	0.4%	1.1%	1.1%
Cumulative Early Values	0.0%	0.0%	0.0%	0.2%	0.4%	0.8%	1.9%	3.0%
Late Values Monthly	0.00%	0.00%	0.01%	0.09%	0.14%	0.16%	0.26%	0.47%
Cumulative Late Values	0.0%	0.0%	0.0%	0.1%	0.2%	0.4%	0.7%	1.1%
Actual Values Monthly	0.00%	0.00%	0.00%	0.00%	0.00%	0.785%	0.765%	0.820%
Cumulative Actual Values	0.00%	0.00%	0.00%	0.00%	0.00%	0.785%	1.55%	2.37%

Months	2009											
	Jan-09	Feb-09	Mar-09	Apr-09	May-09	Jun-09	Jul-09	Aug-09	Sep-09	Oct-09	Nov-09	Dec-09
Early Values Monthly	1.7%	1.5%	1.9%	2.1%	2.7%	2.9%	2.8%	2.6%	2.8%	3.4%	4.1%	5.4%
Cumulative Early Values	4.7%	6.2%	8.1%	10.3%	13.0%	15.9%	18.7%	21.4%	24.2%	27.6%	31.7%	37.1%
Late Values Monthly	0.52%	0.55%	0.49%	0.46%	0.75%	0.69%	0.81%	1.17%	1.23%	2.82%	2.64%	3.91%
Cumulative Late Values	1.6%	2.2%	2.7%	3.1%	3.9%	4.6%	5.4%	6.6%	7.8%	10.6%	13.2%	17.2%
Actual Values Monthly	0.780%	0.800%	1.100%	0.930%	1.210%	1.530%	0.850%	0.730%	1.210%	1.270%	1.09%	1.70%
Cumulative Actual Values	3.15%	3.95%	5.05%	5.98%	7.19%	8.72%	9.57%	10.30%	11.51%	12.78%	13.87%	15.57%

Months	2010											
	Jan-10	Feb-10	Mar-10	Apr-10	May-10	Jun-10	Jul-10	Aug-10	Sep-10	Oct-10	Nov-10	Dec-10
Early Values Monthly	3.6%	4.3%	6.6%	4.8%	3.7%	5.2%	5.4%	7.1%	5.3%	3.7%	3.2%	2.4%
Cumulative Early Values	40.7%	45.0%	51.6%	56.3%	60.1%	65.3%	70.7%	77.8%	83.1%	86.8%	90.0%	92.4%
Late Values Monthly	3.12%	2.79%	3.49%	3.49%	4.27%	4.72%	4.31%	5.06%	5.06%	5.50%	4.53%	7.03%
Cumulative Late Values	20.3%	23.1%	26.6%	30.0%	34.3%	39.0%	43.3%	48.4%	53.5%	59.0%	63.5%	70.5%
Actual Values Monthly	0.92%	1.21%	1.80%	2.06%	2.94%							
Cumulative Actual Values	16.49%	17.70%	19.50%	21.56%	24.50%							

Months	2011										
	Jan-11	Feb-11	Mar-11	Apr-11	May-11	Jun-11	Jul-11	Aug-11	Sep-11	Oct-11	Nov-11
Early Values Monthly	2.7%	1.1%	3.0%	0.9%	0.0%	0.0%	0.0%	0.0%	0.0%	0.0%	0.0%
Cumulative Early Values	95.1%	96.1%	99.1%	100.0%	100.0%	100.0%	100.0%	100.0%	100.0%	100.0%	100.0%
Late Values Monthly	10.21%	6.87%	4.44%	4.04%	3.91%	0.00%	0.00%	0.00%	0.00%	0.00%	0.00%
Cumulative Late Values	80.7%	87.6%	92.0%	96.1%	100.0%	100.0%	100.0%	100.0%	100.0%	100.0%	100.0%
Actual Values Monthly											
Cumulative Actual Values											

Note: 1-Planned Values are based on contractor's cost loaded schedule submitted on 5th January 09.
2-The actual % progress up to 31st May 2010 are based on value of work done.

FIGURE 4.43
Progress S-curve.

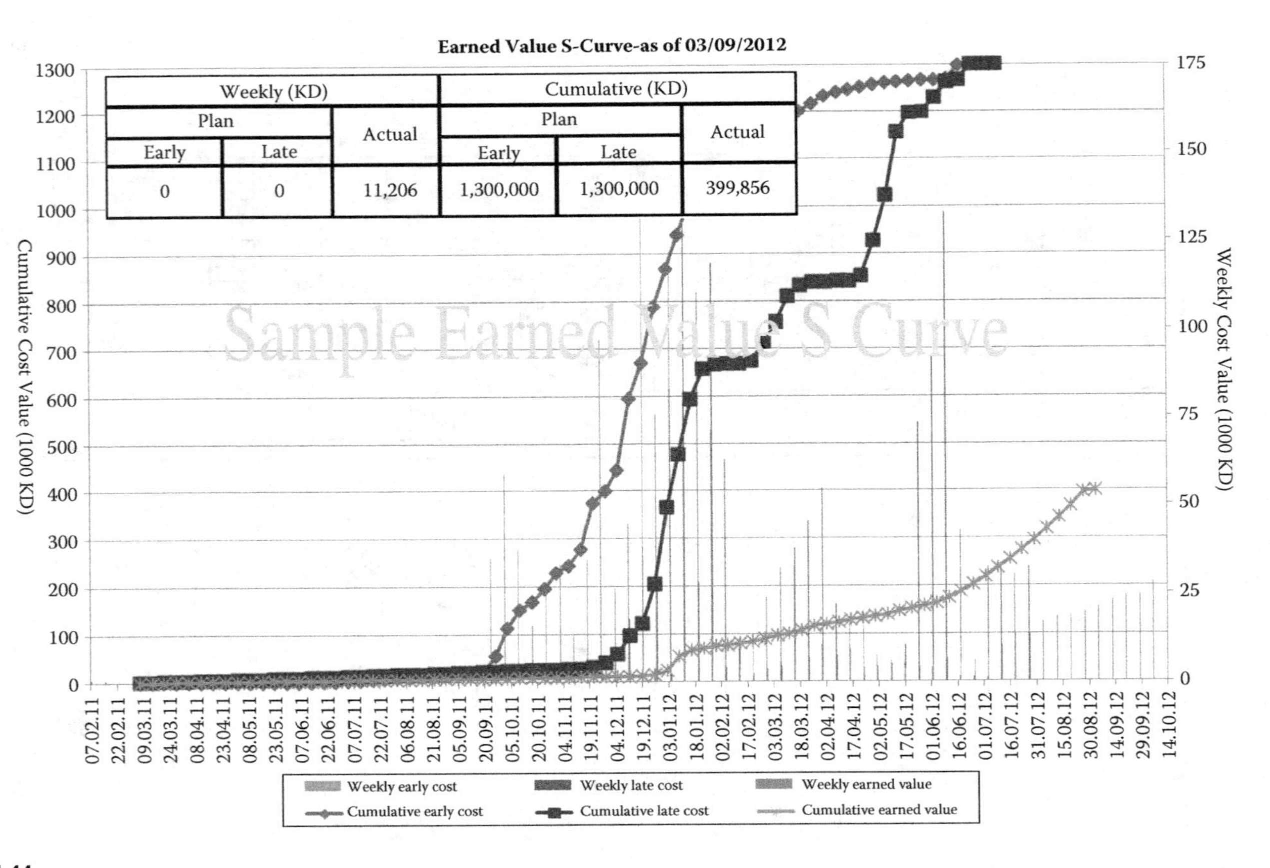

FIGURE 4.44
Diagram for earned value method.

TABLE 4.14

Earned Value Management Terms

Serial Number	Term	Description	Interpretation
1	BCWS (PV)	Planned Value	Planned cost of the total amount of work scheduled to be performed by the milestone date.
2	BCWP (EV)	Earned Value	Actual cost incurred to accomplish the work that has been done to date.
3	ACWP (AC)	Actual Cost	Planned cost to complete the work that has been done.
4	BAC	Budget at Completion	Estimated total cost of project when completed.
5	EAC	Estimate at Completion	Expected total cost of project when completed.
6	ETC	Estimate to Completion	Expected additional cost needed to complete the project.
7	VAC	Variance at Completion	Amount over budget or under budget expected at the end of the project.

TABLE 4.15

Earned Value Management Formulas

Serial Number	Name	Formula	Interpretation
1	Cost Variance (CV)	BCWP–ACWP (EV–AC)	Positive result means under budget. Negative result means over budget.
2	Schedule Variance	BCWP–BCWS (EV–PV)	Positive result means ahead of schedule. Negative result means behind schedule.
3	Cost Performance Index (CPI)	EV/AC BCWP/ACWP	Greater than 1 means work is being produced for less than planned. Less than 1 means the work is costing more than planned.
4	Schedule Performance Index (SPI)	EV/PV BCWP/BCWS	Greater than 1 means project is ahead of schedule. Less than 1 means project has accomplished less than planned and is behind schedule.
5	Estimate at Completion (EAC)	AC+ETC	Expected total cost of project when completed.
6	Estimate to Complete (ETC)	EAC–AC	Expected additional cost needed to complete the project.
7	Variance at Completion (VAC)	BAC–EAC	Amount over budget or under budget expected at the end of the project.

TABLE 4.16

Causes of Variations (Changes)

Serial Number	Causes	
I	Owner	
	I-1	Delay in making the site available on time
	I-2	Change of plans
	I-3	Financial problems/payment delays
	I-4	Change of schedule
	I-5	Addition of work
	I-6	Omission of work
	I-7	Project objectives are not well defined
	I-8	Different site conditions
	I-9	Value engineering
II	Designer (Consultant)	
	II-1	Inadequate Specifications a. Design errors b. Omissions
	II-2	Scope of work not well defined
	II-3	Conflict between contract documents
	II-4	Coordination among different trades and services
	II-5	Design changes/modifications
	II-6	Introduction of latest technology
III	Contractor	
	III-1	Process/Methodology
	III-2	Substitution of material
	III-3	Nonavailability of specified material
	III-4	Charges payable to outside party due to cancellation of certain items/products
	III-5	Delay in approval
	III-6	Contractor's financial difficulties
	III-7	Unavailability of manpower
	III-8	Unavailability of equipment
	III-9	Material not meeting the specifications
	III-10	Workmanship not to the mark
IV	Miscellaneous	
	IV-1	New regulations
	IV-2	Safety considerations
	IV-3	Weather conditions
	IV-4	Unforeseen circumstances

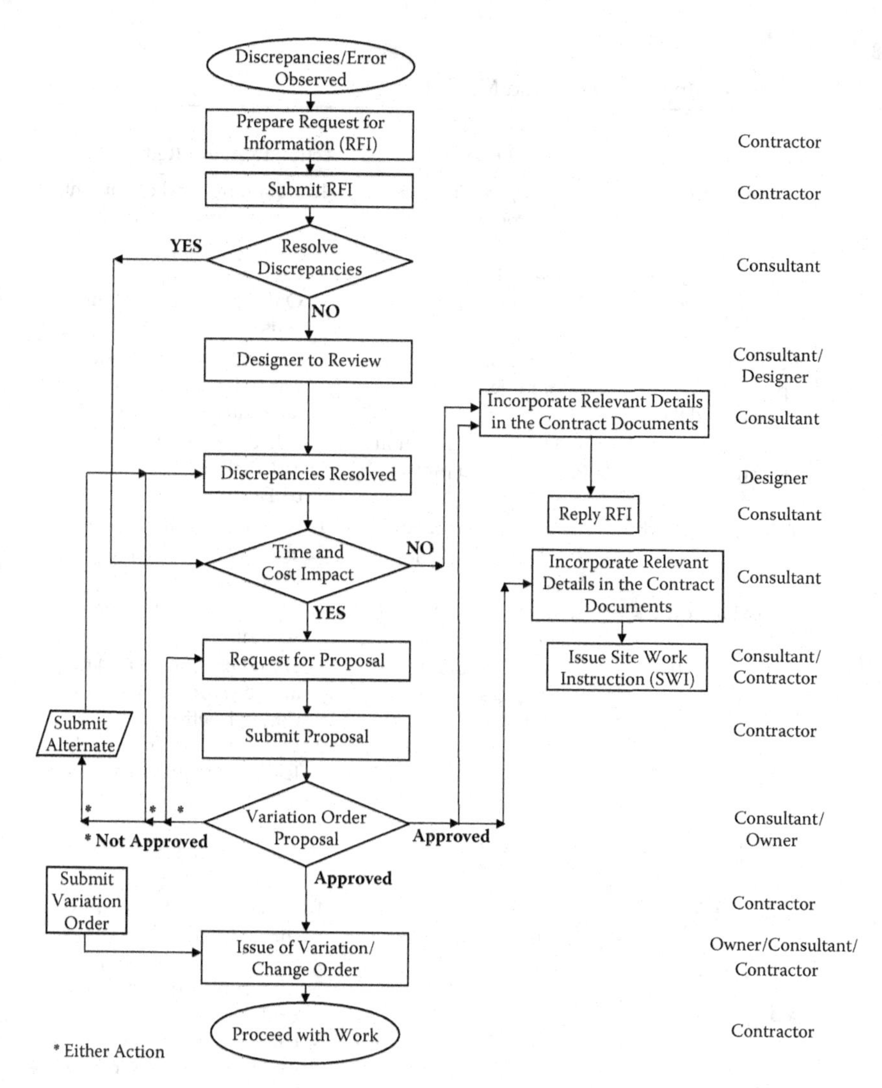

FIGURE 4.45
Flow chart for implementing change requests.

disruption of work and its consequent impact on cost and time. Figure 4.45 shows a flowchart to implement change requests, and Table 4.17 shows the effects of changes and their impact on construction projects.

4.8.6.2 Work Progress

The approved contractor's construction schedule is the performance baseline for construction projects and is achieved by collecting information through

TABLE 4.17

Major Causes for Changes, Effects, and Mitigation

Serial Number	Cause	Effect	Action to Avoid/Reduce Effects
1	Delay in making the site available on time	1. Delay in completion of project	1. Obtaining legal documents and title deeds in time
2	Change of plans	1. Delay in completion 2. Changes in project cost	1. Clearly defining the Change Order procedure for such cause
3	Financial problems/payment delays	1. Delay in completion of project	1. Owner to anticipate such risk and should have alternative solutions
4	Change of schedule	1. Delay in completion 2. Changes in project cost	1. Schedule compression: a. Crashing b. Fast tracking
5	Addition of work	1. Changes in schedule 2. Changes in project cost	1. Resource leveling 2. Team effort by all parties to keep to schedule
6	Omission of work	1. Changes in project cost	1. Practically no effect on schedule
7	Project objectives are not well defined	1. Delay in completion 2. Changes in project cost	1. Designer (Consultant) to get all the clarification during design briefing. 2. Designer to include all the technical requirements using their expertise 3. Clearly defining Change Order procedure for such cause
8	Different site conditions	1. Delay in completion 2. Changes in project cost	1. Conduct pre-design study through specialist consultant/contractor 2. Collect historical data prior to start of design 3. Clearly defining Change Order procedure for such cause
9	Value engineering	1. Delay in completion 2. Changes in project cost	1. It will optimize the value
10	Inadequate specifications a. Design errors b. Omissions	1. Delay in completion 2. Changes in project cost	1. Designer (Consultant) to ensure accuracy of design drawings, specifications, documents, and Bill of Quantity
11	Scope of work not well defined	1. Delay in completion 2. Changes in project cost	1. Designer to ensure accuracy in design documents 2. Clearly defining Change Order procedure for such cause

TABLE 4.17 (*Continued*)

Major Causes for Changes, Effects, and Mitigation

Serial Number	Cause	Effect	Action to Avoid/Reduce Effects
12	Conflict between contract documents	1. Changes in schedule	1. Clearly defining Change Order procedure for such cause
13	Coordination among different trades and services	1. Changes in schedule	1. All participants to resolve issue amicably
14	Design changes/ modifications	1. Delay in completion 2. Changes in project cost	1. Resolve issue amicably without any effect on schedule
15	Introduction of latest technology	1. Delay in completion 2. Changes in project cost	1. Resolve issue amicably without any effect on schedule 2. Clearly defining Change Order procedure for such cause 3. Include provisional amount for such changes
16	Process/ Methodology	1. Changes in cost	1. Resolve issue amicably without any effect on schedule and cost
17	Substitution of material	1. Changes in cost	1. Resolve issue amicably without any effect on schedule 2. Clearly defining Change Order procedure for cost effect
18	Nonavailability of specified material	1. Changes in cost	1. Designer to consider availability of specified material during construction phase. 2. Contractor can propose alternate equal material
19	Charges payable to outside party due to cancellation of certain items/ products	1. Changes in cost	1. Clearly defining Change Order procedure for such cause
20	Delay in approval	1. Changes in schedule	1. Resolve issue amicably without any effect on schedule

(*Continued*)

TABLE 4.17 (*Continued*)

Major Causes for Changes, Effects, and Mitigation

Serial Number	Cause	Effect	Action to Avoid/Reduce Effects
21	Contractor's financial difficulties	1. Delay in completion	1. Contractor to make work progress as scheduled to claim progress payment 2. Contractor should anticipate work delays and reserve funds for contingency
22	Unavailability of manpower	1. Delay in completion	1. Utilize available workforce by working overtime
23	Unavailability of equipment	1. Delay in completion	1. Utilize available equipment by working extra time 2. Try to hire equipment from other sources
24	Material not meeting the specifications	1. Changes in cost + OR –	1. Propose alternative/ substitute 2. Clearly defining Change Order procedure for such cause
25	Workmanship not up to the mark	1. Delay in completion 2. Cost of rework	1. Provide training
26	New regulations	1. Changes in schedule 2. Changes in cost	1. Clearly defining Change Order procedure for such cause
27	Safety considerations	1. Delay in completion	1. Provide training
28	Weather conditions	1. Delay in completion 2. Cost impact	1. Clearly defining Change Order procedure for such cause
29	Unforeseen circumstances	1. Delay in completion 2. Cost impact	1. Clearly defining Change Order procedure for such cause

different methods. Work progress is monitored through various types of logs and S-curves.

4.8.6.2.1 Logs

There are various types of logs used in construction projects to monitoring and control construction activities. The six main logs used in a construction project are as follows:

1. Subcontractors Submittal & Approval Log
2. Submittal Status Log
3. Shop Drawings and Materials Logs—E1
4. Procurement Log—E2

5. Equipment Log
6. Manpower Logs

These logs provide necessary information about the status of subcontractors, materials, shop drawings, procurement, and availability of contractor's resources and help determine its effects on project schedule and project completion.

4.8.6.2.2 S-Curves

Figure 4.46 illustrates the S-curve for actual work versus planned work.

4.8.6.2.3 Reports

4.8.6.2.3.1 Progress Reports Apart from different types of logs and submittals, progress curves, and time control charts, the contractor's progress is monitored through various types of reports and meetings:

- Daily report
- Weekly report
- Monthly report

A monthly report giving details of all the site activities along with photographs is submitted by the contractor to the consultant/owner to update them about the progress of work during the month. Table 4.18 shows a sample monthly progress report.

4.8.6.2.3.2 Safety Report The contractor submits a safety report every month listing important activities with photographs.

4.8.6.2.3.3 Risk Report The contractor identifies occurrence of new risks and report them if they will have any effect on project progress and performance.

4.8.6.2.4 Meetings

4.8.6.2.4.1 Progress Meetings Progress meetings are conducted at agreed-upon intervals to review the progress of work and discuss the impediments, if any, to smooth progress of construction activities. The contractor submits a pre-meeting submittal to the project manager/consultant normally two days in advance of the scheduled meeting date. The submittal consists of

1. List of completed activities
2. List of current activities
3. Two-week look ahead
4. Critical activities
5. Materials submittal log

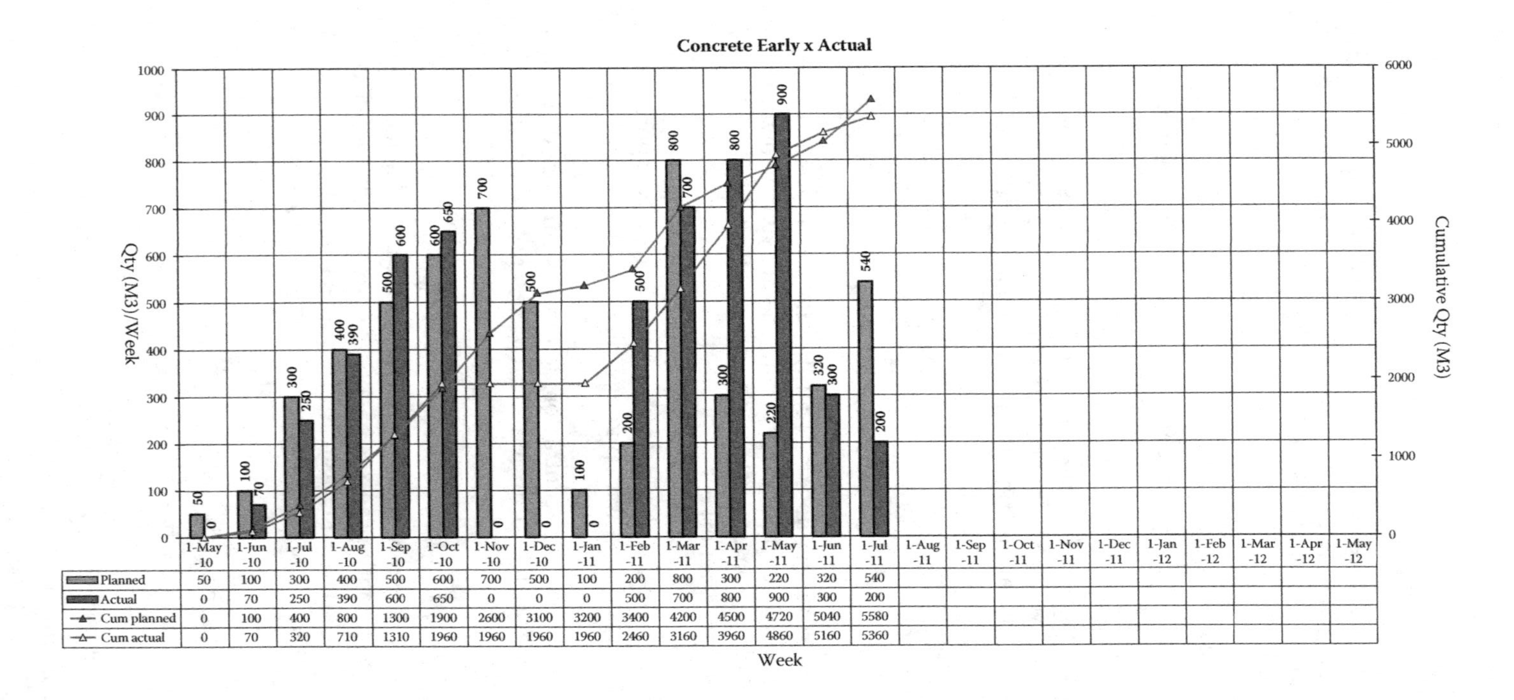

FIGURE 4.46
S-curve for actual versus planned.

TABLE 4.18

Monthly Progress Report

Contents of Monthly Report			
Serial Number		**Contents**	**Description**
1	Executive Summary—Tabular		
	1.1	Summary Status Report	Brief description of the Project Status to date, i.e., manpower, cash, activities
	1.2	NOC Report	No Objection Certificate Report
	1.3	Project Manager's Narrative	Narrative description of Project Status to date
2	Progress Layouts		
	2.1	Updated Milestone Table	Comparison of planned vs. actual for contractual milestones per construction unit\design. Delays for major trades tracked through using color theme.
	2.2	Updated Major of Events Table	Comparison of planned vs. actual for major trades per construction unit\design. Delays for major trades tracked through color theme.
	2.3	Updated Layouts	Same as (2.1) above but presentation per milestone phase (drawing)
3	Updated Execution Program		
	3.1	Updated Milestone Schedule Roll up "Update versus latest Target"	To indicate the status of the Control & Key Milestones comparing the current status with the baseline
	3.2	Updated Detailed Schedule	All activities in details
	3.3	One Month Look Ahead Program	Same as (3.2), but includes only the detailed activities for the coming month but in MS Excel format
4	Submittal Status Report (E1 Log)		Updated status of submittals
	4.1	Submittal Status Report	Briefly describe the Project Submittal Status to date
	4.2	Detailed E1 Log	Detailed description of the Project Submittal Status to date
5	Procurement Status Report (E2 Log)		Updated procurement Status
6	Status of Information Requested		
	6.1	RFI Status report	Request for Information (RFI)
	6.2	NCR Report	Nonconformance Request (NCR)
	6.3	PCO & NOV Log	Potential Change Order & Notice of Variation Summary

(*Continued*)

TABLE 4.18 (*Continued*)

Monthly Progress Report

Contents of Monthly Report		
Serial Number	**Contents**	**Description**
7	Updated Cost Loaded Program	
	7.1 T-7 Updated Status Report	Money Progress = Physical * Budgeted Cost for each running or completed activity
	7.2 Updated Cost Loaded Schedule	Same as T-7 but money values, not percentages
	7.3 Updated Work In Place (%) Report	Histogram & Cumulative Curve
	7.4 Updated Cash Flow Report	Histogram & Cumulative Curve
8	Updated Manpower Histogram	Histogram & Cumulative Curve
9	Updated Schedule of Construction Equipment & Vehicles	Tabular Report
10	Updated Critical Indicators	
	10.1 Shop Drawings Status Report	Histogram & Cumulative Curve
	10.2 Material Status Report	Histogram & Cumulative Curve
	10.3 Construction Leading Indicators	Each trade alone
	10.4 Line of Balance Diagram	Indicates all progress of the major trades as line cumulative chart
11	Updated Progress Photographs	—
12	Updated Safety Inspection Checklist	Tabular Report
13	Contractor Information	Organization Chart, Tabular Report

6. Shop drawings submittal log
7. Procurement log

Apart from the issues related to progress of works and programs, site safety and quality related matters are also discussed in these meetings. These meetings are normally attended by the owner's representative, designer/consultant staff, contractor's representative, and subcontractors' responsible personnel.

4.8.6.2.4.2 Coordination Meetings Coordination meetings are held from time to time to resolve coordination matters among various trades.

4.8.6.2.4.3 Quality Meetings Quality meetings are conducted to discuss on-site quality issues and improvements to the construction process to avoid/reduce rejection and rework.

4.8.6.2.4.4 Safety Meetings Safety meetings are held to discuss safety related matters.

Table 4.19 lists the points to be discussed during safety meetings.

The frequency of meetings is agreed to by all the parties at the beginning of the construction phase. Normally, the resident engineer prepares the agenda for the meeting and circulates it among all the participants. The contractor informs the resident engineer in advance about the points the contractor would like to discuss, which are included in the agenda.

The following are the main items included in the agenda:

- Type of meeting
- Date of meeting
- Time of meeting
- Place of meeting
- Points to be discussed

TABLE 4.19

Points to be Reviewed during Monthly Safety Meeting

Serial Number	Points to be Reviewed	Yes/No
1	Whether meetings were held as planned and attended by all the concerned persons	
2	Whether all the items from the previous meetings were addressed and appropriate actions were taken	
3	Whether all the accidents/incidents/near misses were recorded and reviewed to identify common issues and learning points	
4	Whether training needs were identified	
5	Whether records of training programs were maintained	
6	Whether actions/feedback from weekly meetings were reviewed and appropriate action was taken	
7	Whether suggestions/comments from safety tours were implemented	
8	Whether safety awareness programs were regularly held	
9	Whether safety warning signs are displayed and operative	
10	Whether sirens and alarm bells are functioning properly	
11	Whether temporary firefighting system is active and all the equipment is up to date	
12	Whether regular check and audits were performed	
13	If any new regulations were introduced by authority, whether all concerned have been informed	
14	Was there any visit by competent authority and were their observations acted upon	
15	Whether escape route and assembly points are displayed	
16	Whether the entire workforce has personal protective equipment	
17	Whether the first aid box has all the required medicines	

- Attendees
- Any other special requirements

The proceedings of meetings are recorded and circulated among all attendees and others per the approved responsiblity matrix. Figure 4.47 illustrates a sample form for minutes of a meeting.

4.8.6.2.5 Time Control

Completion of the construction project within the defined schedule is critical. The time control status is prepared in different formats to monitor the project completion time. Figure 4.48 illustrates the project progress status of a building construction project. This chart presents the overall picture of the elapsed period of the project, the remaining period of the scheduled project duration, and actual progress versus planned progress.

Project Name
Consultant Name
MINUTES OF MEETING

Contract No.:	**Date:**
Contractor:	

Meeting Type		MOM No.	
		Date	
Meeting Location		Time	

ITEM	DESCRIPTION OF DISCUSSION	STATUS	PRIORITY	ACTION			
				By	Due	Started on	Closed on

Prepared by:

Distribution Owner Contractor A/E

FIGURE 4.47
Minutes of meeting form.

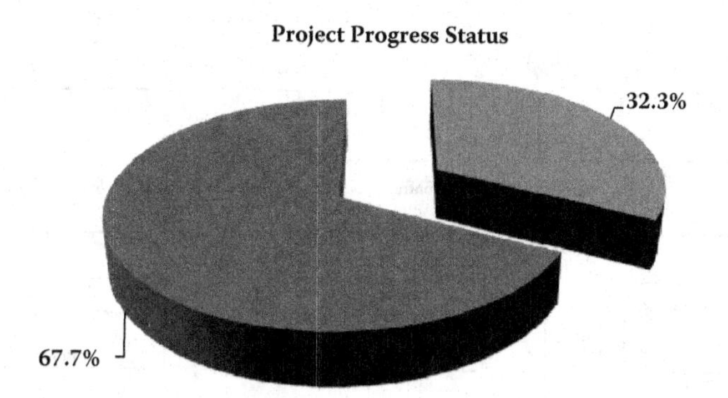

Time Elapsed-Days	413	32.30%
Time Remaining-Days	867	67.70%

Current Project Status			
	June 15, 2013		
Completion Period (Days)	1280		
Duration Elapsed (Days)	413		
% Elapsed Duration	32.30%		
Remaining Duration (Days)	867		
% Remaining Duration	67.70%		
Total Contract Amount (KD)	142,061,073.44		
Earned Value in KD to date	**Early**	**Late**	**Actual**
	31,494,939.98	25,613,611.54	19,422,904
Project Completion % till date	22.17%	18.03%	13.67%

FIGURE 4.48
Project progress status.

4.8.6.2.6 Digitized Monitoring of Work Progress

The work progress is monitored through daily and monthly progress reports. The monthly progress report consists of progress photographs to document the physical progress of work. These photographs are used to compare compliance with the planned activities and actual performance.

With the advent of technology, it is possible to monitor and evaluate construction activities using cameras and related software technologies. In this process, digital images are captured with cameras. These photographs are processed using a photo modeler software, and a 3D model view of the

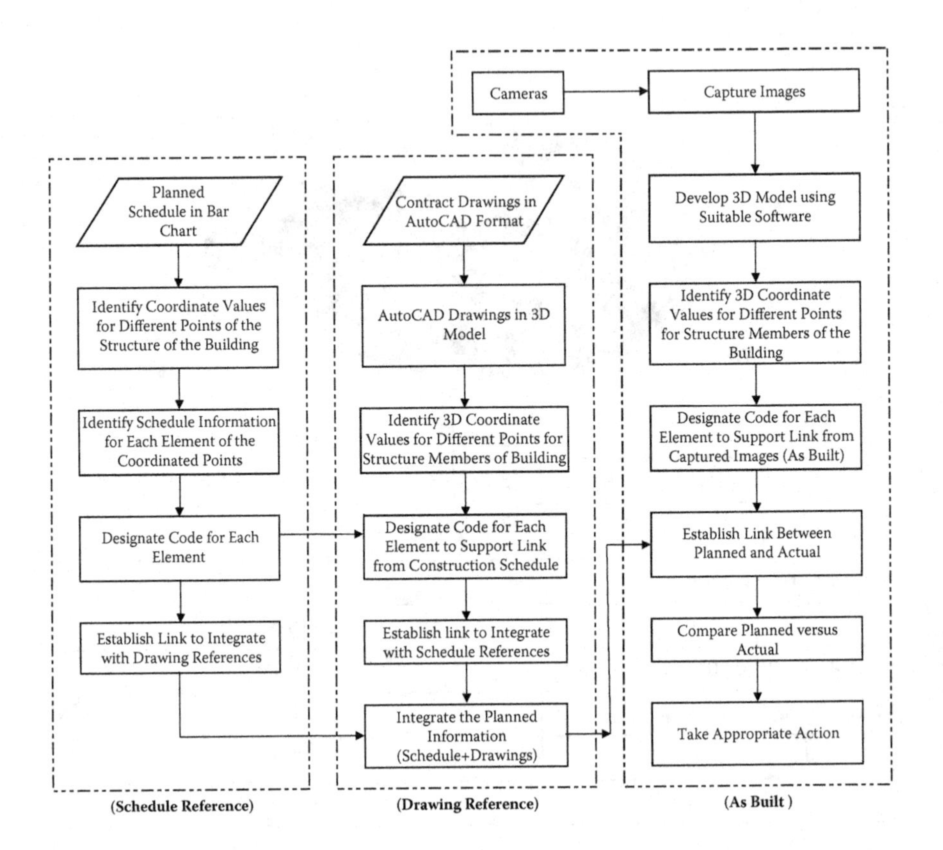

FIGURE 4.49
Digitized monitoring of progress.

digital picture captured from the site is developed. The captured as-built data is compared to the planned activities by interfacing through the Integrated Information Modeling System. The system is used to

- Improve the accuracy of information
- Avoid delays in getting the information
- Improve communication among all parties
- Improve effective control of the project
- Improve document recording
- Help reduce claims

Figure 4.49 illustrates a schematic for a digitized monitoring system.

4.8.6.2.7 Monitoring of Submittals

It is required that the contractor's submittals be processed and response sent within the specified period mentioned in the contract. The consultant

Project Name
Consultant Name

Contractor Name:
Week Start Date:

Contract No.:

Transmittal No.	Rev. No.	Transmittal Date	Date Received	Description	Document Reference	Responsible Engineer	Action Code	Reply Date	Action Pending	Remark/Reason for Pending

FIGURE 4.50
Submittal monitoring.

is required to maintain the log for material submittals, shop drawing submittals, and also all the correspondence between the contractor and client. Figure 4.50 illustrates an example log for monitoring material/shop drawing submittals, and Figure 4.51 illustrates the process for progress payment approval.

4.8.6.3 Quality Control

The construction project quality control process is a part of the contract documents, which provide details about specific quality practices, resources, and activities relevant to the project. The purpose of quality control during construction is to ensure that the work is accomplished in accordance with the requirements specified in the contract. Inspection of construction works is carried out throughout the construction period either by the construction supervision team (consultant) or the appointed inspection/auditing agency. Quality is an important aspect of construction projects. The quality of construction projects must meet the requirements specified in the contract documents. Normally, the contractor provides on-site inspection and testing facilities at the construction site. On a construction site, inspection and testing is carried out at three stages during the construction period to ensure quality compliance:

1. During the construction process. This is carried out with the checklist request submitted by the contractor for testing of ongoing works before proceeding to the next step.
2. Receipt of subcontracted or purchased material or services. This is triggered by a material Inspection Request submitted by the contractor to the consultant upon receipt of material. In certain cases,

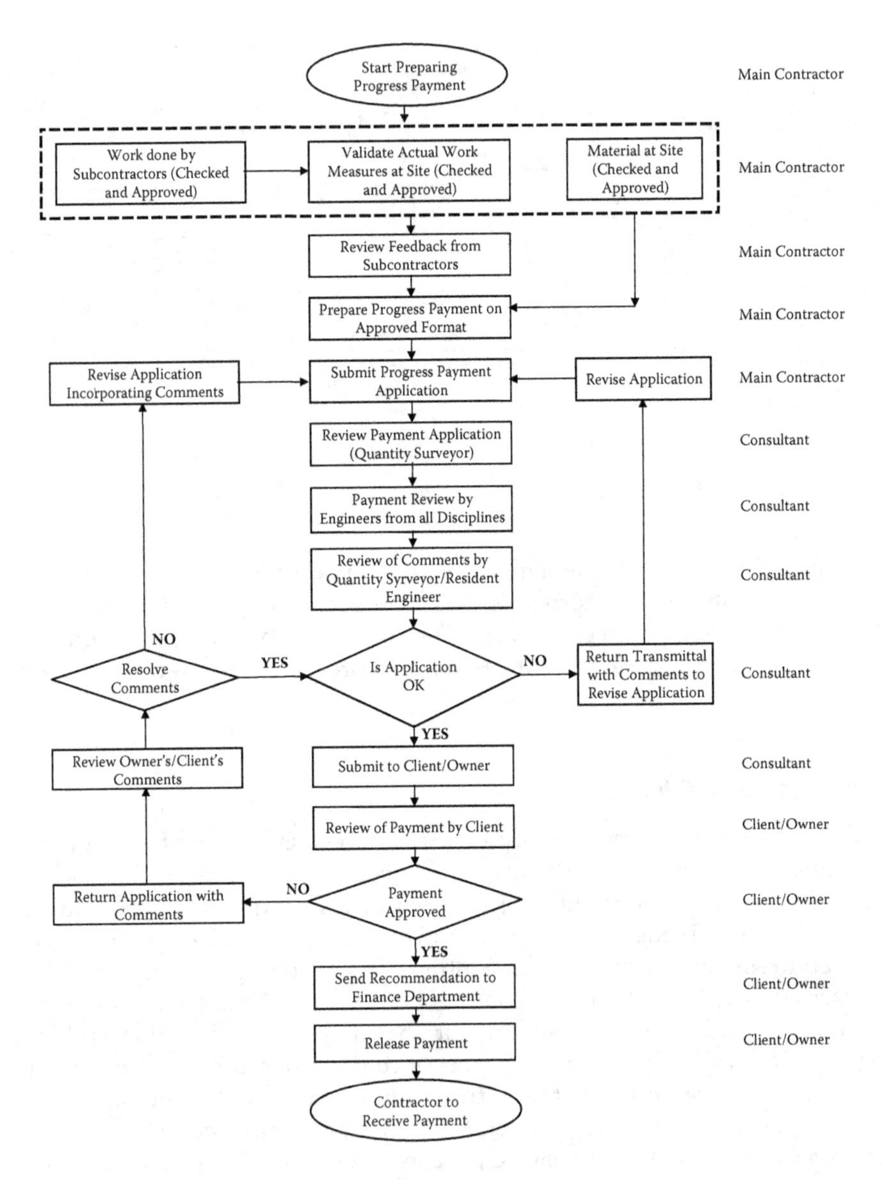

FIGURE 4.51
Progress payment approval process.

the inspection is carried out at the manufacturing premises/location of the product/system prior to dispatch of the material.

3. Before final delivery or commissioning and handover.

Quality management in construction is a management function. In general, quality assurance and control programs are used to monitor design and construction conformance to established requirements as determined by the contract specifications. Instituting quality management programs reduces costs while producing the specified facility.

Table 4.20 illustrates the responsibilities matrix for QC-related personnel.

Construction project documents usually specify details of the materials and workmanship required and also the type of inspection and tests to be performed to prove compliance. The contactor submits a quality control plan to achieve the required quality standards and objectives of the project.

With the commencement of the project, the activities related to mobilization such as setting up of site offices, stores, etc., starts, followed by excavation and installation of dewatering and shoring systems.

Quality in construction is achieved through proper control at every stage of execution and installation of works. The contractor should have complete knowledge of the project they have been contracted to construct. Safe and reliable construction should be the objective of all working on the project. The contractor has to submit shop drawings, materials, products, equipment, and systems to the owner/consultant for their review and approval before they can be deployed in the project. The procedure and requirements to submit shop drawings, materials, and samples are specified in detail under the Submittal section of contract specifications.

Quality of construction has mainly three elements:

1. Defined Scope
2. Budget
3. Schedule

The contractor has to comply with all the requirements specified in the contract documents and execute the works per the approved Quality Control Plan (QCP), which is the contractor's everyday tool to ensure quality.

During the construction process, the contractor has to submit the checklists to the consultant to inspect the works. Submission of checklists or requests for inspection are ongoing activities during the construction process to ensure proper quality control of construction. Concrete work is one of the most important components of building construction. The concrete work has to be inspected and checked at all the stages to avoid rejection or rework. Necessary care has to be taken right from the control of the design mix of the concrete until the casting is complete and cured. The contractor has to submit checklists at different stages of concrete work and has to perform certain tests, specified in the contract, during casting of concrete.

In certain projects, the owner asks the contractor to bring in an independent testing agency for quality control of concrete. The agency is responsible for quality control of materials, performing tests, and submitting test reports to the owner.

TABLE 4.20

Responsibility for Site Quality Control

		Linear Responsibility Chart							
		Owner	Consultant	Contractor					
Sr. No.	Description	Owner/ Project Manager	Consultant/ Designer	Contractor Manager	Quality Incharge	Quality Engineers	Site Engineers	Safety Officer	Head Office
1	Specify Quality Standards	□	■						
2	Prepare Quality Control Plan			□	■	□			□
3	Control Distribution of Plans and Specifications			□	■	□			□
4	Submittals			■	□		■		
5	Prepare Procurement Documents			□			■		■
6	Prepare Construction Method Procedures			□	□	□	■		
7	Inspect Work in Progress		■	■		□	■		
8	Accept Work in Progress		■						

9	Stop Work in Progress	■	□					
10	Inspect Materials Upon Receipt		■	□	■		■	
11	Monitor and Evaluate Quality of Works		■	□	■	■	■	
12	Maintain Quality Records			□	■			
13	Determine Disposition of Nonconforming Items	□	□	■				
14	Investigate Failures	□	□	■	■	□	■	
15	Site Safety			□				■
16	Testing and Commissioning	□	■	□			■	
17	Acceptance of Completed Works	■	□	□				

■ Primary Responsibility

□ Advise/Assist

Source: Abdul Razzak Rumane (2010), *Quality Management in Construction Projects*, CRC Press, Boca Raton, FL. Reprinted with permission from Taylor & Francis Group.

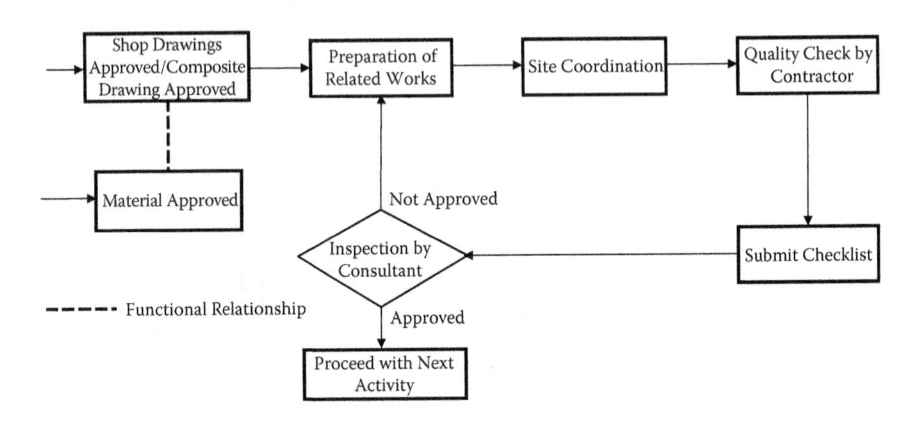

FIGURE 4.52
Sequence of execution of work. (*Source:* Abdul Razzak Rumane (2010). *Quality Management in Construction Projects.* CRC Press, Boca Raton, FL. With permission.)

The contractor has to submit checklists for all the installations and works executed per the contract documents and per the approved shop drawings and materials.

In order to achieve quality in construction projects, all the works have to be executed per the approved shop drawings using approved material and fully coordinating with different trades. Figure 4.52 illustrates the sequence of execution of work.

Following this sequence will help the contractor avoid rejection of work. Rejection of checklists will result in rework, which will consume the contractor's time and money. Frequent rejection of work may delay the project, ultimately affecting the overall completion schedule.

Table 4.21 shows the matrix of quality control areas of contractor and subcontractors.

4.8.6.3.1 *Cost of Quality during Execution of Work*

There are certain hidden costs which may not directly affect the overall cost of the project, however they may cost the contractor in terms of money and may affect the completion schedule of the project.

Rejection/nonapproval of executed/installed works by the supervisor owing to noncompliance with specifications will cause the contractor loss in terms of

- Material
- Manpower
- Time
- The contractor will have to rework or rectify the work, which will need additional resources and extra time to do the work as specified.

TABLE 4.21

Contractor's Quality Control Responsibilities

		Areas of Quality Control							
		Main Contractor		Subcontractors					
Sr. No.	Activity	Head Office/ Quality Manager	Project Site/ Project Manager	Structural	Interior	Mechanical (HVAC+Phff)	Electrical	Landscape	External
1	Prepare Quality Control Plan	□	■	□	□	□	□	□	□
2	Construction Schedule	□	■	□	□	□	□	□	□
3	Mobilization	□	■	□	□	□	□	□	□
4	Staff Approval	□	■	□	□	□	□	□	□
5	Prepare Material Submittal		■	■	■	■	■	■	■
6	Submit Material Transmittal		■	□	□	□	□	□	□
7	Prepare Shop Drawings		■	■	■	■	■	■	■
8	Submit Shop Drawing Transmittal		■	□	□	□	□	□	□
9	Material Sample	□	■	■	■	■	■	■	■
10	Receiving Material Inspection		■	■	■	■	■	■	■
11	Material Testing		■	■	■	■	■	■	■

(Continued)

TABLE 4.21 (*Continued*)

Contractor's Quality Control Responsibilities

		Areas of Quality Control							
		Main Contractor		Subcontractors					
Sr. No.	Activity	Head Office/ Quality Manager	Project Site/ Project Manager	Structural	Interior	Mechanical (HVAC+Phff)	Electrical	Landscape	External
12	Mockup		■	■	■	■	■	■	■
13	Site Work Inspection		■	■	■	■	■	■	■
14	Quality of Work		■	■	■	■	■	■	■
15	Prepare Checklist		■	■	■	■	■	■	■
16	Submit Checklist		■	□	□	□	□	□	□
17	Corrective/ Preventive Action		■	■	■	■	■	■	■
18	Daily Report		■	□	□	□	□	□	□
19	Monthly Progress Report		■	□	□	□	□	□	□

20	Progress Payment	□	■	□	□	□	□	□	□
21	Site Safety		■	■	■	■	■	■	■
22	Safety Report		■	□	□	□	□	□	□
23	Waste Disposal		■	■	■	■	■	■	■
24	Reply to Job Site Instruction		■	□	□	□	□	□	□
25	Reply to Nonconformance Report		■	□	□	□	□	□	□
26	Documentation		■	□	□	□	□	□	□
27	Testing and Commissioning		■	■	■	■	■	■	■
28	Project Closeout Documents	□	■	□	□	□	□	□	□
29	Punch List		■	■	■	■	■	■	■
30	Request for Issuance of Substantial Completion Letter	□	■	□	□	□	□	□	□

■ Primary Responsibility
□ Advise/Assist

- This may disturb the contractor's work schedule and affect execution of other activities. The contractor has to emphasize a zero defect policy, particularly for concrete works.

To avoid rejection of works, the contractor has to take the following measures:

1. Execution of works per approved shop drawings using approved material
2. Following approved method of statement or manufacturers' recommended method of installation
3. Conducting continuous inspection during the construction/installation process
4. Employing a properly trained workforce
5. Maintaining good workmanship
6. Identifying and correcting deficiencies before submitting the checklist for inspection and approval of work
7. Coordinating requirements of other trades, for example, if any opening is required in the concrete beam for crossing of a services pipe

Timely completion of the project is one of the objectives to be achieved. To avoid delay, proper planning and scheduling of construction activities is necessary. Since construction projects involve many participants, it is essential that requirements of all the participants be fully coordinated. This will ensure execution of activities as planned, resulting in timely completion of the project. Normally the construction budget is fixed at the inception of project; therefore, it is required to avoid variations during the construction process as it may take time to get approval for additional budget, resulting in project delay.

Categories of costs related to the construction phase can be summarized as follows:

Internal Failure Cost

- Rework
- Rectification
- Rejection of checklist
- Corrective action

External Failure Cost

- Breakdown of system
- Repairs
- Maintenance
- Warranty

Appraisal Cost

- Design review/preparation of shop drawings
- Preparation of composite/coordination drawings
- On-site material inspection/test
- Off-site material inspection/test
- Mockup
- Pre-checklist inspection
- Functionality tests

Prevention Cost

- Preventive action
- Training
- Work procedures
- Method statement
- Calibration of instruments/equipment
- Planning and managing the quality system

PDCA is a well-known tool for continual process improvement. The contractor may use principle of the PDCA cycle (Deming Wheel) to improve the construction/installation process and avoid rejection of work by the supervision team (consultant).

- PLAN
- Prepare Shop Drawings
- Submit Materials
- Get Shop Drawings Approved
- Get Materials Approved
- DO
- Follow Approved Shop Drawings
- Use Approved Material
- Follow Method of Statement
- Follow Manufacturer's Recommendation
- CHECK
- Check all Conformities against Specifications
- Check Work Executed per Approved Shop Drawings
- Check Approved Material Installed
- Check Functionality of Installed Works/Systems

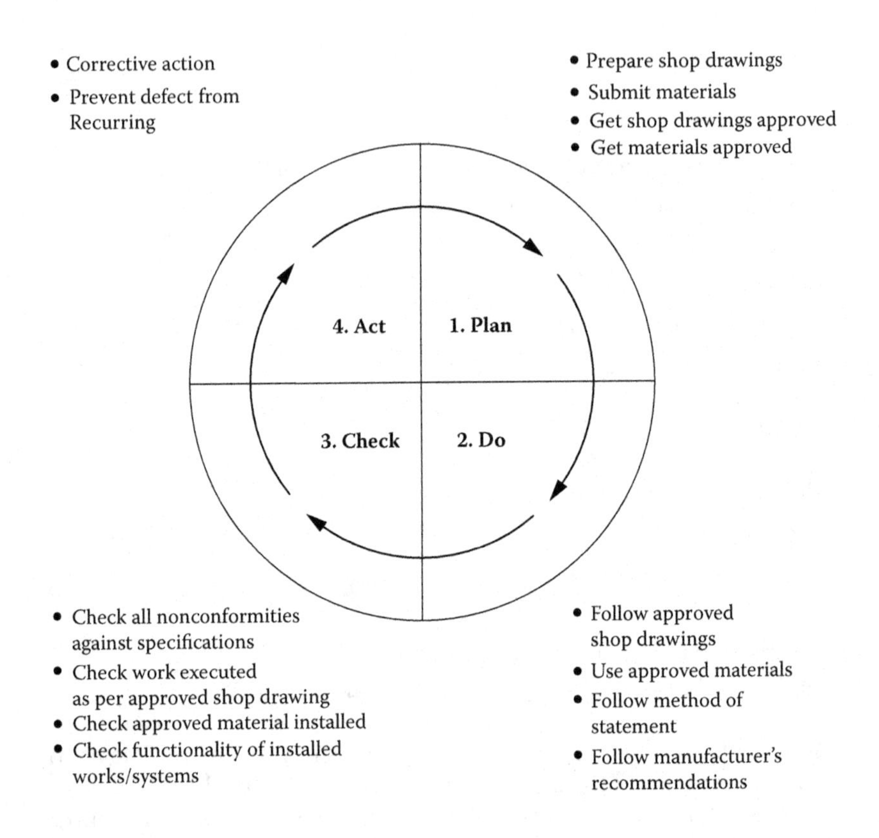

FIGURE 4.53
PDCA for execution of work.

- ACT
- Corrective Action
- Prevent Defect from Recurring

Figure 4.53 shows the PDCA cycle for execution/installation of works.

4.8.6.3.2 Six Sigma DMAIC Tool for Concrete Works

Concrete work is one of the most important components of building construction. In order to have appropriate process control of concrete, the contractor has to perform various tests at different stages:

1. Slump test
2. Air content and unit weight
3. Water/cementation ratio
4. Compressive strength test
5. Depth of concrete over uppermost reinforcement bar of top layer

The frequency of carrying out these tests has to be decided in advance per the contract document/standards being followed for the project. Necessary care has to be taken right from the control of the design mix of the concrete until the casting is complete and cured. The contractor has to inspect and check the concrete work at all the stages to avoid rejection or rework. The contractor has to submit checklists at different stages of concrete work and has to make certain that tests specified in the contract are performed during casting of concrete. The following are the seven checklists that are part of the contract documents which the contractor submits at different stages of the concreting process:

1. Checklist for Quality Control of Form Work
2. Notice for Daily Concrete Casting
3. Checklist for Concrete Casting
4. Checklist for Quality Control of Concreting
5. Report on Concrete Casting
6. Notice for Testing at Lab
7. Concrete Quality Control Form

In certain projects, the owner asks the contractor to involve an independent testing agency for quality control of concrete. The agency is responsible for quality control of materials, performs tests, and submits test reports to the owner.

Six Sigma is, basically, a process quality goal. It is a process quality technique that focuses on reducing variation in processes and preventing deficiencies in product. It is basically an improvement tool that uses a large amount of collected data for rigorous analysis to minimize variation and prevent deficiencies in product. Quality in construction is different from quality in the manufacturing or services industries as the product (project) is not repetitive but a unique piece of work with specific requirements. Construction projects do not pass through a series of processes where the output can be repeatedly monitored by inspection and tested at various stages of production and analyzed as is the case in manufacturing. The Six Sigma methodology involves utilizing rigorous data analysis to minimize variation in the process. It focuses on reducing process variation and preventing product defects.

Appendix C discusses the systematic approach of the Six Sigma methodology concept of team work, team roles, and the analytical tool DMAIC (Define, Measure, Analyze, Improve, Control) to improve the concrete casting process in construction projects.

4.8.6.4 Site Safety

The construction industry has been considered dangerous for a long time. The nature of work at the site always presents some dangers and hazards. There is a relatively high number of injuries and accidents at construction sites. Safety represents an important aspect of construction projects. Every project manager tries to ensure that the project is completed without major accidents on the site.

The construction site should be a safe place for those who are working there. Necessary measures are always required to ensure the safety of all those working at the construction site. Effective risk control strategies are necessary to reduce and prevent accidents.

Contract documents normally stipulate that the contractor, upon signing the contract, has to implement a safety and accident prevention program. It emphasizes that all the personnel have to put in efforts to prevent injuries and accidents. In the program, the contractor has to incorporate requirements of safety and health requirements of local authorities, manuals of accident prevention in construction, and all other local codes and regulations. The contractor has to also prepare an Emergency Evacuation Plan (EEP). The EEP is required to protect personnel and to reduce the number of fatalities in case of major accidents at the site. The evacuation routes have to be displayed at various locations in a prominent manner. Transfer points and gathering points have to be designated and sign boards are to be displayed all the time. Evacuation sirens have to be sounded on a regular basis in order to ensure smooth functioning of the evacuation plan.

A safety violation notice is issued to the contractor/employee if the contractor or any of his employees are not complying with safety requirements. Penalties are also imposed on the contractor for noncompliance with the site safety program. Figure 4.54 shows a sample disciplinary notice form for breach of safety rules. Different-colored cards may be issued along with the notice. Figure 4.55 depicts the principles underlying the issuance of different-colored cards.

The safety program shall encompass the prevention of accidents, injury, occupational illness, and property damage. The contract specifies that a safety officer is engaged by the contractor to follow safety measures. The safety officer is normally responsible for

1. Conducting safety meetings
2. Monitoring on-the-job safety
3. Inspecting the works and identifying hazardous areas
4. Initiating a safety awareness program
5. Ensuring the availability of first aid and emergency medical services per local codes and regulations
6. Ensuring that the personnel are using protective equipment such as hard hat, safety shoes, protective clothing, life belt, and protective eye coverings
7. Ensuring that the temporary firefighting system is working

PROJECT NAME	
CONSULTANT NAME	
Safety Disciplinary Notice	
Notice No.:	Date:
Name of Employee: Contractor Name: Area/Floor: Date & Time of Observance:	
Type of Notice	
☐	**First/Verbal Warning** (White Card)
☐	**Second/Written Warning** (Yellow Card)
☐	**Suspension from Site** (Red Card)
Reason for Issuance of Notice:	
Action Required by Recipient:	
Date by Which Action is Required:	
Issued by: Signature:	Date:
Reveived by:	Date:

CC: Owner ☐ Resident Engineer ☐ Project Manager ☐

FIGURE 4.54
Disciplinary notice form.

8. Ensuring that work areas and access areas are free from trash and hazardous material
9. Housekeeping

Figure 4.56 is a summary procedure to be followed once an accident takes place at the site.

Serial Number	Card Color	Type of Disciplinary Action	Warning Validity	Reasons for Disciplinary Action
1	White	Verbal followed by Safety Discipline Notice	One month to three months	1. Failure to use personal protective equipment 2. Failure to use defined access 3. Working on plant, crane, vehicle without license 4. Working with unsafe scaffolding 5. Working on unsafe platform 6. Using unsafe sling or ropes for lifting 7. Working on unsafe ladders
2	Yellow	Issuance of Safety Discipline Notice and suspension from work for rest of the day	Six months	1. Repetition of activities listed under "White Card" within one month of issuance of first notice 2. Failure to observe HSE-related instructions 3. Failure to work as per instructed method of work, as per given task 4. Failure to follow storage principles about hazardous materials
3	Red	Issuance of Safety Disciplinary Notice and suspension from site for one month	One year	1. Breach of safety rules where there is risk to life 2. Removal of safety devices, interlocks, guard rails, barriers without any authority 3. Deliberately exposing public to danger by not complying with agreed safe methods of work 4. Disposal of hazardous material in unsafe area

FIGURE 4.55
Concept of safety disciplinary action.

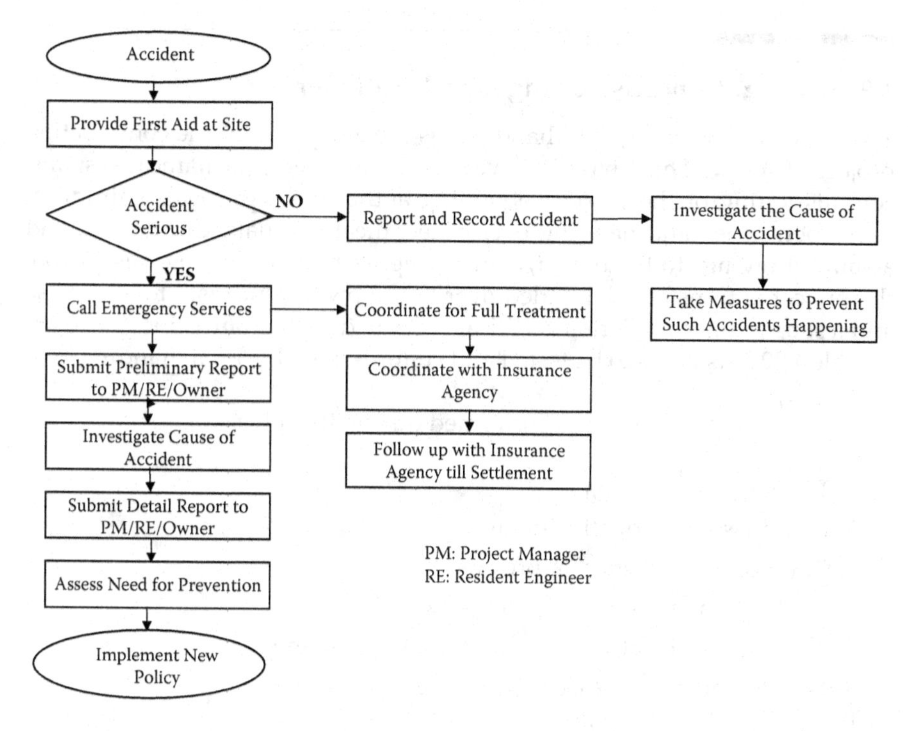

FIGURE 4.56
Summary procedure for actions after accident.

Construction sites have many hazards that can cause serious injuries and accidents. The contractor has to identify these areas and ensure that all site personnel and subcontractor employees working at the site are aware of unsafe acts, potential and actual hazards, and the immediate corrective action to be taken, and that they adhere to the safety plan and procedures.

4.8.7 Inspection

The inspection of construction works is performed throughout the execution of the project. Inspection is an ongoing activity to physically check the installed works. Checklists are submitted by the contractor to the consultant, who inspects the executed works/installations. If the work is not carried out as specified, then it is rejected, and the contractor has to rework or rectify the same to ensure compliance with the specifications. During construction, all the physical and mechanical activities are accomplished on site. The contractor carries out the final inspection of the works to ensure full compliance with the contract documents.

4.9 Testing, Commissioning, and Handover

Testing, commissioning, and handover is the last phase of the construction project life cycle. This phase involves testing of electromechanical systems, commissioning of the project, obtaining authorities' approval, training of user's personnel, and handing over of technical manuals, documents, and as-built drawings to the owner/owner's representative. During this period, the project is transferred/handed over to the owner/end user for their use and a substantial completion certificate is issued to the contractor.

Table 4.22 lists items to be tested and commissioned prior to handing over of the project.

The following activities are performed during this phase:

1. Testing of all systems
2. Commissioning of all systems
3. Obtaining authorities' approvals
4. Submission of as-built drawings
5. Submission of technical manuals and documents
6. Training of owner's/user's personnel
7. Handover of spare parts
8. Handover of facility to owner/end user
9. Occupancy of new facility
10. Preparation of punch list
11. Issuance of substantial certificate
12. Lessons learned

Table 4.23 shows a sample form for preparation of the punch list, and Figure 4.57 lists project closeout documents to be submitted for project closeout. Table 4.24 lists the responsibilities of the consultant during the closeout phase.

4.10 Operation and Maintenance

Once the project is completed and fully functional, it then becomes the part of the operation system. The operation and maintenance (O&M) of a completed project is associated with the quality of the constructed project.

The owner may employ the same contractors who built the facility or engage another specialist operation and maintenance organization to manage the operation and maintenance. It is likely that the owner will have his

TABLE 4.22

Major Items for Testing and Commissioning of Equipment

Serial Number	Discipline	Items
1	HVAC	1. Pipe cleaning and flushing 2. Chemical treatment 3. Pumps 4. Duct work 5. Air handling unit 6. Heat recovery unit 7. Split units 8. Chillers 9. Cooling towers 10. Heating system (controls, piping, pumps) 11. Fans (ventilation, exhaust) 12. Humidifiers 13. Starters 14. Variable frequency drive 15. Motor control centers (MCC panels) 16. Chiller control panels 17. Building management system (BMS) 18. Interface with fire alarm system 19. Thermostat 20. Air Balancing
2.1	Mechanical (Public Health)	1. Piping 2. Pipe Flushing and Cleaning 3. Pumps 4. Boilers 5. Hot Water System 6. Water Supply and Purity 7. Fixtures 8. Power Supply for Equipment 9. Controls 10. Drainage System 11. Irrigation System
2.2	Mechanical (Fire Suppression)	1. Sprinklers 2. Piping 3. Fire Pumps 4. Power Supply and Controls 5. Emergency Power Supply for Fire Pumps 6. Hydrants 7. Hose Reels 8. Water Storage Facility 9. Gaseous Protection System for Communication Rooms 10. Fire Protection System for Diesel Generator Room 11. Interface with Fire Alarm System 12. Interface with BMS
3.1	Electrical (Power)	1. Lighting Illumination Levels 2. Working of Photo Cells and Controls 3. Wiring Devices (Power Sockets)

(*Continued*)

TABLE 4.22 (*Continued*)

Major Items for Testing and Commissioning of Equipment

Serial Number	Discipline	Items
		4. Lighting Control Panels 5. Electrical Distribution Boards 6. Electrical Bus Duct System 7. Main Switch Boards/Sub Main Switch Boards 8. Main Low Tension Panels 9. Isolators 10. Emergency Switch Boards 11. Motor Control Centers (MCC Panels) 12. Audio Visual Alarm Panel 13. Diesel Generator 14. Automatic Transfer Switch (ATS) 15. UPS (Uninterrupted Power Supply) 16. Earthing (Grounding) System 17. Lightning Protection System 18. Surge Protection System 19. Power Supply to Equipment (HVAC, Mechanical, Elevators, Others) 20. IP Rating of Outdoor Switches, Isolators, Switch Boards 21. Interface with BMS
3.2	Electrical (Low Voltage)	1. Fire Alarm System 2. Communication System 3. CCTV System 4. Access Control System 5. Public Address System 6. Audio Visual System 7. Master Satellite Antenna System
4	Elevator	1. Power Supply 2. Speed 3. Capacity to Carry Design Load 4. Number of Stops 5. Emergency Landing 6. Emergency Call System 7. Elevator Management System 8. Interface with Fire Alarm System
5	External Works	1. Lighting Poles 2. Boundary Wall Lighting 3. Lighting Bollards 4. Irrigation System 5. Electrical Distribution Boards
6	General	1. Power Supply for Gate Barriers 2. Automatic Gates 3. Rolling Shutters 4. Window Cleaning System 5. Gas Detection System 6. Water/Fluid Leak Detection System

TABLE 4.23

Punch List

OWNER NAME		
PROJECT NAME		
PUNCH LIST		
Punch List Number:		Date:
Building Name		Floor
Area/Zone		Room Number
Serial Number	**Item**	**Remark**
1	Flooring	
2	Skirting	
3	Walls	
4	Ceiling	
5	Door	
6	Windows	
7	Door Hardware	
8	Window Hardware	
9	Electrical Light Fixture	
10	Wiring Devices (Sockets)	
11	HVAC Grill	
12	HVAC Diffuser	
13	Thermostat	
14	Sprinklers	
15	Fire Alarm Detectors	
16	Speakers	
17	Communication System Devices	
18	TV Outlets	
19	Any Other Item	

own team to operate and maintain the facility. The following documents are required to operate and maintain the completed facility. These documents are normally submitted by the contractor during the testing, commissioning, and handover phase of the construction project.

- As-built drawings
- Operation and maintenance manuals
- Operating procedures
- Source information for availability of spare parts of the installed works

The cost of operation and maintenance should be considered in the feasibility studies of the project. Post-construction costs have a great impact on the overall cost of the life-cycle cost of the project.

Description	As-Built Drawings	Operation and Maintenance	Government Authorities	Guarantees	Warranties	Spare Parts	Test Certificates	Samples	Record Documents	Testing and Commissioning	Punch Lists	Final Cleaning	Training	Taking Over Certificate	Remarks
ARCHITECTURAL WORKS															
CIVIL WORKS															
MECHANICAL WORKS															
HVAC WORKS															
ELECTRICAL WORKS (LIGHT & POWER)															
ELECTRICAL WORKS (LOW VOLTAGE)															
FINISHES															
EXTERNAL WORKS															

FIGURE 4.57
Project closeout documents.

4.10.1 Categories of Maintenance

The following are the different types of maintenance normally being carried out, depending on the nature of the activity, system, or equipment, to avoid interference/interruption with essential installation operation, danger to life or property, or involving high cost for replacement of an item.

4.10.1.1 Preventive Maintenance

Preventive maintenance is the planned, scheduled periodic inspection, adjustment, cleaning, lubrication, and/or parts replacement of equipment and systems in operation. Its aim is prevention of breakdown and failure. Preventive maintenance consists of many activities required to be checked and performed to ensure that the operations are safe and can be performed without any danger to life or facility, and that the operations will not warrant high cost or long lead time for replacement. Preventive maintenance is the cornerstone of any good maintenance program.

4.10.1.2 Scheduled Maintenance

Scheduled maintenance includes those maintenance tasks whose cycle exceeds a specified period. This type of maintenance relates mainly to

TABLE 4.24

Typical Responsibilities of Consultant during Project Closeout Phase

Serial Number	Responsibilities
1	Ensure that occupancy permit from respective authorities is obtained
2	Ensure that all the systems are functioning and operative
3	Ensure that Job Site Instruction (JSI) and Nonconformance Report (NCR) are closed
4	Ensure that site is cleaned and all the temporary facilities and utilities are removed
5	Ensure that master keys are handed over to the owner/end user
6	Ensure that guarantees, warrantees, bonds are handed over to the client
7	Ensure that operation and maintenance manuals are handed over to the client
8	Ensure that test reports, test certificates, inspection reports are handed over to the client
9	Ensure as-built drawings are handed over to the client/end user
10	Ensure that spare parts are handed over to the client
11	Ensure that the snag list is prepared and handed over to the client
12	Ensure that training for client/end user personnel is completed
13	Ensure that all the dues of suppliers, subcontractors, contractor are paid
14	Ensure that retention money is released
15	Ensure that substantial completion certificate is issued and maintenance period commissioned
16	Ensure that the supervision completion certificate from the owner is obtained
17	Ensure that lessons learned are documented

finishing items, such as paintings, polishing of woodwork, parking and road markings, roof maintenance, and testing of the fire alarm system.

4.10.1.3 Breakdown Maintenance

Breakdown maintenance for systems or equipment is unscheduled and unanticipated because of system or component failure. If the problems relate to an essential service or involve any hazardous effect, then an emergency response is necessary. If the problem is not critical, then the routine response is adequate.

4.10.1.4 Routine Maintenance

Routine maintenance is an action taken by the maintenance team to restore a system or piece of equipment to its original performance, capacity, efficiency, or capability.

4.10.1.5 Replacement of Obsolete Items

Replacement of obsolete items refers to the work undertaken by the maintenance team to bring a system or component into compliance with new codes of safety regulations or to replace an item that is unacceptable, inefficient, for which spare parts are no longer available, or that has become obsolete because of technological changes. Early detection of problems may reduce repair and replacement costs, prevent malfunctioning, and minimize downtime. For example, communication system, audio visual system, and equipment need regular checks, replacement of obsolete items, and a system update.

4.10.1.6 *Predictive Testing and Inspection*

Predictive testing and inspection refers to testing and inspection activities that involve the use of sophisticated means to identify maintenance requirements. For example, specialized tests are to be carried out

- To locate wear problems on bus bar bends of an electrical main low tension panel
- To detect insulation cracks
- To locate thinning of pipe walls and cracks
- To identify vibration problems for equipment such as chillers, diesel generator sets, and pumps
- To locate heat buildup (rise in temperature) in an electrical bus duct

4.10.2 O&M Program

Addressing operation and maintenance (O&M) considerations at the start of the project contributes greatly to improved working environments, higher productivity, and reduced energy and resource costs. During the design phase of the project and up to handover of the facility, O&M personnel should be involved in identifying maintenance requirements for inclusion in the design. The goal of effective O&M is to realize the intent of the original design team, that is, the systems, equipment deliver services to the user to feel comfortable, healthy, and safe. O&M should also include long-term goals such as economy, energy conservation, and environmental protection. To create an effective O&M program, the following procedures should be followed:

- Ensure that up-to-date as-built drawings for all the systems are available.
- Ensure that operational procedures and manuals for the installed equipment are available.
- Prepare a master schedule for operation and preventive/predictive maintenance.
- Implement preventive maintenance programs complete with maintenance schedules and records of all maintenance performed for all the equipment and systems installed in the project/facility.

- Follow the manufacturers' recommendations for maintenance of equipment and systems.
- Ensure that the maintenance personnel have full knowledge of the installed equipment and systems that he or she is responsible for operating and maintaining.
- Ensure that O&M personnel are provided training during the testing, commissioning, and handover phase.
- Offer training and professional development opportunities for each O&M team member.
- Implement a monitoring program that tracks and documents equipment performance to identify and diagnose potential problems.
- Implement a monitoring program that tracks and documents systems performance to identify and diagnose potential problems.
- Perform predictive testing and inspection for critical and important items.
- Use preventive maintenance and standbys, etc., so that the failed components can be isolated and repaired without interrupting system performance, thus minimizing equipment failures.

For an effective O&M program, the following seven specific items should be considered:

1. HVAC system and equipment
2. Indoor air quality system and equipment
3. Electrical systems and equipment
4. Water fixtures and system
5. Life safety (fire suppression systems)
6. Cleaning equipment and products
7. Landscape maintenance

4.11 Assessment of Quality

In today's global competitive environment characterized by an ever-increasing customer demand for higher performance requirements, organizations are facing many challenges. They are finding that their survival in the competitive market is increasingly in doubt. To achieve competitive advantage, effective quality improvement is critical for an organization's growth. This can be achieved through development of long-range strategies for quality. The assessment of an existing quality system that is being implemented provides the factual basis for developing long-term organizational strategies for quality.

4.11.1 Assessment Categories

Assessment of quality comprises

- Cost of poor quality
- The organization's standing in the marketplace
- The quality culture in the organization
- Operation of the quality management system

4.11.1.1 Cost of Poor Quality

This assessment is mainly to find out

a. Customer dissatisfaction due to product failure after sales
b. Annual monetary loss due to products and processes that are not achieving their objectives

In order to measure customer satisfaction, an independent product survey can be carried out to discover the reasons and improvement measures to be adopted. Such an assessment can be done through a questionnaire related to the quality of work (workmanship) and functioning, and getting the customer's opinion/reaction.

The management can use the the information from the questionnaire for quality improvement. All necessary improvements, if required, for proper functioning of the installed system shall be carried out to maintain customer satisfaction. Normally, the contractor is required to maintain the completed project for a certain period under the maintenance contact, and therefore necessary modifications can be carried out during this period. Such defects/modifications have to be recorded by the organization as feedback information, and necessary measures can be taken to avoid a repetition in other projects. The cost of carrying out such repairs/modifications is borne by the contractor, which helps improve customer satisfaction.

Poor quality results in higher maintenance and warranty charges. In construction projects, these costs are mainly due to

- Rejection of works during construction: This results in rework by the contractor to meet the specification requirements. If any extra time and cost is required to redo the work, the contractor is not compensated for the same.

- Rejection of shop drawings: Repeated rejection of shop drawings results in the contractor reproducing the same by spending extra time and cost and at the same time delaying the start of relevant activities.

Preventive measures are required for overcoming these problems. Necessary data can be collected to analyze the reasons for rejections. With a cause and effect diagram, reasons for rejection can be analyzed, and necessary improvements can be implemented.

4.11.1.2 Organization's Standing in the Marketplace

The company's standing in the marketplace determines the business opportunities for the organization. The organization needs to know where it stands regarding quality in the marketplace. This can be done through a market study. A number of questions related to the product should be considered for such studies, which can be carried out by the marketing department of the company. The input from the study should be considered for quality improvement. Various quality tools can be used to analyze the situation, and system can be improved taking into consideration the analysis results.

4.11.1.3 Quality Cultures in the Organization

The concept of quality culture has changed during the past three decades. The quality approach is centered on organizationwide participation aimed at long-term success through customer satisfaction. It is focused on participative management and strong operational accountability at the individual contribution level.

Employees' opinions and understanding of the company's products, as well as implementation of a quality system, help in developing a long-term strategy based on companywide assessment of quality. This can be achieved through focus group discussions, by using questionnaires to survey employees' opinions, or by doing both.

Listed below are examples of seven questions that can be used by building construction organizations seeking feedback from their employees:

1. Do you know what is a good quality level for this project?
2. Are you aware of the company's emphasis on quality?
3. Are you familiar with project quality?
4. Do you know the inspection procedures for this project?
5. Do you know how you can provide high-quality workmanship?
6. How you will do the remedial work?
7. How can you improve project quality?

4.11.1.4 Operation of Quality Management System

This element of assessment is related to the evaluation of present quality management system-related activities. Assessment of present quality activities can be evaluated from two perspectives:

1. Assessment that focuses on customer satisfaction results but includes an evaluation of the current quality system
2. Assessment that focuses on evaluation of the current quality system with little emphasis on customer satisfaction results

In either case, the assessment can be performed by the organization itself or by an external body. The assessment performed by the organization is known as self-assessment, whereas, an assessment performed by an external body is referred to as a quality audit. A quality audit is a formal or methodical examination, review, and investigation of the existing system to determine whether agreed-upon requirements are being met.

Audits are mainly classifieds as

First Party—Audit your own organization (internal audit)

Second Party—Customer audits the supplier

Third Party—Audits performed by an independent audit organization

Third party audits may result in independent certification of a product, process, or system such as ISO9000 quality management system certification. Third party certification enhances an organization's image in business circles. The assessment provides feedback to the management on the adequate implementation and effectiveness of the quality system.

4.11.2 Self-Assessment

Self-assessment is a comprehensive, systematic, and regular review of an organization's activities, the results of which are referenced against a specific model. The goal of this assessment is to identify:

- What your organization is doing well
- What it is not doing well
- What it is not doing at all
- Where and how it can make measurable improvements

Appendix A: Application of Six Sigma DMADV Tool to Develop Designs for Construction Projects (Buildings)

A1.0 Introduction

Six Sigma is, basically, a process quality goal. It is a process quality technique that focuses on reducing variation in processes and preventing deficiencies in a product. In a process that has achieved Six Sigma capability, the variation is small compared to the specification limits. Six Sigma started as a defect reduction effort in manufacturing and was then applied to other business processes for the same purpose.

Six Sigma is a measurement of "goodness" using a universal measurement scale. Sigma provides a relative way to measure improvement. Universal implies that sigma can measure anything from coffee mug defects to missed chances of closing a sales deal. It simply measures how many times a customer's requirements were not met (a defect), given a million opportunities. Sigma is measured in defects per million opportunities (DPMO).

Six Sigma methodology focuses on

- Leadership principles
- Integrated approach to improvement
- Engaged teams
- Analytic tools
- Hard-coded improvements

Six Sigma methodology has four leadership principles. These are

1. Align
2. Mobilize
3. Accelerate
4. Govern

Teamwork is absolutely vital for complex Six Sigma projects. For teams to be effective, they must be engaged, involved, focused, and committed to meeting their goals. Engaged teams must have leadership support.

There are four types of teams. These are

1. Black Belts
2. Green Belts
3. Breakthrough
4. Blitz

For typical Six Sigma projects, four critical roles exist

1. Sponsor
2. Champion
3. Team leader
4. Team member

The following are the analytic tools used in Six Sigma projects:

1. Ford Global 8D Tool
   - Ford Global 8D is primarily used to bring performance back to a previous level.
2. DMADV
   - DMADV is primarily used for the invention and innovation of modified or new products, services, or processes.
3. DMAIC
   - DMAIC is primarily used for the improvement of an existing product, service, or process.
4. DMADDD
   - DMADDD is primarily used to drive the cost out of a process by digitization improvements.

A2.0 Construction Projects

Construction projects are unique and nonrepetitive in nature and need specific attention to maintain quality. To a great extent, each project has to be designed and built to serve a specific need. It is the owner's or designer's (consultant) responsibility to prepare the overall project documentation for converting the owner's conception and need into a specific facility with detailed direction through drawings and specifications adhering to the economic objectives.

In order to achieve a Zero Defect policy during the construction phase, the designer (consultant) has to develop project documents to ensure

- Conformance to owner's requirements
- Compliance to standards and codes
- Compliance with regulatory requirements
- Great accuracy to avoid any disruption/stoppage/delay of work during construction
- Completion within the stipulated time to avoid delay in starting construction

A3.0 Six Sigma in Construction Projects

The following is an example procedure to develop design for construction projects (buildings) using the Six Sigma DMADV analytic tool set.

DMADV stands for:

D → Define

M → Measure

A → Analyze

D → Design

V → Verify

DMADV Tool

1. Define → What is important?
2. Measure → What is needed?
3. Analyze → How will we fulfill?
4. Design → How do we build it?
5. Verify → How do we know it will work?

The DMADV tool has the following four uses:

1. The DMADV tool is used primarily for the invention and innovation of modified or new products, services, or processes.
2. The DMADV tool is used when a product, service, or process is required but does not exist.
3. DMADV is proactive, solving problems before they start. This tool is also called DFSS (Design for Six Sigma).
4. Using this toolset, Black Belts optimize performance before production begins.

Table A.1 illustrates the fundamental objectives of the DMADV tool.

TABLE A.1

Fundamental Objectives of Six Sigma DMADV Tool

DMADV	Phase	Fundamental Objective
1	**Define**—What is important?	Define the project goals and customer deliverables (internal and external)
2	**Measure**—What is needed?	Measure and determine customer needs and specifications
3	**Analyze**—How do we fulfill?	Analyze process options and prioritize based on capabilities to satisfy customer requirements
4	**Design**—How do we build it?	Design detailed processes capable of satisfying customer requirements
5	**Verify**—How do we know it will work?	Verify design performance capability

A4.0 Application of DMADV Tool in Design of Construction Projects

A4.0.1 Define Phase

What is important? (Define the project goals and customer deliverables.)

The key deliverables of this phase are

- Establish the goal.
- Identify the benefits.
- Select the project team.
- Develop the project plan.
- Develop the project charter.

A4.0.1.1 Goal

Develop construction project design using Six Sigma tool to meet the owner's needs and satisfy the project quality requirements.

A4.0.1.2 Benefits

The measurable benefits of adopting this process are:

- Minimize design errors.
- Minimize omissions.
- Reduce design rework.

- Reduce risk and liabilities.
- Construction project (building) design that will meet the owner's needs without rejection of documents by the owner's review team (Project Manager, Construction Manager) at the first submission itself and without affecting the construction process, thus reducing external failure cost.
- It will not affect the construction process, thus reducing external failure cost.
- Reduce construction overruns.
- Increase customer satisfaction.
- Improve reputation.
- Increase profits.

A4.0.1.3 Selection of Team

The team shall consist of

1. Sponsor—Project Manager
2. Champion:
   - Design Manager (Civil/Structural)
   - Quality Manager
3. Team Leader—Principal Engineer (Civil/Structural)
4. Team Members—Structural Design Engineer, CAD Technician, Quality Control Engineer, Quantity Surveyor, Cost Engineer, Planner, Owner's Representative, and End User

Similar teams can be organized for other trades such as architectural, HVAC, mechanical (P&FF), and electrical including low-voltage systems, landscape, and external works.

Figure A.1 illustrates the design management team and their major responsibilities.

A4.0.1.4 Project Plan

Time frame in the form of a Gantt chart shall be prepared to meet the target dates for

- Data collection and analysis
- Development of concept design
- Submission and approval of concept design for client review and comments

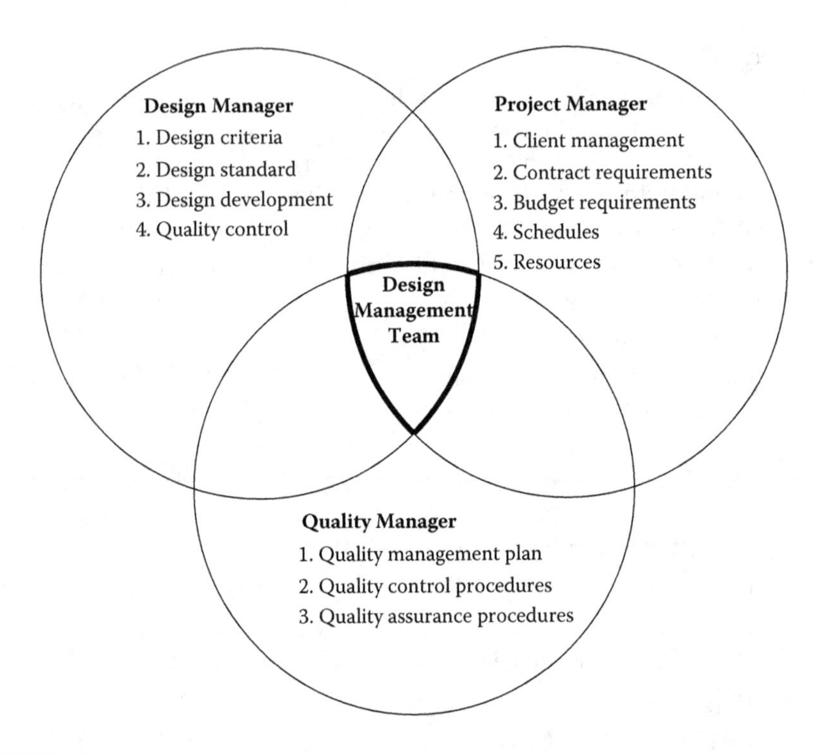

FIGURE A.1
Design management team.

- Development of schematic design
- Submission and approval of schematic design for client review and comments
- Development of detail design (design development)
- Regulatory/authority approvals
- Preparation of contract documents
- Submission of detail design and documents for client review and comments
- Submission of the project design and documents (final)
- Review and coordination meetings (both internal and with client)
- Monitoring of design schedule
- Monitoring and controlling resources

A4.0.1.5 Project Charter

In this example, the project objective is to create a construction project (building) design within the stipulated schedule that

- Will meet the owner's need.
- Will meet relevant standards, codes, regulatory requirements.
- Will be approved by the owner/owner's representative at the first submission without any major comments on the design and specification documents.
- The documents shall be accurate and shall not affect construction quality.

A4.0.2 Measure Phase

What is needed? (Measure and determine customer needs and specifications.)
The key deliverable in this phase are

1. Matrix of owner's requirements (Terms of Reference or TOR)—Scope of work
2. Codes and standards to be followed
3. Regulatory requirements
4. Sustainability (environmental, social, economical)
5. Implementation of LEED requirements
6. Energy conservation requirements
7. Maintainability
8. Constructability
9. HSE requirements
10. Fire protection requirements
11. Project schedule
12. Project budget
13. Design review procedure by the owner/client
14. Other disciplines requirements for coordination purpose
15. Procedure to incorporate changes/revisions requested by the client/owner
16. Number of drawings to be produced
17. Duration of completion of project design and submission of documents at different stages/phases of the project
18. Quality system requirements within the organization or portion of organization
19. Any special conditions of the client

A4.0.3 Analyze Phase

How we fulfill? (Analyze process options and prioritize based on capability to satisfy customer requirements.)

The key deliverables in this phase are

- Data collection
- Prioritization of data under major variables
- Compliance to organization's procedures and guidance

A4.0.3.1 Data Collection

Data shall be collected at different phases of the construction project life cycle, for design considerations, to meet requirements for development of the following three phases:

1. Concept design
2. Schematic/preliminary design
3. Detailed design/design development

The following data shall be collected to develop construction project design at different phases:

- Identification of need by the owner
- Identify number of floors
- Identify building usage
- Identify technical and functional capability requirements
- Soil profile and laboratory test of soil
- Identify topography of the project site
- Identify wind load, seismic load, dead load, and live load
- Identify existing services passing through the project site
- Identify existing roads, structure surrounding the project site
- Identify environmental compatibility requirements
- Identify energy conservation requirements
- Identify sustainability requirements
- Identify regulatory/authority requirements
- Identify codes and standards to be followed
- Identify social responsibility requirements
- Identify health and safety features
- Identify fire protection requirements
- Identify aesthetics requirements
- Identify zoning requirements
- Identify project constraints
- Identify ease of constructability

- Identify critical activities during construction
- Identify method statement requirements
- Identify requirements of other disciplines for coordination purposes
- Identification of team members
- Project time schedule
- Financial implications and resources for the project
- Identify cost effectiveness of the project
- Identify 3D information areas
- Identify suitable software program
- Identify number of drawings to be produced
- Identify milestone for development of each phase of design

A4.0.3.2 Arrangement of Data

The generated data shall be prioritized in an orderly arrangement under the following major variables during concept, schematic, and design development phases.

- Owner's need
- Regulatory compliance
- Sustainability
- Safety
- Constructability

A4.0.4 Design Phase

How we build it? (Design detailed processes capable of satisfying customer requirements.)

The key deliverables in this phase are

- Development of concept design
- Development of schematic design
- Development of detail design

A4.0.4.1 Development of Concept Design

While developing the concept design, the designer shall consider the following:

- Project goals/owner need
- Number of floors
- Usage of building
- Technical and functional capability

- Regulatory requirements
- Authorities' requirements
- Environmental compatibility
- Constructability
- Health and safety
- Cost effectiveness over the entire life cycle of the project

A4.0.4.2 Development of Schematic Design

While developing the schematic design, the designer shall consider the following:

- Concept design deliverables
- Site location in relation to the existing environment
- Building structure
- Floor grade and system
- Tentative size of columns, beams
- Stairs
- Roof
- Authorities' requirements
- Energy conservation issues
- Available resources
- Environmental issues
- Sustainability
- Requirements of all stakeholders
- Optimized life-cycle cost (value engineering)
- Constructability
- Functional/aesthetic aspect
- Services requirements
- Project schedule
- Project budget
- Preliminary contract documents (outline specifications)

A4.0.4.3 Development of Detail Design

While developing the detail design, the designer shall consider the following:

- Schematic design deliverables
- Authorities' requirements

- Energy conservation issues
- Environmental issues
- Sustainability
- Requirements of all stakeholders
- Environmental compatibility
- Available resources
- Number of floors
- Property limits/surrounding areas
- Excavation
- Dewatering
- Shoring
- Backfilling
- Substructure
- Design of foundation based on field and laboratory test of soil investigation
- Subsurface profiles, subsurface conditions, and subsurface drainage
- Coefficient of sliding on foundation
- Degree of difficulty for excavation
- Method of protection of below-grade concrete members against impact of soil and groundwater
- Geotechnical design parameters
- Design load such as dead load, live load, and seismic load
- Grade and type of concrete
- Type of footings
- Type of foundation
- Energy-efficient foundation
- Size of bars for reinforcement and the characteristic strength of bars
- Clear cover for reinforcement
- Reinforcement bar schedule, stirrup spacing
- Superstructure
- Columns
- Walls
- Stairs
- Beams
- Slab
- Parapet wall

- Height of each floor
- Beam size and height of beam
- Location of columns in coordination with architectural requirements
- Openings for services
- Deflection that may cause fatigue of structural elements; crack or failure of fixtures, fittings or partitions; or discomfort of occupants
- Movement and forces due to temperature
- Equipment vibration criteria
- Expansion joints
- Insulation
- Concrete tanks (water storage)
- Services requirements (shafts, pits)
- Shafts and pits for conveying system
- Building services to fit in the building
- Coordination with other trades and conflict resolution
- Calculations required as per contract requirements

A4.0.4.4 Preparation of Contract Documents

Contract documents shall be prepared as per MasterFormat™ (2012 edition).

A4.0.5 Verify Phase

How do we know it will work? (Verify design performance capability.)

The key deliverables in this phase are

- Review and check the design for quality assurance using thorough itemized review checklists to ensure that design drawings fully meet the owner's objectives/goal.
- Review and check contract documents (design drawings, specifications, and contract documents).
- Check for accuracy.
- Check calculations.
- Review studies and reports.
- Review discipline requirements.
- Review interdiscipline requirements and conflict.

- Review constructability.
- Management review.

After verification, the documents can be released for submission.

In case of any comments from the client/client's representative, the design shall be reviewed and modified accordingly.

Appendix B: Application of Six Sigma DMADV Tool in Development of Contractor's Construction Schedule

B1.0 Introduction

Six Sigma is, basically, a process quality goal. It is a process quality technique that focuses on reducing variation in process and preventing deficiencies in a product. In a process that has achieved Six Sigma capability, the variation is small compared to the specification limits.

Six Sigma started as a defect reduction effort in manufacturing and was then applied to other business processes for the same purpose.

Six Sigma methodology focuses on

- Leadership principles
- Integrated approach to improvement
- Engaged teams
- Analytic tools
- Hard-coded improvements

Six Sigma methodology has four leadership principles. These are

1. Align
2. Mobilize
3. Accelerate
4. Govern

Teamwork is absolutely vital for complex Six Sigma projects. For teams to be effective, they must be engaged, involved, focused, and committed to meeting their goals. Engaged teams must have leadership support. There are four types of teams. These are

1. Black Belts
2. Green Belts

3. Breakthrough
4. Blitz

For typical Six Sigma projects, four critical roles exist. These are

1. Sponsor
2. Champion
3. Team leader
4. Team member

The following are the analytic tools used in Six Sigma projects:

1. Ford Global 8D Tool
   - Ford Global 8D is primarily used to bring performance back to a previous level.
2. DMADV
   - DMADV is primarily used for the invention and innovation of modified or new products, services, or process.
3. DMAIC
   - DMAIC is primarily used for the improvement of an existing product, service, or process.
4. DMADDD
   - DMADDD is primarily used to drive the cost out of a process by digitization improvements

B2.0 Contractor's Construction Schedule

Construction projects are unique and nonrepetitive in nature, and need specific attention to maintain quality. The contractor's construction schedule is an important document used during the construction phase. It is used to plan, monitor, and control project activities and resources. The document is voluminous and important. It has to be prepared accurately in order to follow up the work progress without deviation from the milestones set up in the contract documents.

The contractor's construction schedule (CCS) is very important in construction projects. In most cases, the contractor experiences problems with approval of the CCS, at the very first submission, from the construction manager/project manager/consultant. The CCS is rejected by the construction manager/project manager/consultant if it does not meet the specification requirements. The contractor is not paid unless the CCS is approved.

B3.0 Six Sigma in Construction Projects

The following is an example procedure to develop the CCS using the Six Sigma DMADV analytic tool set.

DMADV stands for:

D → Define

M → Measure

A → Analyze

D → Design

V → Verify

DMADV Tool

1. Define → What is Important?
2. Measure → What is needed?
3. Analyze → How will we fulfill?
4. Design → How do we build it?
5. Verify → How do we know it will work?

DMADV tool has the following uses:

1. DMADV tool is used primarily for the invention and innovation of modified or new products, services, or process.
2. DMADV tool is used when a product, service, or process is required but does not exist.
3. DMADV is proactive, solving problems before they starts. This tool is also called as DFSS (Design for Six Sigma).
4. Using this toolset, Black Belts optimize performance before production begins.

Table B.1 illustrates the fundamental objectives of DMADV.

TABLE B.1

Fundamental Objectives of Six Sigma DMADV Tool

DMADV	Phase	Fundamental Objective
1	**Define**—What is important?	Define the project goals and customer deliverables (internal and external)
2	**Measure**—What is needed?	Measure and determine customer needs and specifications
3	**Analyze**—How do we fulfill?	Analyze process options and prioritize based on capabilities to satisfy customer requirements
4	**Design**—How do we build it?	Design detailed processes capable of satisfying customer requirements
5	**Verify**—How do we know it will work?	Verify design performance capability

B4.0 Application of DMADV in Development of the CCS

B4.0.1 Define Phase: What Is Important? (Define the Project Goals and Customer Deliverables)

The key deliverables of this phase are

- Establish the goal
- Identify the benefits
- Select the project team
- Develop the project plan
- Project charter

Goal: Develop the CCS using Six Sigma tools

Benefits: The measurable benefits of adopting this process will result in a CCS that will meet all the requirements of the specifications and will be approved by the construction manager/project manager/consultant at the first submission, thus reducing the repetitive work, which will help with implementation of the schedule right from the early stages of the project.

Selection of Team: The team shall consist of

1. Sponsor—Project Manager
2. Champion—Construction Manager
3. Team leader—Planning and Control Manager
4. Team members—Planning engineer, cost engineer, and one representative from each subcontractor

Project Plan: Time frame in the form of a Gantt chart shall be prepared to meet the target dates for submission of the CCS.

B4.0.2 Measure Phase: What Is Needed? (Measure and Determine Customer Needs and Specifications)

The key deliverable in this phase is

- Identify specification requirements

B4.0.2.1 Identify Specification Requirements

The following are the specification requirements listed in most contract documents.

The contractor has to submit the construction schedule in a bar-chart time-scaled format, to show the sequence and interdependence of activities required for complete performance of all items of work under the contract. The contractor shall use a computerized precedence diagram CPM technique to prepare the CCS. The schedule shall include but not be limited to the following:

- Project site layout
- Concise description of works
- Milestones (contractual milestones or constraints)
- Number of working days
- WBS activities shall consist of all those activities that take time to carry out the execution/installation and on which resources are expended
- Construction network of project's phases (if any) including various subphases
- Construction network of the project arrangements (activities) and sequence
- Time schedules for various activities in a bar-chart format
- The minimum work activities to be included in the program shall include items stated in the BOQ
- Early and late finish dates
- Time schedule for critical path
- Schedule text report showing activity, start and finish dates, total float, and relationship with other activities
- Summary schedule report showing number of activities, project start, project finish, number of relations, open ends, constraints, and milestones
- Total float of each activity
- Cost loading
- Expected progress cash flow S-curve
- Resource loaded S-curve

- Manpower loading
- Labor and crew movement and distribution
- Resource productivity schedule
- The number of hours per shift
- Average weekly usage of manpower for each trade
- Resource histogram showing the manpower required for different trades per time period for each trade (weekly or monthly)
- Equipment and machinery loading
- Schedule of mobilization and general requirements
- Schedule of subcontractor's and suppliers submittal and approval
- Schedule of materials submittals and approvals
- Schedule of long lead materials
- Schedule of procurement
- Schedule of shop drawings submittals and approvals
- Regulatory/authorities requirements
- Schedule of testing, commissioning, and handover
- Expected cash flow for executed work (during progress of work)

B4.0.3 Analyze Phase: How We Fulfill? (Analyze Process Options and Prioritize based on Capability to Satisfy Customer Requirements)

The key deliverables in this phase is

- Data collection
- Prioritization of data under major variables

B4.0.3.1 Data Collection

- Identify milestone dates
- Identify project calendar
- Identify resource calendar
- Identify project constraints such as access, logistics, delivery, seasonal, national, safety, existing work flow discontinuity, proximity of adjacent concurrent work
- Identify WBS (work break down) activities using BOQ
- Relate WBS activities with BOQ (Bill of Quantity/Bill of Material) and contract drawings
- Identify zoning/phasing
- Identify codes for all activities as per contract document divisions/sections using CSI format

- Identify volume of work for each activity
- Identify logical relationship
- Identify sequencing of activities
- Identify critical activities and their effect on critical path
- Identify duration/time schedule of each activity
- Identify early and late finish dates
- Review contract conditions and technical specifications
- Identify mobilization requirements
- Identify regulatory/authorities' requirements
- Identify subcontractors/suppliers
- Identify close out requirements
- Identify suitable software program
- Identify submittal requirements
- Identify project method statement
- Identify materials requirements
- Identify long lead items
- Identify procurement schedule
- Identify shop drawing requirements
- Identify manpower resources with productivity rate
- Identify equipment and machinery
- Identify testing, commissioning, and handover requirements
- Identify special inspection requirements
- Determine cost loading schedule of values/cost estimates of activities
- Identify project progress cash flow
- Identify and include items not listed in the specifications but that are important for the project scheduling

B4.0.3.2 Arrangement of Data

The generated data can be prioritized in an orderly arrangement under the following seven major variables:

1. Milestones
2. WBS activities
3. Time schedule
4. General requirements

5. Resources
6. Engineering
7. Cost loading

B4.0.3.2.1 Milestones

- Contractual milestone
- Project calendar
- Resource calendar
- Constraints
- Access, logistics, delivery, seasonal, national
- Existing workflow discontinuity, proximity of adjacent concurrent work, safety

B4.0.3.2.2 WBS Activities

- Activities related to BOQ
- Activities coding
- Activities sequencing
- Durations
- Logical relationship
- Critical activities
- Zoning/staging

B4.0.3.2.3 Time Schedule

- Schedule text report
- Time schedule for various activities
- Summary schedule
- Time schedule for critical paths
- Early and late finish dates

B4.0.3.2.4 General Requirements

- Mobilization
- Regulatory approval
- Subcontractor approval
- Closeout
- Contract conditions
- Submittals

- Softwares
- Updates

B4.0.3.2.5 Resources

- Manpower
- Productivity rate
- Hours per shift
- Work volume per activity
- Resource histogram
- Equipment and machinery
- Resource loaded S-curve

B4.0.3.2.6 Engineering

- Method statement
- Materials
- Procurement
- Shop drawing
- Long lead materials
- Inspection
- Factory inspection
- Third-party inspection
- Testing and commissioning

B4.0.3.2.7 Cost Loading

- Planned work S-curve
- Cost loading
- Schedule of values

These variables, along with related subvariables arranged in the form of an Ishikawa diagram, are shown in Figure B.1.

B4.0.4 Design Phase: How We Build It? (Design-Detailed Process(es) Capable of Satisfying Customer Requirements) See Figure B.2

The key deliverable in this phase is

- Preparation of program using suitable (specified) software program.
- Planning and Control Manager can prepare the CCS based on the collected data and sequence of activities.

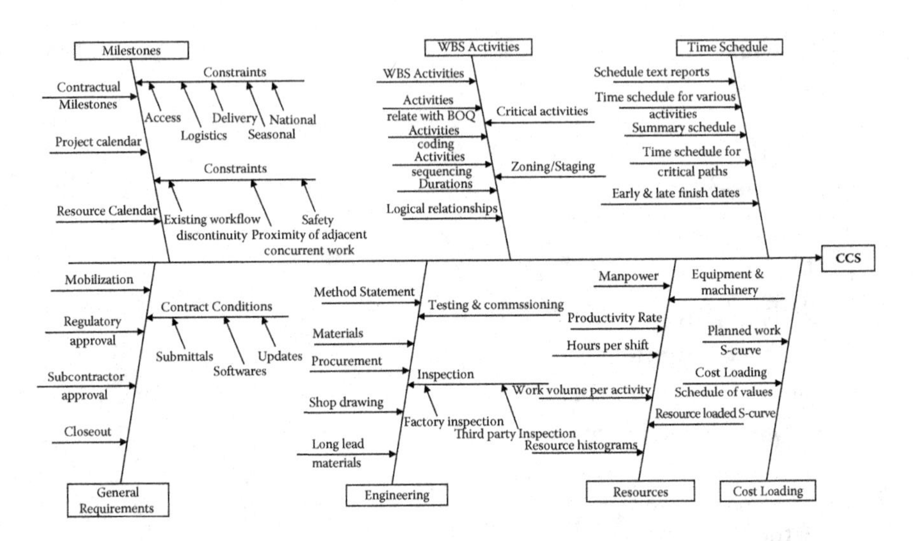

FIGURE B.1
Ishikawa diagram for CCS.

Project Name—Construction Project (Sample)									
Activity ID	Activity Name	Duration	Remaining Duration	Start	Finish	Successors	Actual Start	Actual Finish
1 SUMMARY SCHEDULE		1349	540	24-Jan-10A	18-Jun-14		24-Jan-10	
1.1 MILESTONES		58	0	24-Jan-10A	31-Mar-10A		24-Jan-10	31-Mar-10
1.1.1 MOBILIZATION		58	0	24-Jan-10A	31-Mar-10A		24-Jan-10	31-Mar-10
A1000	SIGNING OF CONTRACT	0	0	24-Jan-10A		A1020	24-Jan-10	
A1020	PRELIMINARY SITE SURVEY	7	0	24-Jan-10A	31-Jan-10A	A1030	24-Jan-101	31-Jan-10
A1030	SITE OFFICE MOBILIZATION & SETUP	45	0	08-Feb-10A	31-Mar-10A	A1170	08-Feb-10	31-Mar-10
1.2 ENGINEERING		449	0	01-Feb-10A	21-Jul-11A		01-Feb-10	21-Jul-11
1.2.1 CIVIL		92	0	01-Feb-10A	18-May-10A		01-Feb-10	18-May-10
1.21.1 LAYOUTS		51	0	01-Feb-10A	31-Mar-10A		01-Feb-10	31-Mar-10
A1170	PREPARATION OF CIVIL SHOP DRAWINGS	28	0	01-Feb-10A	04-Mar-10A	A1180	01-Feb-10	04-Mar-10
A1180	SUBMISSION OF CIVIL SHOP DRAWINGS	3	0	04-Mar-10A	07-Mar-10A	A1190	04-Mar-10	07-Mar-10
A1190	APPROVAL OF CIVIL SHOP DRAWINGS	21	0	08-Mar-10A	31-Mar-10A	A1290	08-Mar-10	31-Mar-10
1.21.2 PIPE SUPPORTS & PITS		41	0	01-Apr-10A	18-May-10A		01-Apr-10	18-May-10
A1290	PREPARATION OF PIPE SUPPORTS AND PITS DRAWINGS	18	0	01-Apr-10A	21-Apr-10A	A1300	01-Apr-10	21-Apr-10
A1300	SUBMISSION OF PIPE SUPPORTS AND PITS DRAWINGS	2	0	22-Apr-10A	24-Apr-10A	A1310	22-Apr-10	24-Apr-10
A1310	APPROVAL OF PIPE SUPPORTS AND PITS DRAWINGS	22	0	24-Apr-10A	18-May-10A	A1350	24-Apr-10	18-May-10
FIGURE B.2		Project Name—Construction Project						Sheet 1 of 8
Construction Schedule		Company Name—Messrs ABC						

FIGURE B.2
Summary CCS.

1.2.2 MECHANICAL		74	0	19-May-10A	14-Aug-10A		19-May-10	14-Aug-10
A1350	PREPARATION OF CAD & ISOMETRIC DRAWINGS	48	0	19-May-10A	14-Jul-10A	A1360	19-May-10	14-Jul-10
A1360	SUBMISSION OF CAD & ISOMETRIC DRAWINGS	5	0	15-Jul-10A	20-Jul-10A	A1370	15-Jul-10	20-Jul-10
A1370	APPROVAL OF CAD & ISOMETRIC DRAWINGS	21	0	21-Jul-10A	14-Aug-10A	A1370	21-Jul-10	14-Aug-10
1.2.3 ELECTRICAL		115	0	15-Aug-10A	03-Jan-11A		15-Aug-10	03-Jan-11
1.23.1 CABLE LAYOUT/ SCHEDULE		88	0	15-Aug-10A	30-Nov-10A		15-Aug-10	30-Nov-10
A1470	PREPARATION OF CABLE LAYOUT & EARTHING DRAWINGS	58	0	15-Aug-10A	23-Oct-10A	A1490	15-Aug-10	23-Oct-10
A1490	SUBMISSION & APPROVAL OF CABLE LAYOUT & EARTHING DRAWINGS	30	0	24-Oct-10A	30-Nov-10A	A1530	24-Oct-10	30-Nov-10
1.23.2 LIGHTING & SMALL POWER		27	0	01-Dec-10A	03-Jan-11A		01-Dec-10	03-Jan-11
A1530	PREPARATION OF LIGHTING & UPS DRAWINGS	7	0	01-Dec-10A	09-Dec-10A	A1540	01-Dec-10	09-Dec-10
A1540	SUBMISSION & APPROVAL OF LIGHTING & UPS DRAWINGS	20	0	11-Dec-10A	03-Jan-11A	A1590	11-Dec-10	03-Jan-11
1.2.4 INSTRUMENTATION		168	0	04-Jan-11A	21-Jul-11A		04-Jan-11	21-Jul-11
1.24.1 INSTRUMENT CABLE LAYOUT		78	0	04-Jan-11A	06-Apr-11A		04-Jan-11	06-Apr-11
A1590	PREPARATION OF INSTRUMENT DRAWINGS	35	0	04-Jan-11A	13-Feb-11A	A1600	04-Jan-11	13-Feb-11
FIGURE B.2		Project Name—Construction Project						Sheet 2 of 8
Construction Schedule		Company Name—Messrs ABC						

FIGURE B.2 ***(Continued)***

Project Name—Construction Project (Sample)								
Activity ID	**Activity Name**	**Duration**	**Remaining Duration**	**Start**	**Finish**	**Successors**	**Actual Start**	**Actual Finish**
A1600	SUBMISSON OF INSTRUMENT DRAWINGS	3	0	14-Feb-11A	17-Feb-11A	A1610	14-Feb-11	17-Feb-11
A1610	APPROVAL OF INSTRUMENT DRAWINGS	40	0	19-Feb-11A	06-Apr-11A	A1710	19-Feb-11	06-Apr-11
1.24.2 STUDIES		90	0	07-Apr-11A	21-Jul-11A		07-Apr-11	21-Jul-11
A1710	ENVIRONMENTAL IMPACT ANALYSIS	30	0	07-Apr-11A	11-May-11A	A1730	07-Apr-11	11-May-11
A1730	HSE ENVIRONMENTAL STUDY	30	0	12-May-11A	15-Jun-11A	A1740	12-May-11	15-Jun-11
A1740	QUANTITATIVE RISK ANALYSIS	30	0	16-Jun-11A	21-Jul-11A	A1750	16-Jun-11	21-Jul-11
1.3 PROCUREMENT		522	46	27-Feb-11A	15-Nov-12		27-Feb-11	
1.3.1 MECHANICAL		254	0	27-Feb-11A	27-Dec-11A		27-Feb-11	27-Dec-11
1.3.1.1 HIGH PRESSURE INJECTION PUMP & VESSELS		174	0	27-Feb-11A	22-Sep-11A		27-Feb-11	22-Sep-11
A1750	PREPARATION, SUBMISSION & APPROVAL OF DATA SHEETS	45	0	27-Feb-11A	20-Apr-11A	A1790	27-Feb-11	20-Apr-11
A1790	PROPOSE & APPROVE MANUFACTURER	15	0	20-Apr-11A	08-May-11A	A1800	20-Apr-11	08-May-11
A1800	PROCUREMENT OF HIGH PRESSURE PUMPS & VESSELS	100	0	08-May-11A	06-Sep-11A	A1810	08-May-11	06-Sep-11
A1810	INSPECTION ON SITE & APPROVAL	14	0	06-Sep-11A	22-Sep-11A	A2170	06-Sep-11	22-Sep-11
1.3.1.2 PIPINGS & VALVES		80	0	22-Sep-11A	27-Dec-11A		22-Sep-11	27-Dec-11
A2170	PROCUREMENT OF PIPING MATERIALS AND VALVES	60	0	22-Sep-11A	04-Dec-11A	A2200	04-Dec-11	25-Jun-12
FIGURE B.2		Project Name—Construction Project						Sheet 3 of 8
Construction Schedule		Company Name—Messrs ABC						

FIGURE B.2 ***(Continued)***

A2200	INSPECTION AND APPROVAL OF PIPING MATERIALS AND VALVES AT SITE	20	0	04-Dec-11A	27-Dec-11A	A2270	27-Dec-11	25-Jun-12
1.3.2 ELECTRICAL		150	0	27-Dec-11A	25-Jun-12A		27-Dec-11	21-Jan-12
1.3.2.1 EARTHING & POWER DISTRIBUTION SYSTEMS		150	0	27-Dec-11A	25-Jun-12A		27-Dec-11	05-Mar-12
A2270	SUBMISSION OF ELECTRICAL TECHNICAL DETAILS	20	0	27-Dec-11A	21-Jan-11A	A2280	27-Dec-11	31-May-12
A2280	APPROVAL OF TECHNICAL SUBMITTALS	35	0	21-Jan-12A	05-May-12A	A2290	21-Jan-11	25-Jun-12
A2290	PROCUREMENT OF EARTHING SYSTEMS, POWER DISTRIBUTION SYSTEMS	75	0	05-Mar-12A	31-Mar-12A	A2300	05-Mar-11	
A2300	INSPECTION ON SITE & APPROVALS	20	0	31-May-12A	25-Jun-12A	A2430	31-May-11	
1.3.3 INSTRUMENTATION		118	46	25-Jun-12A	15-Nov-12		25-Jun-12	
1.3.3.1 CONTROLS AND AUTOMATION		118	46	25-Jun-12A	15-Nov-12		25-Jun-12	
A2430	PROCUREMENT OF PLC CONTROLLER	118	46	25-Jun-12A	15-Nov-12	A2440	25-Jun-12	
A2440	PROCUREMENT OF HM WORK STATIONS	118	46	25-Jun-12A	15-Nov-12	A2450	25-Jun-12	
1.3.3.2 FIELD INSTRUMENTS		111	45	02-Jul-12A	14-Nov-12		02-Jul-12	
A2450	PROCUREMENT OF FIELD INSTRUMENTS	98	32	02-Jul-12A	30-Oct-12	A2460	02-Jul-12	
A2460	INSPECTION AT SITE & APPROVAL	13	13	31-Oct-12s	14-Nov-12	A2490		
FIGURE B.2		Project Name—Construction Project						Sheet 4 of 8
Construction Schedule		Company Name—Messrs ABC						

FIGURE B.2 ***(Continued)***

Project Name—Construction Project (Sample)									
Activity ID	Activity Name	Duration	Remaining Duration	Start	Finish	Successors	Actual Start	Actual Finish
1.4 CONSTRUCTION		486	486	14-Nov-12	08-Jun-14			
1.4.1 CIVIL		264	264	14-Nov-12	22-Sep-13			
1.4.1.1 LAYOUTS & FOUNDATIONS		88	88	14-Nov-12	28-Feb-13			
A2490	EXCAVATION AND EARTH WORKS	78	78	14-Nov-12	14-Feb-13	A2500		
A2500	INSTALL UNDERGROUND WATER LINES AND ELECTRIC LINES	5	5	14-Feb-13	20-Feb-13	A2510		
A2510	FORM/POUR & BACKFILLING	5	5	20-Feb-13	28-Feb-13	A2660		
1.4.1.2 CONTROL ROOM & SHEDS		100	100	28-Feb-13	25-Jun-13			
A2660	FABRICATION/CONSTRUCTION OF CONTROL ROOMS & SHEDS	100	100	28-Feb-13	25-Jun-13	A2750		
1.4.1.3 PIPE SUPPORTS & SLEEPERS		24	24	25-Jun-13	23-Jul-13			
A2750	CONCRETE WORKS FOR PIPE SUPPORTS	8	8	25-Jun-13	04-Jul-13	A2760		
A2760	STRUCTURAL STEEL WORKS FOR PIPE SUPPORTS	11	11	04-Jul-13	17-Jul-13	A2780		
A2780	INSTALLATIONS OF PIPE SLEEPERS	5	5	18-Jul-13	23-Jul-13	A2790		
1.4.1.4 PITS		30	30	23-Jul-13	28-Aug-13			
A2790	CONSTRUCTION OF PITS FOR DRAIN VESSEL	30	30	23-Jul-13	28-Aug-13	A2830		
1.4.1.5 MISCELLANEOUS WORKS		22	22	28-Jul-13	22-Sep-13			
FIGURE B.2		Project Name—Construction Project						Sheet 5 of 8
Construction Schedule		Company Name—Messrs ABC						

FIGURE B.2 ***(Continued)***

A2830	INSTALLATION OF CRASH BARRIERS	12	12	28-Aug-13	11-Sep-13	A2850		
A2850	CONCRETE FOOTING FOR INSTRUMENT SUPPORTS	10	10	11-Sep-13	22-Sep-13	A2870		
1.4.2 MECHANICAL		137	137	22-Sep-13	01-Mar-14			
1.4.2.1 HIGH PRESSURE INJECTION PUMPS & VESSELS		52	52	22-Sep-13	21-Nov-13			
A2870	INSTALLATION OF HIGH PRESSURE INJECTION PUMP	20	20	22-Sep-13	15-Oct-13	A2880		
A2880	INSTALLATION OF FILTER FEED PUMP	16	16	15-Oct-13	03-Nov-13	A2890		
A2890	INSTALLATION OF VESSELS	9	9	12-Nov-13	21-Nov-13	A2930		
1.4.2.2 PIPINGS		85	85	21-Nov-13	01-Mar-14			
A2930	LAYING OF RTRPPIPINGTILL PLANT AREA	16	16	21-Nov-13	10-Dec-13	A2940		
A2940	PIPING INSIDE FENCE-PLANT AREA	45	45	10-Dec-13	101-Feb-14	A2950		
A2950	PIPING FROM PLANT AREA TO INJECTION WELL	24	24	01-Feb-14	01-Mar-14	A2960		
1.4.3 ELECTRICAL		16	16	19-Mar-14	07-Apr-14			
A2960	INSTALLATION OF EARTHING & POWER DISTRIBUTION SYSTEMS	8	8	19-Mar-14	29-Mar-14	A2970		
A2970	LIGHTING & SMALL POWER SYSTEM INSTALLATION	8	8	29-Mar-14	07-Apr-14	A2990		
FIGURE B.2		Project Name—Construction Project						Sheet 6 of 8
Construction Schedule		Company Name—Messrs ABC						

FIGURE B.2 ***(Continued)***

Project Name—Construction Project (Sample)								
Activity ID	Activity Name	Duration	Remaining Duration	Start	Finish	Successors	Actual Start	Actual Finish
1.4.4 INSTRUMENTATION		53	53	07-Apr-14	08-Jun-14			
1.4.4.1 CONTROL & AUTOMATION		8	8	07-Apr-14	16-Apr-14			
A2990	INSTALLATION OF PLC CONTROL SYSTEM	8	8	07-Apr-14	16-Apr-14	A3000		
A3000	INSTALLATION OF HMI WORKS STATION	8	8	07-Apr-14	16-Apr-14	A3010		
1.4.4.2 FIELD INSTRUMENTS		45	45	16-Apr-14	08-Jun-14			
A3010	INSTALLATION OF FIELD INSTRUMENTS AND CABLES	45	45	16-Apr-14	08-Jun-14	A3030		
1.5 TESTING & COMMISSIONNING		9	9	08-Jun-14	18-Jun-14			
1.5.1 PRECOMMISSIONING		**6**	**6**	**08-Jun-14**	**15-Jun-14**			
A3030	PRECOMMISSIONING OF HIGH PRESSURE INJECTION PUMPS	2	2	08-Jun-14	10-Jun-14	A3040		
A3040	PRECOMMISSIONING OF FEED PUMPS	2	2	11-Jun-14	14-Jun-14	A3050		
A3050	PRECOMMISSIONING OF VESSELS	2	2	14-Jun-14	15-Jun-14	A3080		
FIGURE B.2		Project Name—Construction Project						Sheet 7 of 8
Construction Schedule		Company Name—Messrs ABC						

FIGURE B.2 (*Continued*)

1.5.2 COMMISSIONING		3	3	16-Jun-14	18-Jun-14			
A3080	COMMISSIONING OF HIGH PRESSURE INJECTION PUMPS	1	1	16-Jun-14	16-Jun-14	A3090		
A3090	COMMISSIONING OF FILTER FEED PUMPS	1	1	17-Jun-14	17-Jun-14	A3100		
A3100	COMMISSIONING OF VESSELS	1	1	17-Jun-14	18-Jun-14	A3140		
A3140	PROJECT COMPLETION TARGET DATE	0	0		18-Jun-14			
FIGURE B.2		Project Name—Construction Project						Sheet 8 of 8
Construction Schedule		Company Name—Messrs ABC						

Source: Abdul Razzak Rumane. (2010). *Quality Management in Construction Projects.* CRC Press, Boca Raton, FL. Reprinted with permission of Taylor & Francis Group.

FIGURE B.2 ***(Continued)***

B4.0.5 Verify Phase: How Do We Know It Will Work? (Verify Design Performance Capability)

The key deliverables in this phase are

- Review of the schedule by the team members to ascertain all the required elements are included for compliance with specification requirements.
- Submit the CCS to the construction manager/project manager/consultant.
- Update the schedule as and when required.

Appendix C: Application of Six Sigma DMAIC Tool in Construction Projects

C1.0 Introduction

Six Sigma is, basically, a process quality goal. It is a process quality technique that focuses on reducing variation in processes and preventing deficiencies in a product. In a process that has achieved Six Sigma capability, the variation is small compared to the specification limits. Six Sigma started as a defect reduction effort in manufacturing and was then applied to other business processes for the same purpose.

Six Sigma is a measurement of "goodness" using a universal measurement scale. Sigma provides a relative way to measure improvement. Universal means sigma can measure anything from coffee mug defects to missed chances of closing a sales deal. It simply measures how many times a customer's requirements were not met (a defect), given a million opportunities. Sigma is measured in defects per million opportunities (DPMO).

Six Sigma methodology focuses on

- Leadership principles
- Integrated approach to improvement
- Engaged teams
- Analytic tools
- Hard coded improvements

Six Sigma methodology has four leadership principles. These are

1. Align
2. Mobilize
3. Accelerate
4. Govern

Team work is absolutely vital for complex Six Sigma projects. For teams to be effective, they must be engaged, involved, focused, and committed to meeting their goals. Engaged teams must have leadership support. There are four types of teams. These are

1. Black Belts
2. Green Belts
3. Breakthrough
4. Blitz

For typical Six Sigma projects, four critical roles exist.

1. Sponsor
2. Champion
3. Team leader
4. Team member

The following are the analytic tools used in Six Sigma projects

1. Ford Global 8D Tool
   - Ford Global 8D is primarily used to bring performance back to a previous level.
2. DMADV
   - DMADV is primarily used for the invention and innovation of modified or new products, services, or processes.
3. DMAIC
   - DMAIC is primarily used for the improvement of an existing product, service, or process.
4. DMADDD
   - DMADDD is primarily used to drive the cost out of a process by digitization improvements.

C2.0 Construction Projects

Construction projects are unique and nonrepetitive in nature and need specific attention to maintain the quality. Construction activities mainly consist of

- Construction of concrete foundations, footings, beams, walls, slab, roofing, finishes, furnishings, conveying systems, electromechanical services, low-voltage systems, landscape works, and external works.

- In construction, an activity may be repeated at various stages, but it is done only once for a specific work. Therefore, it has to be done right from the outset.

In most cases, contractors experience problems with approval of executed works at the very first submission of the checklist. In order to achieve a zero defect policy during the construction process, and to avoid rejection of executed works, the contractor has to take the following measures:

- Execution of works per approved shop drawings
- Use of approved material
- Follow approved method of statement
- Conduct continuous inspection
- Employ trained workforce
- Coordinate requirements of other trades

C3.0 Six Sigma in Construction Projects

The following is an example procedure to develop a quality management system for the execution of concrete structural works in construction projects using the Six Sigma DMAIC tool.

The DMAIC tool is primarily used

- For improvement of an existing product, service, or process
- When a product, service, or process is failing to meet customer requirements or is not performing adequately

Using this toolset, Black Belts, Green Belts approach their projects driving process performance to never-before-seen levels.

The Six Sigma DMAIC tool can be applied at various stages in construction projects:

- Detailed Design Stage—To enhance coordination in order to reduce repetitive work
- Construction Stage—Preparation of builder's workshop drawings and composite drawings, as this needs a lot of coordination among different trades
- Construction Stage—Preparation of the contractor's construction schedule
- Execution of Works

The DMAIC process contains five distinct steps that provide a disciplined approach to improving existing process and processes and products through the effective integration of project management, problem solving, and statistical tools. Each step is designed to ensure that

- Companies apply the breakthrough strategy in a systematic way
- Six Sigma projects are systematically designed and executed
- Incorporating the results of these projects to running the day-to-day business

DMAIC stands for

D → Define
M → Measure
A → Analyze
I → Improve
C → Control

More about the DMAIC tool:

1. Define → What is important?
2. Measure → How we are doing?
3. Analyze → What is wrong?
4. Improve → What needs to be done?
5. Control → How do we guarantee performance?

Table C.1 illustrates the fundamental objectives of the DMAIC tool.

TABLE C.1
Fundamental Objectives of Six Sigma DMAIC Tool

DMAIC	Phase	Fundamental Objective
1	Define—What is important?	Define the project goals and customer deliverables (internal and external)
2	Measure—How are we doing?	Measure the process to determine current performance
3	Analyze—What is wrong?	Analyze and determine the root causes of the defects
4	Improve—What needs to be done?	Improve the process by permanently removing the defects
5	Control—How do we guarantee performance?	Control the improved process's performance to ensure sustainable results

C4.0 Application of DMAIC Tool to Development of Design for Construction

C4.0.1 Define Opportunities: What Is Important?

The objective of this phase is

- To identify and/or validate the improvement opportunities that will achieve the organization's goals and provide largest payoff, develop the business process
- To define critical customer requirements, and prepare to function as an effective project team

The key deliverables of this phase are

- Team charter
- Action plan
- Process map
- Quick win opportunities
- Critical customer requirements
- Develop project plan
- Prepared team

C4.0.1.1 Team Charter

The scope of the team's charter is to

- Ensure that structural concrete works are executed without any defects and concrete strength is as specified.
- Reduce rejection of executed works by the supervisor/independent testing agency.

C4.0.1.2 Action Plan

The objective is to ensure that each activity during execution of structural concrete works is carried out as specified and per standards and codes by ensuring that

- The work follows approved shop drawings
- Concrete mix is per approved sample
- Proper transportation of concrete ready mix from batching plant
- Follow casting procedures and method statement

C4.0.1.3 Process Map: (Sequence of Execution of Work)

The structural concrete work mainly consists of the following activities:

- Form work
- Reinforcement work
- Concrete
- Concrete pouring/casting
- Curing

Figures C.1 and C.2 illustrate the process maps.

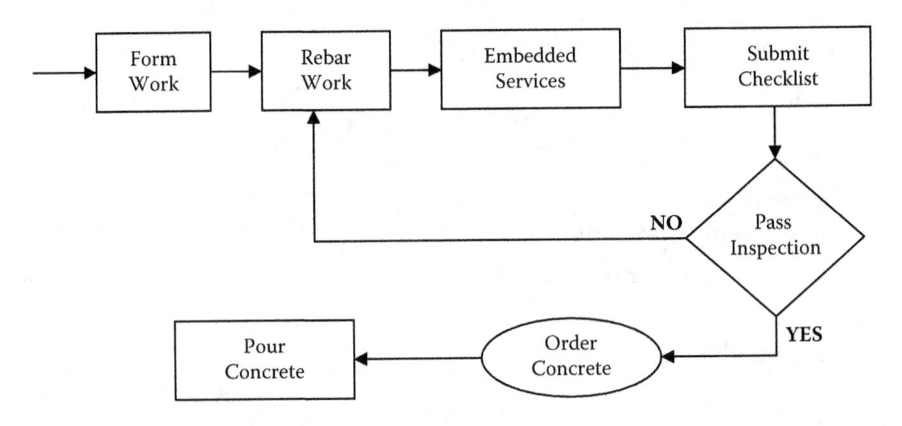

FIGURE C.1
Process map.

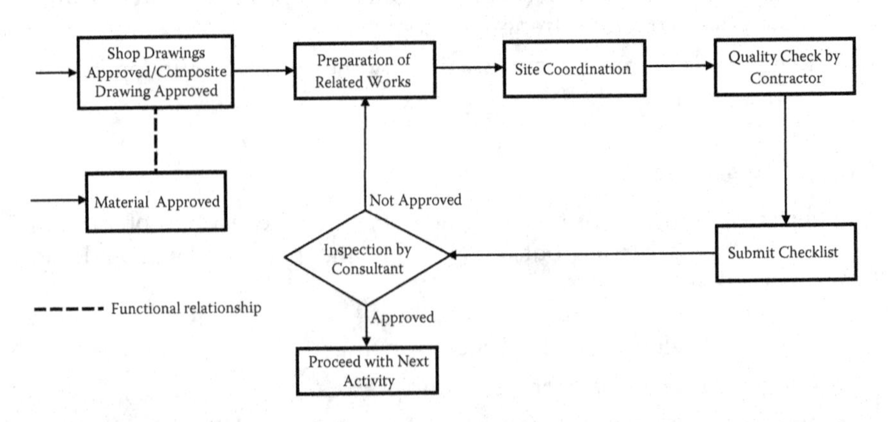

FIGURE C.2
Process map.

C4.0.1.4 Critical Customer Requirements

- Concrete quality as specified
- Control of the design mix per approved sample
- Concrete casting to comply with specified strength
- Concrete thickness and level as specified
- Concrete finishes as specified
- Temperature during casting
- Concrete curing
- Laboratory testing of sample cubes during casting
- Construction joints

C4.0.1.5 Develop Project Plan

A project plan in the bar-chart format (time frame) shall be prepared to complete the process within three to five concrete castings to ensure that executed works are not rejected and rework is not required for the remaining works.

C4.0.1.6 Prepared Team

The team shall be selected taking into consideration the qualifications of team members, background, and experience and shall consist of

1. Sponsor—Contractor's project manager
2. Champion—Site structural engineer
3. Team leader—Quality control engineer
4. Team members—Coordination engineer, MEP cocoordinator, civil/structural foreman (supervisor), lab technician, batching plant technician

Each member shall be notified about his or her role, responsibilities, and authority to perform the job.

C4.0.2 Measure Performance: How Are We Doing?

The objectives of this phase are

- To identify critical measures that are necessary to evaluate success or failure and meet customer requirements
- To establish a baseline for the process the team is analyzing

- To measure the process in order to determine current performance, how exactly the process operates
- To quantify the problems

The key deliverables in this phase are

- Input, process, and output indicators
- Operational definitions
- Data collection format and plans
- Baseline performance
- Productive atmosphere

C4.0.2.1 Input, Process, and Output Indicators

There are mainly three areas that may exhibit variations in the structural concrete works. These are

1. Form work
2. Reinforcement
3. Concrete casting

Create a process map in detail exactly as it exists for these areas and identify

- What are key input variables?
- What are key process variables?
- What are key output variables?

The process map would include time, people, and material elements to ensure that the current state is clearly understood.

C4.0.2.2 Operational Definitions

An operational definition is a specific description of the defect, process, product, or service to be measured.

Define the specific process or set of validation tests to be used to identify the characteristics you would like to measure.

Operational definitions are a very clear and precise explanation of the items being measured.

There are four criteria used in validating an operational definition:

1. The requirements to be measured must be agreed upon.
2. The method of statement must be agreed upon.

3. There must be agreement on what the definition will not include.
4. The customer must agree with the team on an appropriate operational definition.

C4.0.2.3 Data Collection Format and Plans

In the definition phase, critical customer requirements are developed.

Data collection should relate both to the problem statement and what the customer considers to be critical to quality. The data will then be graphed or charted to obtain a visual representation of the data.

If the team was collecting error data, then a Pareto chart would be a likely graphical choice. If a trend chart is needed to show how the process behaves over time, histograms are another way to observe the process data. Another widely utilized tool is the control chart. Other quality tools can be used depending on the need.

This data shall be used both as baseline data for improvement efforts and to estimate the current state process sigma. This will be a relative indicator of how close the current process is to delivering zero defects.

The following are the main sources of variations:

- Material
- Method
- Manpower
- Machine
- Measurement
- Environment

The causes of variations are

1. Chance/inherent causes
2. Special/assignable causes

C4.0.2.4 Baseline Performance

Measure baseline process performance (process capability, yield, sigma level).

C4.0.3 Analyze Opportunity: What Is Wrong?

The objectives of this phase are

- Stratify and analyze the opportunity to identify a specific problem and define an easily understood problem statement.
- Determine true sources of variation and potential failure modes that lead to customer dissatisfaction.

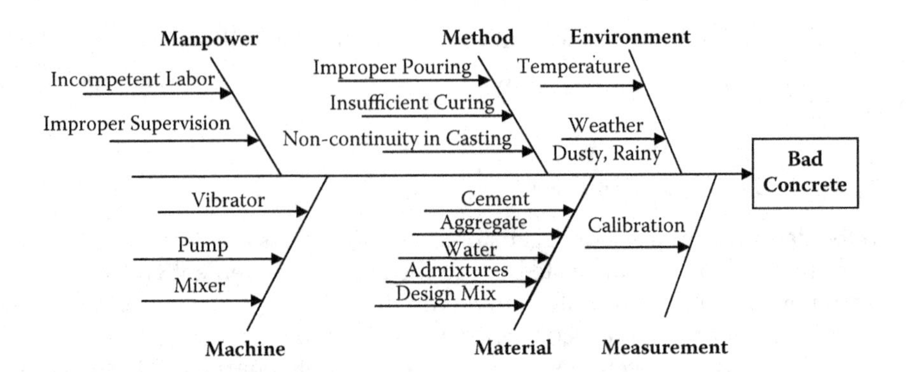

FIGURE C.3
Root cause analysis diagram.

The key deliverables in this phase are

- Data analysis
- Validated root causes
- Source of variation
- Failure modes and effects analysis (FMEA)
- Problem statement
- Potential solutions

Figure C.3 illustrates the root cause analysis diagram.

C4.0.3.1 Validated Root Causes

Identify the source of variation.

C4.0.3.2 Problem Statement

To achieve concrete strength as specified without any rework and rejection by the testing agency/consultant.

C4.0.3.3 Potential Solutions

Identify how problems will be eliminated.

C4.0.4 Improve Performance: What Needs to be Done?

The objectives of this phase are

- To identify, evaluate, and select the right improvement solutions
- To develop a change management approach to assist the organization in adapting to the changes introduced through solution implementation

The key deliverable in this phase are

- Solutions
- Process maps and documentation
- Pilot results
- Implementation milestones
- Improvement impacts and benefits
- Storyboard
- Change plans

C4.0.4.1 *Solutions*

- List out solutions.

C4.0.4.2 *Process Maps and Documentation*

- Establish a revised process map.

Figure C.4 is the revised process for casting.

C4.0.4.3 *Pilot Results*

- Run the process per the revised process map, conduct a pilot study, and note the results.

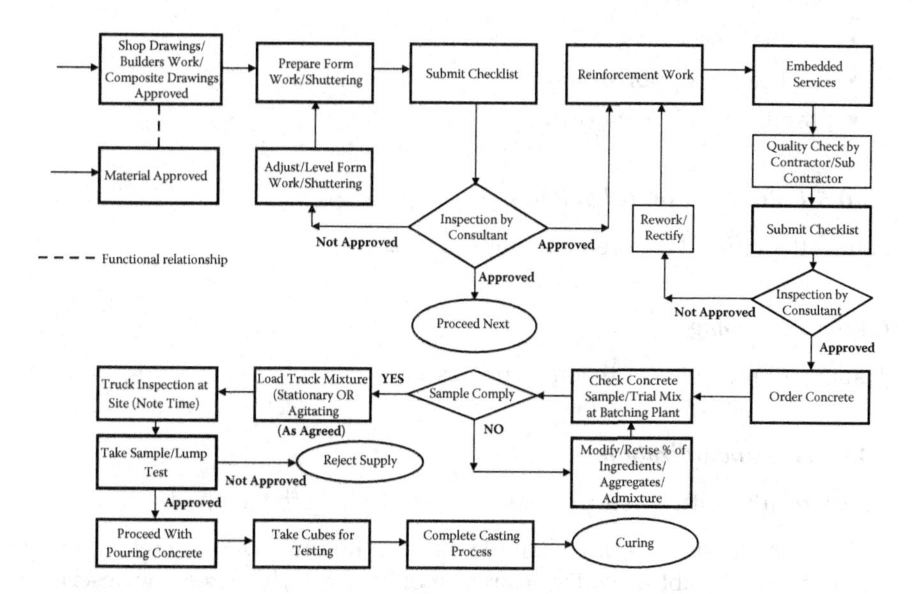

FIGURE C.4
Process map.

C4.0.4.4 Improvement Impacts and Benefits

- Identify the impact on the process due to application of the revised process.

C4.0.5 Control Performance: How Do We Guarantee Performance?

The objectives of this phase are

- Understand the importance of planning and executing against the plan
- Understand how to disseminate lessons learned

The key deliverables in this phase are

- Process control systems
- Standards and procedure
- Training
- Team evaluation
- Change implementation plan
- Potential problem analysis
- Solution results
- Success stories
- Trained associates
- Replication opportunities
- Standardization opportunities

C4.0.5.1 Process Control Systems

Follow the revised concrete casting process.

C4.0.5.2 Training

Train team members with a new process.

C4.0.5.3 Standardization

Check results with a new process, and standardize the method.

NOTE: The system is an example system to improve concrete quality. It may not be possible to collect a large amount of data; however, the described system will definitely improve the quality and reduce rework.

Bibliography

Abdul Razzak Rumane (2010), *Quality Management in Construction Projects*, CRC Press, Boca Raton, FL (A Taylor & Francis Group Company).

Index

N

O

P

Printed in the United States
by Baker & Taylor Publisher Services